Inductively Coupled Plasma–Mass Spectrometry

Practices and Techniques

Inductively Coupled Plasma–Mass Spectrometry

Practices and Techniques

Howard E. Taylor
Boulder, Colorado

ACADEMIC PRESS
A Harcourt Science and Technology Company

San Diego   San Francisco   New York   Boston   London   Sydney   Tokyo

Front cover photograph: An inductively coupled plasma. (For more details, see Chapter 3.)

This book is printed on acid-free paper. ∞

Academic Press
A Harcourt Science and Technology Company
525 B Street, Suite 1900, San Diego, California 92101-4495, USA
http://www.academicpress.com

Academic Press
Harcourt Place, 32 Jamestown Road, London NW1 7BY, UK
http://www.academicpress.com

Library of Congress Catalog Card Number: 00-108830

International Standard Book Number: 0-12-683865-8

PRINTED IN THE UNITED STATES OF AMERICA
00 01 02 03 04 05   SB   9 8 7 6 5 4 3 2 1

The author expresses his deepest gratitude to all the members of his family for their support during the preparation of the manuscript for this book. I am especially grateful for my wife Cathe for all of her love, support, and understanding and the sacrifices she made to make this possible.

Contents

5   *Sample Introduction*

6 Special Techniques

7 Quantitation Techniques

8 Interferences

9 Optimization

About the Author

Howard E. Taylor is a research chemist with the National Research Program, Water Resources Division, U.S. Geological Survey located in Boulder, Colorado. Dr. Taylor has played a major role over the past 25 years in the development of plasma spectrometric techniques in analytical chemistry, as reflected in his more than 150 technical publications and the presentation of numerous papers at national and international technical meetings. He has served as faculty affiliate at Colorado State University and has taught American Chemical Society Short Courses for more than 15 years.

CHAPTER 1

Introduction

Inductively coupled plasma–mass spectrometry (referred to in this book as ICP-MS), a technique for the elemental chemical characterization of virtually any material, evolved during the late 1990s into a mature analytical procedure. This technique has a variety of characteristics that make it uniquely suited for the solution of chemical analysis problems in many applications. These characteristics include the ability to precisely identify and measure (quantitate) all elements in the periodic table including the often difficult to analyze refractory elements. In addition to this wide scope of elemental analysis, the technique has the inherent capability to perform these determinations in a multielement analysis mode, efficiently providing comprehensive elemental compositional characterization. The technique also has the powerful ability to measure individual isotopes of the analyte elements, providing a capability that has many useful applications ranging from isotope dilution quantitation to stable isotope tracer studies.

Another equally important characteristic is the ability to detect and measure concentrations of analyte elements at very low levels. State-of-the-art instrumentation is currently capable of quantitatively measuring analytes down to 1–10 ng of analyte element per liter, in solution. This direct high measurement sensitivity across the periodic table exceeds the capability of most other modern instrumental methods of analysis. Indeed, even non-metallic elements can be determined with sufficient sensitivity to serve as a useful trace analytical tool. Additional advantageous characteristics, including a large linear dynamic working range, high accuracy and precision of measurement, and minimal interferences, make this a powerful and useful trace analysis tool.

The wide availability of commercial instrumentation, utilizing a variety of approaches to the ICP-MS technology, provides the chemist with several alternatives for the solution of analytical problems. The effective combination of types of spectrometers (or ion measurement technology) coupled with many approaches to sample introduction makes techniques available that can be customized for a specific sample type or form of analyte. This book summarizes the operational characteristics of this technology, the various configurations and types of instrumentation currently available, characteristics and relative advantages of different types of sample introduction, procedures and techniques for attaining and interpreting results, and potential problems with data acquisition and interpretation for specific applications.

1.1 HISTORY

The roots of ICP-MS began in the mid-1960s with the advent of a technique called inductively couple plasma–atomic emission spectrometry (ICP-AES). For decades, prior to this, atomic emission spectrometry (flame, direct current-arc, and controlled-waveform spark) was the predominant method used for elemental analysis. The work of Greenfield *et al.* (1964) and work done essentially simultaneously by Wendt and Fassel (1965) introduced an emission spectrometric technique that provided high sensitivity trace element analysis with a multielement detection capability. This technique is still widely used today and can be studied in publications by Boumans (1987) and Montaser and Golightly (1992).

The most advantageous and unique feature of this technique is the use of the atmospheric pressure argon inductively coupled plasma (ICP) for sample atomization and efficient atomic excitation. This plasma is a highly energetic media consisting of inert ionized gas. Its high equivalent temperature (7000–10,000 K) provides exceptional atomization (i.e., decomposition of complex materials in a sample to individual atoms) followed by highly efficient atomic excitation. This characteristic of the plasma results in the ability to analyze very refractory materials (those difficult to atomize) and those with high excitation potentials that are difficult to excite by other atomic emission sources. The result is a very satisfactory trace elemental analytical technique.

Houk *et al.* (1980) demonstrated the advantages that could be realized by using the ICP as an ionization source. They coupled the ICP with a mass spectrometer and identified the ion species found in the plasma and measured their quantitative concentrations. They also discovered that the result was a much simpler spectrum than that obtained in atomic emission spectrometry, potentially reducing interelement interferences, while providing significantly higher quantitative sensitivity. It soon became apparent that measuring ion currents by this instrumental approach produced much higher signal-to-background ratios, yielding significantly higher analytical sensitivity, which resulted in much improved detection limits.

Soon commercial instrumentation became available, with the first equipment being offered by a Canadian company called Sciex (which subsequently merged with Perkin Elmer). Sciex delivered an instrument called the *Elan* to customer laboratories in 1984. As with most highly sophisticated instrumentation, the early models often exhibited design deficiencies that were corrected by an evolutionary process as new models were introduced into the marketplace and as users developed techniques and methodologies that required specialized apparatus. Models of instrumentation that were available in 2000 are listed in Table 1.1.

1.2 OVERVIEW

The first ICP-MS instruments employed a conventional crystal-controlled (27-MHz) radio-frequency ICP operating at between 1 and 2 kW of incident power. They, as well as subsequent models, utilize atmospheric pressure argon gas flowing through a concentric quartz torch. A sample, usually in the form of an aqueous solution, is converted to an aerosol by a nebulization process and transported to the plasma by an argon gas stream. In the plasma the analyte elements are atomized, followed immediately by ionization. The composition of the ion population in the plasma is proportional to the concentration of the analyte species in the original sample solution. Ions that are produced by the ICP are representatively sampled and extracted from the plasma, where they are separated and measured by a quadrupole mass spectrometer. The key to the

TABLE 1.1   Commercial ICP-MS Instrumentation Available in 2000

Manufacturer	Model	Type
Finnigan MAT Corp.	Element2	Magnetic
Finnigan MAT Corp.	Neptune	Multiple collector
GBC Scientific Equip.	Optimass 8000	Time-of-flight
Hewlett Packard Corp.[a]	HP4500 Series	Quadrupole
JEOL	JMS-Plasmax 2	Magnetic
LECO Corp.	Renaissance	Time-of-flight
Micromass Inc.	Platform ICP	Collision cell quadrupole
Micromass Inc.	Plasma Trace 2	Magnetic
Micromass Inc.	Iso-Plasma Trace	Multicollector
Nu Instruments	Nu Plasma	Multicollector
Perkin Elmer/Sciex	Elan 6100	Quadrupole
Perkin Elmer/Sciex	Elan 6100 DRC	Reaction cell quadrupole
Seiko Instruments	SPQ 9000	Quadrupole
Spectro Analytical Inst.	SpectroMass 2000	Quadrupole
Thermo-Jarrell Ash	POEMS II	Combination AES/MS
Varian Corp.	Ultramass 700	Quadrupole
VG Elemental	PlasmaQuad 3	Quadrupole
VG Elemental	PQ Excell	Collision cell quadrupole
VG Elemental	Axiom SC	Magnetic
VG Elemental	Axiom MC	Multicollector

[a]Yokogawa Anal. Inst.

technology is a cleverly engineered interface to sample ions from the atmospheric pressure argon plasma and transport them into the low-operating pressure ($10^{-4} - 10^{-6}$ torr) of the mass spectrometer without modifying their compositional distribution. This interface allows the marriage of two well-developed technologies in the field of analytical chemistry—the inductively coupled plasma from the atomic emission spectrometry community, and the quadrupole mass spectrometer from the organic compound analysis field.

As the nature of the technique evolved, modification and improvements in both technologies increased the versatility and performance of the system. These improvements in the ICP included changes in the radio-frequency (RF) power generator, utilizing various frequencies other than 27 MHz, improved tuning and RF coupling networks between the generator and the plasma torch, and the utilization of mixed-gas plasmas (N_2, O_2, etc.). At the spectrometer end, high-resolution magnetic sector mass spectrometers and, more recently, time-of-flight mass spectrometers, have greatly enhanced the performance characteristics of the instrumentation.

However, the conventional quadrupole argon ICP-MS still provides the operating analytical chemist with sufficient utility to solve the majority of elemental analysis problems at a reasonable equipment investment cost.

Several reference sources are available that summarize the characteristics of this technique. These references should be consulted for additional details regarding specific topics. Table 1.2 lists these references in alphabetical order.

TABLE 1.2   Selected References

Author/editor	Title	Publisher	Year
Adams, F, Gijbels, R. and Van Grieken, R.	Inorganic Mass Spectrometry	Wiley	1988
Date, A.R. and Gray, A.L.	Applications of Inductively Coupled Plasma Mass Spectrometry	Blackie	1989
Evans, E.H., Giglio, J.J., Castillano, T.M. and Caruso, J.A.	Inductively Coupled and Microwave Induced Plasma Sources for Mass Spectrometry	Royal Society of Chemistry	1995
Hill, Steve J.	Inductively Coupled Plasma Spectrometry and Its Applications	Sheffield Academic Press	1999
Holland, G. and Eaton, A.N.	Applications of Plasma Source Mass Spectrometry	Royal Society of Chemistry	1991
Holland, G. and Eaton, A.N.	Applications of Plasma Source Mass Spectrometry II	Royal Society of Chemistry	1993
Jarvis, K.E., Gray, A.L. and Houk, R.S.	Handbook of Inductively Coupled Plasma Mass Spectrometry	Blackie	1992
Montaser, A. and Golightly, D.W.	Inductively Coupled Plasmas in Analytical Atomic Spectroscopy, 2nd ed.	VCH	1992
Montaser, A.	Inductively Coupled Plasma Mass Spectrometry	Wiley-VCH	1998

CHAPTER 2

Atomic Structure

The existence and properties of individual molecules, atoms, and ions in the gaseous state provide a fundamental basis for elemental spectrometric analytical determinations. Classical techniques based on these principles have been utilized for decades for the major, minor, and trace analysis of substances in virtually every field of science. The enhancement of techniques based on these properties has recently resulted in technology that exhibits ultratrace analysis sensitivity (sub–part-per–million detectability) and allows expansion of the scope of analysis to the rarer and more exotic elements. A basic understanding of some of the fundamental atomic properties of the elements that are used for mass spectrometric analysis is required to fully utilize this technology.

Dalton, in 1803, introduced the concept of atoms and molecules and established the following postulates regarding the laws of chemical composition:

1. Each chemical element is composed of discrete atoms, which are identical in all respects.

2. Atoms cannot be broken down by known chemical processes, and they maintain their unique identity after undergoing these processes.
3. Atoms of different elements may associate or bond together to form chemical compounds which have properties different from those of the atoms.

These principles were used to establish a theory regarding the properties, structure, and behavior of atoms. This chapter provides a basic introduction to these principles.

Atoms are the most fundamental building blocks of matter. Composed of subatomic components including electrons, protons, and neutrons, atoms are the smallest units having characteristics and properties that are unique and identifiable to individual elements. Table 2.1 lists the mass and electrical charge of the most common subatomic particles. Electrical charge is defined here, by convention, as a positive or negative integer.

2.1 BOHR MODEL

Models are often used to describe the structure of atoms, the configuration of their subatomic particles, and their interactive behavior. Of the several models for atomic structure, including those of J. J. Thomson in 1907 and Lord Rutherford in 1911, the Bohr model, proposed in 1913, is used here to illustrate atomic principles. Two assumptions made by Bohr are critical to using his model: (1) Stationary energy states exist such that an atom in one of these states is stable and the atom in this state is populated for a finite period of time and (2) the emission or absorption of radiation from an atom is exactly equal to the difference between two of the discreet energy states.

Although the Bohr model employs a simplified concept that is insufficient to explain the chemical behavior of an element, it provides an illustration to

TABLE 2.1   Physical Properties of Subatomic Particles

Particle	Rest mass (g)	Charge[a]
Electron	9.110×10^{-31}	-1
Neutron	1.675×10^{-27}	0
Proton	1.673×10^{-27}	$+1$

[a] Equivalent to the electrostatic charge of one electron (1.6×10^{-19} C).

assist in the elementary understanding of their atomic characteristics. The atom appears to consist of an extremely small nucleus surrounded by one or more electrons, which in themselves are very small. The nucleus of the atom is made up of both protons and neutrons. As can be seen from Table 2.1, by far the bulk of the mass of the atom is made up by the nucleus (the mass of the electrons is insignificant compared to that of the protons or neutrons).

Because each electron, which is negatively charged, resides a substantial relative distance from the positively charged nucleus, the atom must have associated with it considerable energy to prevent the electron from being attracted to the nucleus. A balance of forces keeps the electrons, which are in constant motion from collapsing into the nucleus. Identification of these forces is a complex matter and is beyond the scope of this discussion. The description of each electron in an atom with respect to its energy is fundamental in understanding the atom's electronic structure.

Figure 2.1 shows a simplified diagram for two elements, the simplest elemental structure, hydrogen, and a more complex element, fluorine. Hydrogen has an atomic number of 1, which means that it has 1 electron and 1 proton (in its nucleus). Because the electron has a charge of -1 and the proton has a charge of $+1$, this hydrogen atom is overall electrically

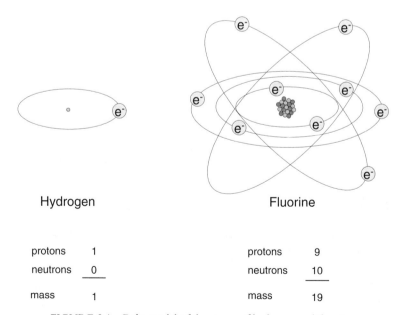

Hydrogen			Fluorine	
protons	1		protons	9
neutrons	0		neutrons	10
mass	1		mass	19

FIGURE 2.1   Bohr model of the atoms of hydrogen and fluorine.

neutral. Also in Figure 2.1, we see that fluorine has an atomic number of 9, which means it consists of 9 electrons and 9 protons in its nucleus. Appendix 2 lists common isotopes of the elements and their abundances.

2.2 ISOTOPES

It is possible for separate atoms of the same chemical element to have different masses. These entities were observed in the study of the transformation of radioactive atoms. In 1913 Soddy named these atoms, which have the same chemical properties but different masses, *isotopes*.

The absolute mass of an isotope can be computed by summing the individual masses of the subatomic particles (see Table 2.1) and reporting the total number of grams that an isotope must weigh; however, it is more convenient to quote masses in relative terms using the atomic mass unit (amu). By definition, the most abundant isotope of carbon is exactly 12.000 amu. All other isotope masses are reported relative to carbon-12. Appendix 1 lists the average atomic mass (i.e, the average of the abundances of the stable isotopes) of each element in atomic mass units.

Some elements such as arsenic, sodium, manganese, and cobalt are monoisotopic, meaning they only have one stable isotope. Other elements, which have multiple isotopes, usually have constant isotopic abundances of each isotope. The exceptions are elements such as lead, which have one or more isotopes that are the decay products of other precursor elements that are radioactive. The isotopic abundances of these elements will vary depending on the concentration and history of the concomitant radioactive elements.

In Figure 2.1 we observe that fluorine, which has 9 electrons and 9 protons, has an atomic mass of 19 (see Appendix 1). Since the relative mass of the proton is approximately 1 amu, then by difference, the nucleus of fluorine is required to have 10 neutrons, which also have an approximate atomic mass of 1 amu each. The various isotopes of a given element have different numbers of neutrons; therefore, their atomic masses vary accordingly.

The atomic structure of the two chlorine stable isotopes is shown in Figure 2.2. Chlorine, which has an atomic number of 17, has 17 electrons and 17 protons. It also has two stable isotopes one with 18 neutrons, resulting in an atomic mass of 35 amu and the other with 20 neutrons resulting in an atomic mass of 37 amu. The relative abundances of these two isotopes are 75.8% for 35 and 24.2% for 37, almost three to one.

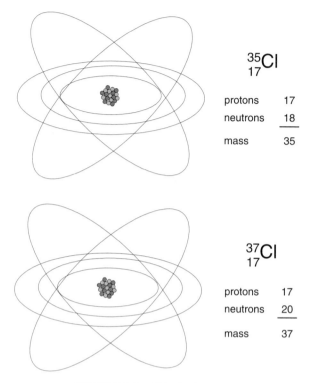

$^{35}_{17}\text{Cl}$

protons	17
neutrons	18
mass	35

$^{37}_{17}\text{Cl}$

protons	17
neutrons	20
mass	37

FIGURE 2.2   Chlorine isotopes.

2.3 IONIZATION

By adding excess external energy, an electron can be removed from a neutral atom. The removed electron has a unit negative charge, resulting in the formation of an ion with a net unit positive charge. Since the relative mass of the electron is negligible compared to the mass of the atomic nucleus, the ion has the same characteristic atomic mass as the original isotope of the element. By application of additional energy, a second electron can be removed from the ion, resulting in a doubly charged species. This ionization process is shown for the element iron in Figure 2.3.

Each element has characteristic first and second ionization potentials that are dependent on the electronic structure of that specific element. The higher the ionization potential, the more externally applied energy is required to induce ionization. First and second ionization potentials for the elements are tabulated in Appendix 1. Energy for ionization can be applied

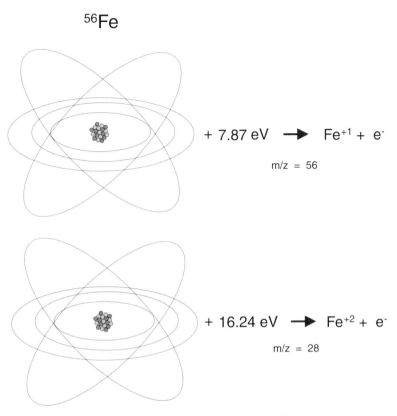

FIGURE 2.3   First and second ionizations of iron-56.

by a thermal radiation process, by collision with other ions or electrons, or by exposure to high-energy photons. By carefully controlling the magnitude of the energy applied, singly charged ions can be efficiently generated with minimal production of doubly charged species.

Molecular species are also formed in the plasma or in the ion beam formed by ions produced in the plasma. These molecular species are predominantly diatomic in structure, although triatomic and larger molecular species are possible. For most triatomic and higher molecular species, the energy required to break them up into simpler molecules is sufficiently small that the probability of them remaining stable in the high-temperature plasma is very low. Therefore, the absence of the majority of these species in the spectrum is expected and observed. A notable exception occurs when the individual components of the polyatomic molecule are present at high concentrations, or their dissociation potential is sufficiently high that the formation of triatomic molecules is observed.

Because the formation of these molecules is an equilibrium process, by the law of mass action, components that are present at high concentration have the higher probability of forming molecular species. Both homogeneous diatomic molecules, such as Ar_2^+, and mixed composition heterogeneous diatomic molecules, such as ArO^+, are formed. A common triatomic molecule that is observed is $ArOH^+$. Molecular ions composed of atoms from the plasma support gas, atoms from solvent that is introduced into the plasma as a carrier for the sample, and atoms from major components of the sample matrix are commonly observed. Polyatomic molecular species usually form single charged ions in the plasma.

CHAPTER 3

Inductively Coupled Plasmas

The definition of a plasma is an electrically neutral gas made up of positive ions and free electrons. Plasmas have sufficiently high energy to atomize, ionize, and excite virtually all elements in the periodic table, which are intentionally introduced into it for the purpose of elemental chemical analysis.

Although there are many types of plasmas (direct current, microwave induced, etc.) the inductively coupled plasma (ICP) has demonstrated the most useful properties as an ion source for analytical mass spectrometry. Gases such as argon, helium, nitrogen, and air have been used to sustain plasmas useful for analytical purposes; however, the inert gases offer some advantages, because of their desirable ionization properties and their availability in relatively pure form. Impurities in the plasma support gas can result in spectral interferences, leading to inaccurate quantitative measurements. Inert gases, specifically argon, also have the advantageous property of minimal chemical reactivity with various analyte species, which can also result in undesirable analytical results.

3.1 PLASMA FORMATION

Inductively coupled plasmas are formed by coupling energy produced by a RF generator to the plasma support gas with an electromagnetic field. The field is produced by applying an RF power (typically 700–1500 W) to an antenna (load coil) constructed from 3-mm-diameter copper tubing wrapped in a two- or three-turn 3-cm-diameter coil, positioned around the quartz torch assembly designed to configure and confine the plasma. An alternating current field is created that oscillates at the frequency of the tuned RF generator. The plasma is initiated by the addition of a few "seed" electrons, generated from the spark of a Tesla coil or a piezoelectric starter, to the flowing support gas in the vicinity of the load coil. After the plasma is initiated it is sustained by a process known as *inductive coupling*. As these seed electrons are accelerated by the electromagnetic RF field, collisions with neutral gas atoms create the ionized medium of the plasma. The mean free path of accelerated electrons in atmospheric pressure argon gas is about 1 μm before a collision occurs with an argon atom. These collisions produce additional electrons. This cascading effect creates and sustains the plasma. Once the gas is ionized, it is self-sustaining as long as RF power is applied to the load coil. The ICP has the appearance of an intensely bright fireball-shaped discharge. Figure 3.1 shows the cross-section of a typical ICP, torch assembly, and load coil. A photograph of a typical ICP is shown in Figure 3.2.

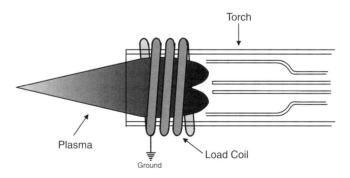

FIGURE 3.1   Plasma with torch assembly and load coil.

FIGURE 3.2   Photograph of an inductively coupled plasma.

3.1.1 Generators

Two basic types of electronic circuits are commonly used to produce the RF energy required to operate an ICP: (1) the fixed-frequency crystal-controlled oscillator and (2) the free-running variable frequency oscillator. Either type of circuit is fully suitable for operating a plasma configured for ion generation. Radio-frequency generators that are used to operate ICPs are basically simple circuits with a limited number of components, which produce an alternating current at a specific frequency. These generators must be capable of operating with up to 2 kW of output power to adequately sustain an atmospheric pressure argon plasma.

Crystal-Controlled Oscillator

The basic frequency of this generator is controlled by a piezoelectric crystal in the feedback circuit of the oscillator (see Figure 3.3). The crystal-controlled oscillators for ICPs typically operate at a frequency of 13.56 MHz; the power supply circuitry includes a frequency doubler to provide a typical plasma operating frequency of 27.12 MHz. Although other frequencies have also been used, this is the most commonly utilized crystal-controlled frequency in plasma systems. At the higher frequency of 27.12 MHz, the coupling between the generator and the plasma is more efficient, which makes the plasma more robust with regard to sample introduction stability.

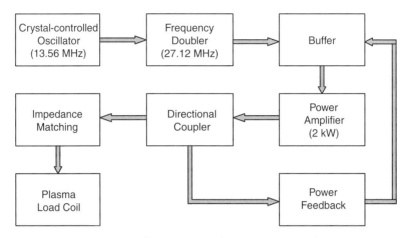

FIGURE 3.3   Block diagram of a crystal-controlled oscillator circuit for a RF power supply.

A thermionic amplifier tube is used to provide the high-power output needed to operate an atmospheric pressure plasma. Crystal-controlled systems usually require a servomotor-operated variable capacitor in an impedance-matching network to maintain tuning of the system and minimize reflected RF power, which extends the functional life of the power tube. A block diagram of this impedance matching circuitry is shown in Figure 3.4.

Free-Running Oscillator

Radio-frequency generators that do not utilize crystal frequency control are defined as *free running*. The frequency of these oscillators is deter-

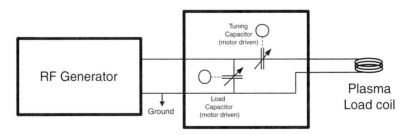

FIGURE 3.4   Block diagram of impedance-matching network with motor-driven tuning capacitors.

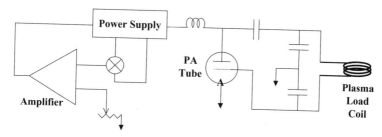

FIGURE 3.5   Circuit diagram of a Colpitts-based free-running RF oscillator.

mined by the combination of values of the components in the circuit. There are several types of free-running generators including the Armstrong, Hartley, Colpitts, and tuned-anode-tuned gate oscillator circuits. A diagram of a typical Colpitts oscillator circuit is shown in Figure 3.5. Since the absolute frequency of the oscillator is dependent on the impedance of the load coil and the plasma coupling, it will vary slightly as the properties of the plasma change, making coupling and tuning of the circuitry much less difficult than with the crystal-controlled systems. As the plasma detunes the coupling circuitry, the frequency changes slightly to compensate.

A desirable feature of the free-running oscillator generator is the ease with which the plasma is initiated, making possible operation with minimal operator adjustment and few moving parts.

Solid-State Generators

In recent years, solid-state semiconductor circuitry has been utilized for the generation of high-power RF energy. Although these generators are significantly less expensive to manufacture, they are usually less efficient at producing higher power levels. The circuitry for these generators is the solid-state analogue for the crystal-controlled oscillator described previously. Some solid-state systems can be operated at either 27.1 or 40.6 MHz at the operator's discretion, with an impedance-matching tuning network functional for either frequency. These types of generators still require a high-power output amplifier tube to produce the energy required to sustain a plasma.

3.1.2 Load Coils

The load coil, which is either an integral part of the RF oscillator circuit (free-running generator) or part of the tuning network in a crystal-controlled oscillator system, usually consists of two to three turns of 3-mm-diameter copper tubing wound in about a 3-cm-diameter spiral. Cooling liquid or gas is circulated through the coil to dissipate thermal energy, which minimizes distortion of the coil from overheating. The coil serves as an antenna to produce an electromagnetic field, which sustains the plasma. It can be thought of as the "primary" winding of a RF transformer, with the "secondary" winding being the plasma itself, thereby transferring energy to maintain the plasma. As seen from Figure 3.1, which is a diagram of the ICP and torch assembly, the plasma is primarily confined within the dimensions of the load coil. Adjacent turns of the coil must not touch each other, but should be wound as close together and be as uniform as possible, to produce a uniform field.

The load coil is usually grounded to earth potential at its front turn (closest to end of torch), rear turn (closest to the sample injector), or center turn. The grounding location influences the formation of a residual secondary discharge at the plasma sampling interface, which can have an significant impact on the formation of molecular oxide and doubly charged ion species in the ion beam of the mass spectrometer. One manufacturer employs the concept of two interlaced coils, one of which is grounded at the front and the other at the rear. They claim that the advantage of this configuration is improved stability and tuning characteristics.

3.1.3 Torches

The torch is a device that is used to contain and assist in configuring the plasma. Torches are typically made of materials that are transparent to the RF radiation. Therefore, they do not attenuate the field generated by the load coil/antenna. Torches can be made of materials such as ceramics or boron nitride; however, by far, most are made from quartz, which has a sufficiently high melting point to allow it to maintain its configuration when operated at temperatures commonly experienced with argon ICPs.

A simple quartz tube centered in the load coil with laminar flow argon gas passing through it at 10–20 L/min will form a simple plasma, when RF power is applied and seed electrons are provided. The plasma or "fireball" that forms will have the shape of a prolate spheroid. If an excessive amount of power is applied, the plasma will reach a sufficiently high temperature to melt the quartz tube, which confines it. This configuration is not fully suitable for analytical spectrometric purposes. It is difficult to efficiently inject a sample aerosol into this type of plasma because of the "skin effect" created by the potential barrier at the surface of the plasma. This barrier tends to deflect aerosol particles around the outside of the spheroid, rather than entrain them in the plasma. Unless appreciable sample can reach the optimal excitation region of the plasma, analytical sensitivity will be significantly limited.

A typical quartz torch suitable for the formation of an atmospheric pressure argon plasma for analytical spectrometric use is shown in Figure 3.6. This torch consists of three concentric quartz tubes. A coolant gas (argon) is introduced into the space between the outer and center tubes at a tangential direction relative to the longitudinal axis of the torch, creating a vortical flow. This gas stream serves two purposes: (1) It isolates the plasma from the internal wall of the outer quartz tube preventing melting, and (2) it encourages the formation of a toroidal (annular)-shaped plasma (see Figure 3.7). The center tube is for the injection of sample aerosol into the plasma. The space between the injector and the intermediate tube is used for the introduction of an auxiliary flow of argon gas to assist in the formation of the plasma and to ensure that the plasma is forced away from the tip of the injector, preventing it from melting. This auxiliary gas flow is usually very low compared to the coolant gas flow and often is not used. Typical argon gas flow rates for the various inputs are listed in Table 3.1. These flow rates can vary depending on the specific dimensions and configuration of the torch being used.

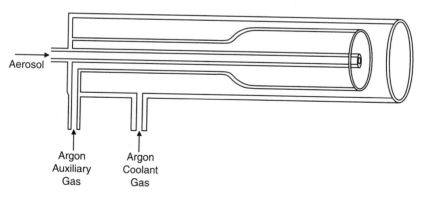

Aerosol

Argon
Auxiliary
Gas

Argon
Coolant
Gas

FIGURE 3.6   Typical quartz torch used for ICP-MS.

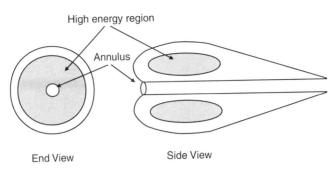

FIGURE 3.7   End and side view of inductively coupled plasma, showing annulus region where aerosol is transported.

The toroidal or annular-shaped plasma offers some highly desirable features for analytical spectrometry. The aerosol injector tube is positioned to direct a stream of sample particles at sufficient velocity to allow their penetration into the bottom of the plasma. The center of the toroid is in effect a tunnel, which facilitates the transport of the aerosol particles into the plasma, greatly reducing the deflecting "skin effect" properties of the spherical-based plasma. This ease of sample introduction greatly enhances the plasma's analytical sensitivity.

The magnitude of each of the gas flows significantly affects the configuration and consequently the temperature and ionization efficiency of the plasma. Although the absolute magnitude of the various flow rates is important for reproducing the operating characteristics of the plasma, precise control of these parameters is critical for maintaining its stability for analytical purposes. The control of these flow rates is maintained by the use of sensitive instruments such as mass flow controllers. A mass flow controller uses an electronic feedback circuit from a flow sensor in the gas stream to regulate gas flow. It precisely regulates the flow rates from the gas supply

TABLE 3.1   Typical Operating
Conditions for an Argon ICP

Power	1–2 kW
Argon flow	
Coolant	15 L/min
Auxiliary	0–2 L/min
Nebulizer	1 L/min
Sampling depth[a]	14–18 mm

[a] Beyond last turn of load coil.

regardless of factors such as gas supply quantity and back pressure. Although it is desirable for mass flow controllers to be used on all gas streams for optimum analytical precision, it is absolutely critical to use this regulation on the injector gas flow rate because this has the most critical impact on analytical performance. For cost consideration reasons, many commercial instruments employ mass flow control on this gas stream only.

3.2 PLASMA CONFIGURATION

The electromagnetic RF field primarily couples with the outer regions of the plasma (i.e., the part that is closest to the outer walls of the torch) thus enhancing the formation of the annular-shaped "doughnut." This region also exhibits the highest temperature, which can be as high as 10,000 K (equivalent temperature based on ionization capacity). In contrast, the central region of the plasma (the center of the annulus, often called the *axial channel*) where the sample is introduced is at a much lower temperature, usually between 5000 and 7000 K. This axial channel contains, in addition to argon gas, the analyte species, components of the sample matrix, and any solvent from the sample introduction apparatus.

Figure 3.8 shows traces that represent lateral profiles of Ar^+ and Mn^+ ion currents sampled across the width of the plasma while Mn is being introduced through the injector tube. The Ar^+ ion current trace shows maxima at two regions equidistant from the central axis. This represents ions being produced by the outside and highest temperature regions of the annular

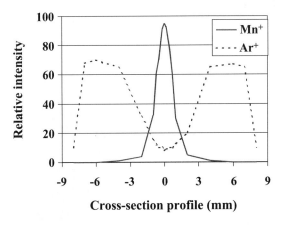

FIGURE 3.8   Cross-sectional distribution of the Mn^+ and Ar^+ ions.

plasma. The minimum in the Ar^+ ion current trace shows the lower population of Ar^+ ions in the central channel. Conversely, when a manganese solution is nebulized into the plasma, the profile for the Mn^+ ion current is observed to have a maximum in the central channel resulting from the direct injection of the analyte aerosol into this region. Absence of Mn^+ in the outer regions of the plasma demonstrates the inefficiency of analyte entering this region of the plasma.

In addition to the high-temperature region of the annular plasma, there are also other operationally defined zones: (1) the initial radiation zone (IRZ), in the axial channel aligned with the RF load coil; (2) the normal analytical zone (NAZ), which is located immediately above the IRZ (usually 10–20 mm); and (3) the preheating zone (PHZ), which is located in the axial channel before reaching the plasma. These zones are shown in Figure 3.9. Solvent vaporization and dry aerosol decomposition to individual molecules or species usually occurs in the PHZ prior to the sample entering the plasma. Atomization or decomposition of crystalline materials and dissociation of molecules occurs in the IRZ. Finally, ionization of atomic species, produced in the IRZ, occurs in the NAZ.

To easily visualize these zones, the nebulization of a solution of yttrium (about 1 mg/L concentration) can be used as an indicator. The bright red color of the central cone defines the IRZ. The red color originates from the atomic emission of Y atoms excited by the plasma. Above the IRZ is a bright blue "bullet," which defines the NAZ. This color is due to excited Y ions, which emit blue visible radiation. If the plasma and sample introduction system are performing satisfactorily, these colored zones should be well defined and reproducible. The use of this

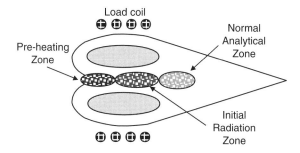

FIGURE 3.9   Various energy zones of the ICP.

approach as an indicator can be effectively used for preliminary alignment of the plasma during setup.

In general, the ICP is not in true thermodynamic equilibrium because the various processes that are occurring in the plasma are individually not in equilibrium. Therefore, the ICP cannot be characterized by a single equilibrium temperature. The Saha equation can be used to make a reasonable estimate of the degree of ionization:

$$\frac{n_i n_e}{n_a} = \frac{2 Z_i(T)}{Z_a} \frac{2\pi m k T}{h^2} e^{-E_i/kT}, \tag{3.1}$$

where n_i, n_a, and n_e are the number densities of ions, atoms, and free electrons, respectively; m is the electron mass; Z_a and Z_i are partition functions; E_i is the ionization potential; and T is the temperature. Using this equation, Houk (1986) calculated the degree of ionization of elements for a plasma temperature of 7500 K and an electron density of 10^{15} cm^{-3}. The results of these calculations for selected elements are shown in Table 3.2. The degree of ionization

TABLE 3.2   Degree of Ionization for Selected Elements (Temperature 7500 K, n_e 1 × 10^{15} cm^{-3})

Element	Degree of ionization (%)
Ar	0.04
As	52
Br	5
Cd	65
Co	93
Cr	98
Cu	90
Fe	96
La	90
Mn	95
Na	100
P	33
Pb	97
Rb	100
Se	33
Sr	96
Tl	100
V	99
Zn	75

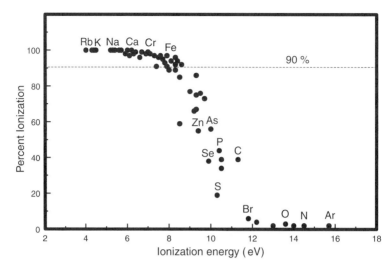

FIGURE 3.10   Degree of ionization as a function of first ionization potential, demonstrating that the majority of elements are greater than 90% ionized.

under these conditions, plotted as a function of ionization energy for all elements, is shown in Figure 3.10. As seen from Appendix 1, most elements have first ionization potentials below 10 eV, which corresponds to >50% ionization, and none of the elements have second ionization potentials below 10 eV.

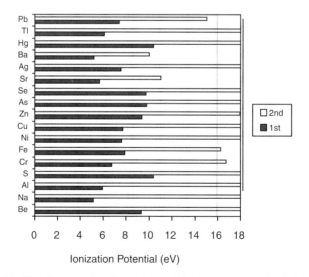

FIGURE 3.11   Bar chart showing first and second ionization potentials of selected elements.

Because the average ionization energy of the atmospheric pressure argon ICP is dominated by the first ionization potential of argon (15.76 eV) and most elements have a first ionization potential below 16 eV, the plasma will efficiently produce singly charged ions for essentially all elements. In addition, few doubly charged ions will be produced because most elements have second ionization potentials greater than that of argon. Notable exceptions are barium and strontium. The bar chart in Figure 3.11 shows the ionization potentials for selected elements, with the dotted line representing the ionization potential for argon (the general ionization capacity of the argon plasma). Those elements with second ionization potential bars that extend beyond the line will have a low probability of forming doubly charged ions. Therefore, for the elements on this chart it is expected that Ba and Sr will have an appreciable probability of forming doubly charged ions, and Pb has only a slight probability of forming doubly charged species.

CHAPTER 4

Instrumentation

All ICP-MS instruments consist of several components including the ICP, a sample introduction system, a mass spectrometer with ion detector, and a data acquisition/readout system. The block diagram in Figure 4.1 shows the relationship and interconnection of these components. Each part of the instrument plays a critical role in the proper overall function of the

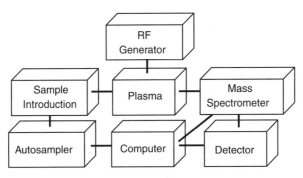

FIGURE 4.1   Block diagram of ICP-MS instrument.

system. This diagram also shows the configuration of the support system for the ICP, including the RF generator and the gas supply and regulation components.

A critical part of the instrument is the interface, which allows ions generated by the ICP to be uniformly transferred to the mass spectrometer for isolation and measurement. This interface is the most innovative aspect of combining the well-known ICP technology, used with atomic emission systems, to commonly used mass spectrometers for ion analysis.

Although commercial instrumentation utilizes several configurations and variations of components, all have basically the same arrangement shown in Figure 4.1.

4.1 INTERFACE

The function of the interface is to representatively sample ions produced in the ICP, export them from the high-temperature atmospheric pressure argon plasma, and facilitate their transport into the mass spectrometer, where they are isolated and their concentrations in the ion beam are measured. The means to perform this action requires the use of two or more concentric water-cooled cones fabricated of metal (commonly Ni or Pt).

A drawing of the interface is shown in Figure 4.2. The outside cone, often called the *sampler* or *extraction* cone, is positioned in the plasma such that the orifice located at its apex is immersed in the NAZ. The diameter of the orifice is approximately 1 mm. Ions produced in the plasma pass through this orifice and form an ion beam. An additional cone called a *skimmer* is positioned immediately behind the sampling cone a few millimeters. The skimmer cone has a much smaller orifice at its apex (<0.5 mm in diameter). This orifice samples the supersonic gas jet expanding through the sampler cone orifice, directing ions into the mass spectrometer.

A dual vacuum pumping system is used to reduce the pressure from the plasma (atmospheric) to the required working pressure of the mass spectrometer. In the first stage, (the space between the sampler and skimmer cones) is evacuated to a pressure of about 1 Torr with a mechanical vacuum pump. The second stage of the interface, the region behind the skimmer cone, is reduced to a pressure of about 10^{-5} Torr, the normal operating pressure of the mass spectrometer, with an oil diffusion or turbomolecular pump. Alternate designs include the generation of the first

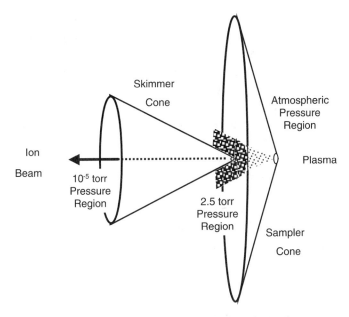

FIGURE 4.2   ICP interface cones, showing the ion beam.

stage at considerably lower pressure than 1 Torr, which is claimed to produce significantly higher sensitivity for the determination of many elements. In addition, some interface designs utilize three stages of pressure
reduction, which is also claimed to improve analytical performance characteristics.

By carefully engineering the size, configuration, and position of these
cones and their orifices in the interface, an ion beam can be produced with
a composition representative of the ion population in the NAZ of the plasma. Therefore, when the mass spectrometer measures the ion abundance in
the transported ion beam, a determination of the analyte components in
the original sample is possible.

4.2 ION LENSES

As the ion beam, passing through the interface shown in Figure 4.2, enters
the mass spectrometer, a negatively charged extraction electrode is used to
attract positive ions and transport them into the electrostatic lens assembly.
This assembly further focuses the ion beam and prepares it for ion analysis
by the mass spectrometer.

Ion lenses are used to assist in the transport of positively charged ions as they leave the interface region. These lenses are usually configured as one or more cylindrical electrodes, through which the ion beam passes. By independently varying the potentials on each of these lenses, the ion beam can be collimated and focused. A schematic diagram of a typical ion lens configuration is shown in Figure 4.3. The first component of an ion lens set is often a metal disk called a *photon stop*, which is mounted in direct alignment with the ion beam behind the skimmer cone of the interface. The purpose of the photon stop is to intercept photons and energetic neutral species produced by the ICP. This prevents them from entering the mass analyzer. It is highly undesirable for photons to enter the mass analyzer because they can scatter and reflect from the analyzer's internal components. If these reflected photons reach the electron multiplier detector, they can significantly increase the background signal produced by the spectrometer. The use of a *photon stop* is important for achieving the sufficiently low background levels required to attain acceptable detection limits for analyte species. Some models of instrumentation use an off-axis deflected beam configuration to accomplish similar results.

The positive analyte ions in the ion beam are directed by positively charged lenses to defect around the photon stop, recombining on its opposite side. In Figure 4.3 the ion beam is shown being diverted around the photon stop and recolimated on the opposite side.

Because all ions have approximately the same velocity (established by the entrained Ar gas flowing through the interface in which they are entrained), their kinetic energy (KE) is dependent on their mass and velocity:

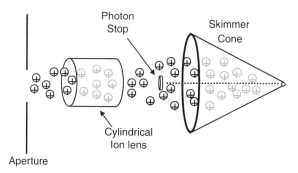

FIGURE 4.3   Typical ion lens configuration in ICP-MS, showing photon stop and cylindrical lens.

$$KE = \frac{mv^2}{2}, \qquad (4.1)$$

where m is the ion mass and v is the ion velocity. Thus the heavier ions will have higher energy than the light ions. This variation in energy, as a function of mass, results in different optimal ion lens voltage settings for ions of each different element.

Maximum measurement sensitivity is obtained when all operating conditions that impact the transport of ions through the mass spectrometer are optimized. Therefore, there is a set of optimal voltages for each ion lens that results in maximum efficiency of transport and hence maximum sensitivity. This variation in optimal lens settings, as a function of mass, means that there is no unique set of conditions that will yield optimal sensitivity for all elements when working in the multielement analysis mode.

A compromise set of conditions is usually selected for the analyte elements being determined. The more variation in the mass of the analytes, the greater the compromise of performance in multielement analysis.

In addition, the process of ion beam divergence when the ion density is relatively high is called the *space charge effect*. The space charge effect impacts all ions in the ion beam. The light ions are deflected more than the heavy ions; which adds an additional factor in optimizing the ion lens setting for analyte elements of different masses. The space charge effect is highly dependent on the magnitude of the ion population of the ion masses of which it is comprised. The ultimate result is that even small changes in the total ion current caused by the addition of sample matrix elements can change the fraction of analyte ions that passes through the lenses.

The previously described space charge effect phenomenon offers an explanation for many of the observed sample matrix interference effects in ICP-MS. The masses of both the analyte element and the matrix are important. The transport of analyte species is suppressed more by heavy matrix ions than light ones, and heavy analyte ions are suppressed less severely than light analyte ions.

Recent commercial instrumentation utilizes computer-controlled real-time adjustment of ion lens voltage settings as different mass analytes are measured. This approach allows close to optimum ion lens settings for each analyte element to be used when operating in the multielement analysis mode, greatly improving the analytical performance (sensitivity, detection limits, and interferences) of the technique.

4.3 MASS SPECTROMETERS

Ions produced by the ICP are measured by the use of a mass spectrometer. The mass spectrometer is essentially a mass filter designed to isolate a specific mass-to-charge ratio (m/z) ion from the multi-ion beam. After separation, the individual ion beams, which are characteristic of specific charged isotopic or molecular species, are sequentially or simultaneously (multicollector spectrometer) directed to a detector devised to measure their individual ion currents. The magnitude of these ion currents is proportional to the population of the analyte ion species in the multicomponent ion beam sampled from the ICP. Therefore, the measurement of the m/z of the ion allows qualitative identification of the isotope or molecule being measured, and the magnitude of the ion current is used to provide quantitation of the amount of the analyte in the original sample.

Several different types of mass spectrometers, each with specific advantageous characteristics, are used for ion isolation and detection. A spectrometer employing a quadrupole mass filter offers simplicity of operation with excellent stability, ease of operation, and relatively low cost of manufacture; however, it is only practical to operate this spectrometer at slightly less than unit mass resolution. Other types of spectrometers, employing magnetic fields to disperse the ion beams, operate at much higher resolutions (up to a resolving power of 1 part in 10,000). These spectrometers require highly stable electromagnets, which are sometimes difficult to scan rapidly, and whose cost of manufacture is much higher than that of the quadrupole spectrometer. Finally, spectrometers that utilize other mass separation principles including time-of-flight and ion-trap principles are discussed in detail later in this section.

4.3.1 Quadrupole Mass Analyzer

A quadrupole mass spectrometer (mass filter) consists of four precisely machined cylindrical rods arranged parallel to each other in a symmetrical configuration (see Figure 4.4). These rods are constructed of highly polished metal or metal-plated (gold) ceramic. The center space between the rods is aligned concentric with the ion beam passing through and configured by the electrostatic ion lenses. The alignment of these rods must be precisely fixed to a symmetrical position with a tolerance of 10 μm or less. As a mixture of ions with different m/z values passes through this center space and travels parallel to the length of the rods, only a single m/z ion

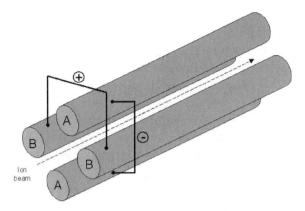

FIGURE 4.4   Quadrupole mass filter.

species is permitted to traverse unimpeded and exit at the opposite end. All other m/z ion species are rejected by the quadrupole. This process is analogous to "filtering" all species except the one m/z species of interest to which the quadrupole spectrometer has been tuned and thus is allowed to pass.

The operation of this quadrupole spectrometer involves the application of both a direct current (dc) potential (E) and a radio-frequency (RF) alternating current potential $(V \cos(\omega t))$ to pairs of the rods. A combined electrical potential of $(E + V \cos(\omega t))$ is applied to two oppositely positioned rods, while simultaneously an applied combined potential of $- (E + V \cos(\omega t))$ is applied to the other two opposing rods such that they oscillate $180°$ out of phase. Figure 4.5 shows an end-view cross-section of these rods with an indication of the applied potentials. As the electrical potentials on

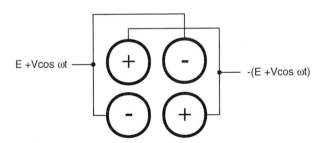

FIGURE 4.5   End-on view of quadrupole rods with electrical connections.

the rods are varied, an electromagnetic field is created that interacts with the beam of mixed ion species. Each ion will be deflected in a spiral path, the magnitude of which is related to the fields created by the applied potentials. All ions, except those with a specific unique m/z will be deflected in such a way as to cause them to travel in a wide spiral and collide with the quadrupole rods. Those ions with the unique m/z value will have a stable path and will be transported through the central axis of the rods, exiting at the opposite end for eventual interaction with the ion detector positioned behind the quadrupole rods. Selection of the ions with a specific m/z that will traverse through the quadruple is determined by the reproducible application of the electrical potentials to the rods.

The measurement of a specific m/z ion current is called single-ion monitoring and can be used for the determination of a single analyte species. Integrating this ion current for a fixed period of time provides a total accumulated charge, which can be related to the concentration of the corresponding analyte in the original sample.

Scanning

By changing the field created by the quadrupole rods, ions with different m/z values can be selectively transported through the spectrometer. This field can be changed in a uniform fashion by altering either the applied dc potential or the frequency of the RF voltage. Thus by continuously varying the applied potentials on the rods as a function of time, the m/z values of the ions passing through the spectrometer are changed in a uniform fashion. This results in a scan of the mass spectrum of the ion beam sampled from the plasma. This mass spectrum is depicted as a plot of the magnitude of the various ion currents as a function of the m/z values of the ions being scanned. A typical scanned mass spectrum of lead and thallium isotopes is shown in Figure 4.6.

Regulated quadrupole rod power supply voltages can be altered under computer control. The rate at which these voltages are changed establishes the scan rate of the spectrometer. Scan rates of up to 3000 amu/s are achieved with most commercially available instrumentation. This means that the entire mass spectrum from an m/z range of 0–300 amu can be accomplished in about 0.1 s, providing a suitable rate for the measurement of transient signals (i.e., those obtained from flow injection, laser ablation, electrothermal vaporization, or chromatographic sample introduction systems). This also provides a mechanism for the rapid repetitive scanning of

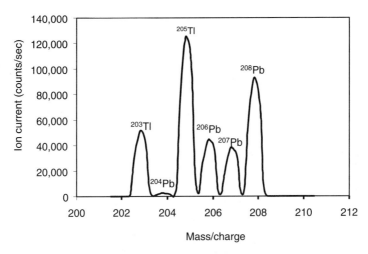

FIGURE 4.6 Scanned mass spectrum of lead and thallium isotopes.

the mass range of interest, which results in an improvement in precision by averaging fluctuations (originating from the ICP) in the measured signal.

Scanning provides a suitable way for rapid multielement qualitative or semiquantitative analysis. Parameters can be established to scan the entire mass spectrum, usually from m/z values of 6 (Li) to 238 (U). At times it is preferable to scan only smaller regions of the mass spectrum. It is possible to specify limited mass ranges that incorporate the m/z values of small groups of elements of interest. These ranges can be established by selecting a series of mass ranges within which the elements of interest would be included. Regions of the mass spectrum that did not include m/z values of elements of interest could be skipped over to save time. For instance, the rod voltages could be set, under computer control, to scan an m/z range of 6 to 11 to measure the major isotopes of Li, Be, and B; from 63 to 67 to measure the isotopes of Cu and Zn; and from 203 to 208 to measure the isotopes of Tl and Pb. Using this strategy, time would not be wasted measuring other mass ranges where no useful ion current information for the elements of interest would be found. In addition, regions of the spectrum containing m/z values for major components in the ion beam, such as $^{40}Ar^{+}$, could be avoided to eliminate the possibility of damage to the detector from ion saturation. The regions from 13 to 22 m/z (containing peaks originating from N, O, and H_2O) and from 32 to 42 m/z (containing peaks originating from O_2 and Ar) must always be avoided to protect the detector.

Peak Hopping

Higher precision and sensitivity can usually be achieved by the *peak hopping* mode of operation. This mode consists of carefully calibrating the quadrupole spectrometer so that, under computer control, voltages are applied so that specific m/z locations in the mass spectrum could be measured in a mode analogous to multiple single-ion monitoring. Ion currents are integrated for specific time periods at the m/z for the isotope of each element of interest. When the integration at each specific location is completed, the quadrupole spectrometer voltages are changed to allow monitoring of the ion current for the m/z of the next analyte isotope in the multielement sequence. The analyte elements of interest and the specific m/z values of the isotopes that are desired to be used for quantitation must be selected in advance. Different integration times for each isotope m/z can be used to accommodate differences in isotopic abundances (see Appendix 2) of the analyte elements, permitting similar sensitivity and precision to be achieved for all analyte elements being measured. Replicate passes are often made through the series of m/z values being used to improve the precision of measurements (i.e., minimizing the effects of fluctuations in the ICP).

The peak hopping approach is generally considered the most desirable for quantitative analysis. It also provides the most time-efficient approach, because time is not wasted making quantitative measurements in regions of the mass spectrum that are not of interest. Peak hopping is also the method of choice when isotope ratio measurements (multiple isotope m/z values are measured for the same element) are being made.

Resolution

Resolution of a mass spectrometer is defined as a measure of the ability to separate adjacent mass regions in the mass spectrum. If one considers two adjacent peaks in the mass spectrum, the separation of these peaks is considered resolved if the magnitude of the valley between the peaks (X_{valley}) is less than 10% of the mean magnitude of the height of the peaks ($\bar{X}$) (see Figure 4.7):

$$\frac{X_1 + X_2}{2} = \bar{X} > 0.1 X_{valley}. \tag{4.2}$$

The distance between the two peaks (ΔM) divided into their mean mass ($\bar{M}$) yields a term R, which is defined as the resolving power at 10% of the valley definition:

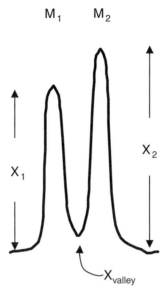

FIGURE 4.7   Resolved mass spectral peaks.

$$R = \frac{\overline{M}}{M_1 - M_2} = \frac{\overline{M}}{\Delta M}.$$ (4.3)

Most commercial quadrupole mass spectrometers operate with a typical resolving power capable of unit mass resolution ($R = 300$). This resolution is capable of resolving most analytical species used in the ICP. However, to resolve isobaric interferences (i.e., species occurring at the same unit mass), a resolution exceeding $R = 3000$ is necessary. Therefore, although the quadrupole is an exceptional spectrometer for accomplishing 90% of common analytical requirements, there are some applications for which simple direct ion current measurements are not sufficient to achieve accurate analytical data. For these applications, either another type of mass spectrometer is required, mathematical corrections must be applied to the data, or in the case of interferences, a chemical separation prior to measurement must be made.

Abundance Sensitivity

The ability to measure a small m/z peak (low concentration analyte) directly adjacent to a large m/z peak (high concentration analyte) is an important consideration for multielement analysis. The abundance sensitivity is the ratio of the net signal at the centroid of a peak to the signal at an m/z at either $m + 1$ or $m - 1$. This ratio defines the ability to measure low concentration analytes in the presence of high concentration analytes or matrix components. The numerically larger this ratio, the better the performance of the quadrupole mass spectrometer. Most commercial instruments have an abundance sensitivity on the low mass side of the peak of up to about 10^6 and about 10^8 on the high mass side. The abundance sensitivity in conjunction with resolution establishes the operating performance of the quadrupole mass spectrometer for quantitative analysis purposes.

4.3.2 Magnetic Sector Mass Analyzer

The classical type of mass spectrometer first demonstrated by Aston (1920), employs the principle of a charged particle passing through a magnetic field, being deflected at an angle proportional to the mass and charge of the particle. The radius of curvature of the deflection is dependent on both the m/z of the ion and the strength of the applied magnetic field. Since then, spectrometers based on this principle of operation have been effectively used in spark source mass spectrometry, for isotope ratio measurements in thermal ionization mass spectrometry, and for accurate mass measurement in organic molecular mass spectrometry.

The primary advantage of magnetic-sector mass spectrometry is its ability to operate at high resolution. A resolving power in excess of 10,000 can be achieved routinely, greatly enhancing the ability to perform interference-free determinations. In addition, because of high measurement sensitivity and low background signals, detection limits in the femtogram per milliliter range (in low-resolution mode) can be made.

Single Focusing

A single-focusing magnetic-sector mass spectrometer consists of a magnetic field establishing the separation of ions according to their m/z. Figure 4.8 shows the arrangement of an ion beam accelerated by an electrical potential (V), collimated by a slit, passing through a curved flight tube positioned between the poles of a high field electromagnet. The flight tube is positioned so that the path of the ion beam is perpendicular to the magnetic field lines of force.

Equation (4.4), which describes the curvature of individual m/z ion beams:

$$r = \frac{(2Vm/z)^{1/2}}{H}, \tag{4.4}$$

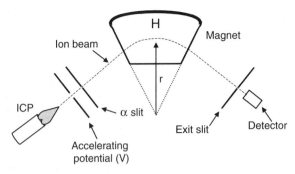

FIGURE 4.8   Magnetic-sector mass spectrometer.

where r is the radius of curvature of the ion path, V is the accelerating voltage, m is the mass of the ion, z is the charge of the ion, and H is the strength of the magnetic field. As seen from the equation, the radius of the ion path is proportional to the square root of the m/z at a constant accelerating voltage and at a constant magnetic field.

Using an ion-sensitive photographic detector, multiple ions can be simultaneously detected. The linear position of lines (images of the entrance slit) on the photoplate is indicative of specific m/z ions separated by the magnetic field. This approach was used for many years because it provided both multielement detection and the high sensitivity achieved by time-integrated measurements. However, this is a cumbersome approach because it requires carefully controlled photographic development techniques and tedious interpretation processing.

A second mode of operation involves the use of an exit slit with an ion collector positioned immediately behind it to measure the intensity of the isolated ion beam. By continuously varying either the accelerating voltage, V, or the magnitude of the magnetic field, H, as a function of time, the separate m/z ion beams can be scanned past the exit slit. When the response of the detector is plotted as a function of time, a spectrum is obtained in which m/z is proportional to H^2. The width of the slit controls the spectral resolution of the measurement.

Double Focusing

By combining a magnetic-sector mass analyzer with an electrostatic analyzer (often termed a *double-sector* or *double-focusing mass spectrometer*), a significant improvement in resolution is realized. Such a double-focusing instrument can achieve a resolution of the order of 100,000. The double-focusing magnetic mass analyzer utilizes two independent sectors. In addition to the magnetic sector, which is used to separate ions according to their m/z, an electrostatic analyzer is used to filter the ions according to their kinetic energy, irrespective of their m/z. The use of two sectors greatly enhances the resolution that can be achieved.

When ions are passed through an electrostatic analyzer (see Figure 4.9), a curved flight tube through an electrostatic field, they travel in a circular path of radius r such that the electrostatic force balances the centrifugal force:

$$\frac{mv^2}{r} = Ez, \tag{4.5}$$

where E is the electrostatic field strength. The radius of curvature is dependent on the ion's energy and not its mass. A broad energy distribution is reduced to a very narrow energy distribution, permitting higher resolutions to be attained.

The electrostatic analyzer can be either positioned before (Nier-Johnson geometry) or after (reverse Nier-Johnson geometry) the magnetic sector. Usually the latter geometry is used in ICP-MS applications because of the better abundance sensitivity that can be achieved in this configuration. Figure 4.10 compares the single-focusing and double-focusing spectrometers (electrostatic analyzer between the magnetic sector and the detector) with an illustration of the comparative resolution of the mass spectra obtained from each.

A spectrum obtained at unit mass resolution (by a quadrupole mass spectrometer from m/z 55 to 57) is compared to one obtained at high resolution (magnetic-sector mass spectrometer at m/z 56) in Figure 4.11. Two species are commonly observed at m/z 56, the major isotope of iron ($^{56}Fe^+$)

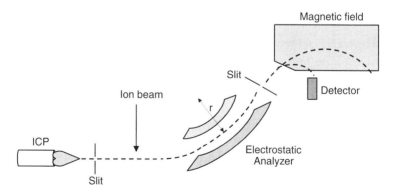

FIGURE 4.9   Nier-Johnson geometry double-focusing magnetic-sector instrument.

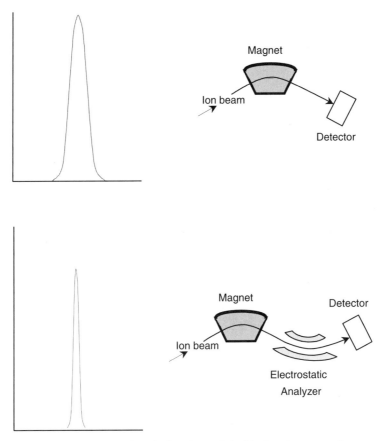

FIGURE 4.10   Comparison of single-focusing and double-focusing magnetic-sector mass spectrometers with an illustration of comparative resolutions.

and the $^{40}Ar^{16}O^+$ diatomic molecule. As seen from Figure 4.11, these two species are not resolved in the unit mass resolution spectrum and, therefore, it is not possible to ascertain the relative contributions from either of these species making a direct determination of iron impossible. The two species are easily resolved by the high-resolution magnetic sector instrument operating at a resolving power of $R = 2550$. While the overall ion transmission is significantly reduced at high resolution, lowering measurement sensitivity, the $^{56}Fe^+$ species can be unambiguously measured.

A double-focusing magnetic-sector mass spectrometer can be scanned either by varying the magnetic field as a function of time or by holding the magnetic field constant, varying the accelerating voltage and the voltage

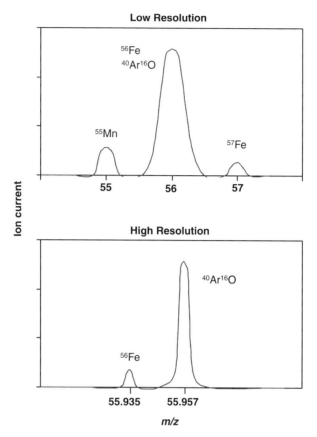

FIGURE 4.11   Comparison of unit mass resolution mass spectrum with a high-resolution mass spectrum.

applied to the electrostatic analyzer. The latter is usually the preferred method because the voltage on the electrostatic analyzer can be scanned much more reproducibly than the magnetic field. A technique called *synchronous scanning*, which is a combination of the two approaches, is often used in ICP–MS to yield optimal performance.

4.3.3 Time-of-Flight Mass Spectrometer

A time-of-flight mass analyzer (TOF–MS) does not depend on a magnetic, electrostatic, or RF field to disperse or filter ions for individual m/z ion detection. A pulsed polyatomic ion beam is accelerated to a constant kinetic energy (zV), where z is the charge on the ion and V is the applied accelerating potential. The ions enter a *drift tube* and acquire a characteristic velocity dependent on their m/z.

Because each m/z ion has the same energy, from the well-known kinetic energy relationship, Eq. (4.1), we observe that ions of different mass travel down the drift tube at different velocities. Therefore, light ions, which are traveling faster, reach the detector at the end of the drift tube in a shorter time than the heavier ions. This time discrimination process can be used to resolve ions of different mass.

To increase the drift path length and thereby improve resolution, an ion reflector is often used to reverse the direction of travel of the ions (see Figure 4.12). In addition, this *reflectron* is used to reduce the kinetic energy distribution of the ions, also resulting in an improvement in resolution.

The pulsing of the ions, which is required for TOF operation, limits the signal integration process that is normally used in continuous ion-beam signal processing obtained with other types of mass spectrometry. A continuous ion beam is generated by the ICP ion source. Pulsing this beam is achieved by the use of an orthogonal interface with a pulsed repeller plate. Background noise is a significant problem with this beam pulsing process. Some manufacturers have introduced instrumentation employing a quadrupole ion lens between the ICP and the TOF-MS with significant success. In some cases this quadrupole is also used to gate the ion beam, producing a pulsed signal.

One of the major advantages of the TOF-MS is the ability to monitor all m/z ions in the mass spectrum simultaneously. It has been reported that more than 30,000 simultaneous mass spectra per second can be obtained by this technique. This is a significant advantage when performing isotope

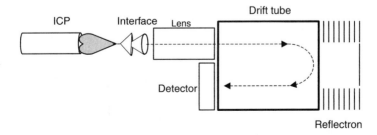

FIGURE 4.12 Time-of-flight mass spectrometer.

ratio measurements and when using transient signal sample introduction systems such as electrothermal vaporization.

4.3.4 Ion–Trap Mass Spectrometer

The operating characteristics of the ion-trap mass spectrometer (IT-MS) are similar to those of the quadrupole system. Instead of symmetrically positioned cylindrical rods characteristic of the quadrupole, the ion trap employs a doughnut-shaped ring electrode and two end-cap electrodes (see Figure 4.13). A RF signal is applied to the ring electrode, which oscillates at a frequency sufficient to cause ions that are introduced into the trap to be stabilized and retained, with the exception of an ion with a specific m/z, which is allowed to leave the trap. This particular m/z ion is directed to the detector for ion current measurement. This method of operation is called the *mass selective instability mode*. Modifications of this approach can be used by applying specific synchronized RF voltages to the end caps, resulting in better performance of the ion trap and in the ability to concentrate the specific ions.

The direct use of an ion-trap mass analyzer for ICP applications has problems associated with collision and scattering of ions due to the abundance of argon species in the ion beam originating from the ICP. Research studies are continuing to develop the coupling of ion traps with other mass analyzers to improve the overall operating characteristics using tandem configurations.

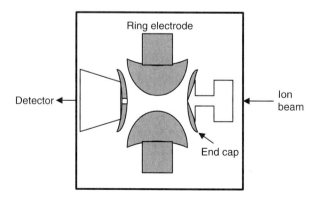

FIGURE 4.13   Cross-section of an ion-trap mass spectrometer.

4.4 DETECTORS

After element-specific ion beams are isolated by an appropriate mass analyzer, a device is needed to measure the magnitude of the ion current produced. Since the magnitude of the ion current is proportional to the concentration of the analyte ions in the original polyatomic ion beam, which is, in turn, proportional to the concentration of the analyte in the original sample prior to atomization and ionization, a means of precisely measuring this ion current is necessary to achieve quantitative analysis. In addition, since ion beams are typically of low magnitude, especially when performing trace analysis, a highly sensitive detection device is required. The most universal approach is the use of a multiplier-type detector, similar in concept to optical photomultiplier tubes.

4.4.1 Continuous Dynode Electron Multiplier

The continuous dynode electron multiplier is a device capable of converting ions into a measurable electrical current. This conversion is accomplished by the ion beam impacting a low-work function metal oxide (PbO) coated surface of the detector. For each ion collision with the surface of the detector, one or more secondary electrons are ejected, which are then multiplied by successive collisions as the ejected electrons are accelerated down the curved tube of the device, resulting in collisions with the opposite wall of the detector. Each accelerated electron ejects multiple secondary electrons when it collides with the surface of the detector, resulting in a multiplication process.

A schematic diagram of the detector is shown in Figure 4.14. It consists of a curved funnel-shaped glass tube with a flared end to accept the ion beam. A large negative potential (to attract positive ions) applied to the front end of the electrically resistive leaded glass construction of the detector provides a continuously increasing positive potential (to accelerate negative charged electrons) down the length of the detector tube. As the electrons cascade down the funnel-shaped tube, they eventually reach the collector where the resulting electrical signal is measured as a direct current.

The most common mode of operation of the continuous dynode multiplier is the pulse-counting mode. In the pulse-counting mode, the high-voltage bias (approximately $-3000\,V$ dc) is set so that a single ion causes a significantly high amplification that the detector rapidly goes into saturation.

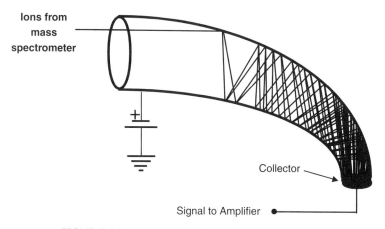

FIGURE 4.14   Continuous dynode electron multiplier.

This results in a large electron pulse (about 100 mV), which is a gain of about 10^8.

When operating in the pulse-counting mode, if the rate of ions impinging on the detector is very high, the period of time between individual ions reaching the detector and the pulse width at the output collector will be similar. Ions arriving during the output pulse generated by the preceding ion will not be measured by the electronic processing circuitry. This process is called *pulse pileup* and requires a detector "dead time" correction to maintain linearity of the pulse counting system. Equation (4.6) is a mathematical expression to correct for this dead time effect:

$$n_{corr} = \frac{n_{meas}}{1 - n_{meas}\tau}, \tag{4.6}$$

where n_{corr} is the dead time corrected count rate, n_{meas} is the measured count rate, and τ is the detector dead time. For example, at a count rate of about 1 MHz, a 1% correction is required. This correction is particularly important when making isotope ratio measurements, especially when there is a significant difference in abundance between the two isotopes being ratioed. Typical dead time corrections on commercial instruments range between 20 and 100 ns. The graph in Figure 4.15 shows the effect of over- or undercorrecting for dead time.

An alternate mode of operation of the electron multiplier is the *analog mode*. In this mode, a lower applied potential is applied to the detector, reducing the gain to 10^3 or less. In this mode, an electrical current is produced at the collector that is proportional to the number of ions striking

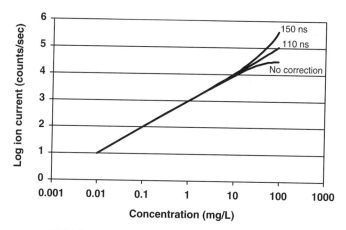

FIGURE 4.15   The effects of dead time corrections.

the entrance of the detector. This current is amplified and digitized by an analog-to-digital (A-D) converter and is subsequently related to the analyte ion concentration. Because the detector is not operated in the saturated mode, dead time corrections are not significant.

A significant drawback of this type of detector is the limited useful life-time before replacement is required. They are particularly susceptible to damage when high-intensity ion beams are measured. They also exhibit the undesirable characteristic of having a shelf life, so spares cannot be kept in reserve for long periods of time.

4.4.2 Discrete Dynode Electron Multiplier

The discrete dynode electron multiplier detector is made up of a series of individual dynode plates arranged in a configuration like that shown in Figure 4.16. Each of these plates has a metal oxide coating to which is applied a successively larger dc potential, such that electrons emitted as a result of ion collision with the first plate will be accelerated toward the second plate. Collisions of each electron with the low-work function surface of the second plate will eject two secondary electrons, which will then be accelerated toward the third plate, where four secondary electrons will be ejected. This cascade process will continue for 15 or 16 stages with the final multiplied electron beam impacting a collector. An electrical current proportional to the number of ions striking the first dynode is measured by

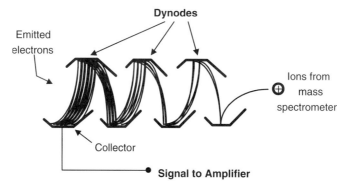

FIGURE 4.16   Discrete dynode electron multiplier.

high-impedance electronic amplification circuitry. Typical amplification or current gain from a detector such as this is about 10^6. Detectors of this type typically have a lifetime of 1–2 years in normal operation.

4.4.3 Faraday Cup

When ion currents are greater that about 10^6 ion/s, it is not practical to use electron multiplier detectors because of their lack of linearity. A Faraday cup detector, which is an analog detector consisting of a metallic cup to collect the ion beam (Figure 4.17), is used to measure high ion currents directly. The signal generated from the ions impinging on the Faraday cup detector is amplified by a low-noise direct current amplifier, capable of measuring ion beams down to about 10^4 ion/s, overlapping the effective working range of the electron multiplier detectors.

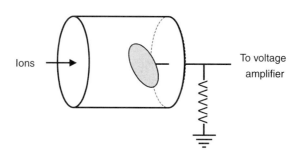

FIGURE 4.17   Faraday cup detector.

The advantage of the Faraday cup detector is high stability, which can be used to measure large ion beams without fear of damaging the detector. In addition, it can be effectively used for making isotope ratio measurements because of its advantageous feature of no mass bias, which results in very high accuracy. However, because it exhibits no appreciable amplification gain, it is not suitable for low-concentration-level trace analysis.

Commercial instrumentation often employs a combination of some form of a Faraday cup along with a high-gain electron multiplier detector. This combination permits both high-intensity ion beams, generated by high-concentration major constituent determinations, along with low-concentration trace element determinations.

CHAPTER 5

Sample Introduction

To use the ICP-MS to perform chemical analysis of virtually any type of material, it is necessary to accurately and reproducibly convert a sample of the material to a form that is suitable for introduction into the plasma, where ionization takes place. The nature of the sample introduction technique is dependent on numerous factors, including the type and form of the sample specimen, the concentration levels, scope and chemical form of the analytes to be determined, and the quantity of sample available for analysis.

Several different approaches are employed, each designed to solve specific analytical problems. Each approach has a common objective, namely, to ultimately convert the sample into an aerosol, which is introduced into the annular opening at the base of the plasma. After entering the plasma, this aerosol is rapidly thermally atomized followed by ionization of the element atoms (see Chapter 3). The block diagram in Figure 4.1 shows the general configuration of the relationship of the sample introduction system to the other components of the ICP-MS. As this figure demonstrates, the sample introduction component is the key part of the system that links the original sample, presented for analysis, to the remainder of the instrument.

Sample introduction systems are available for gas, liquid, and solid samples. To use these systems, often some preliminary form of sample preparation is required to modify the sample prior to introduction. These preliminary steps can consist of dissolution, digestion, filtration, grinding, fusion, polishing, or other forms of sample modification, which assists in the reproducible presentation of the sample to the plasma for ionization prior to measurement.

During these preliminary preparation steps, additional modification of the sample such as preconcentration of the analyte or separation from potentially interfering components in the sample can be achieved. Therefore, the sample introduction technique of choice can play an important role in the design of an optimum strategy for analysis. A summary of proven sample introduction techniques for use with ICP-MS is given in Table 5.1. This table shows the types of techniques that are useful for the determination of analyte elements in gas, liquid, or solid samples and mixtures thereof.

TABLE 5.1   Summary of Common Sample Introduction Techniques for ICP-MS Elemental Analysis

Gaseous samples	Liquid samples	Solid samples
• Hydride generation	• Cross-flow pneumatic nebulizer	• Direct insertion
• Cold-vapor generation	• Concentric pneumatic nebulizer	• Spark atomization
(mercury)	• Babington-type nebulizer	• Laser ablation
• Chromatographic	Conespray	• Slurry nebulization
Gas chromatography	Platinum grid	• Field flow fractionation
Supercritical fluid	Fritted disk	
chromatography	USGS	
	V groove (slot)	
	• Ultrasonic nebulizers	
	• Thermospray nebulizer	
	• Microvolume nebulizer	
	Direct injection	
	High efficiency	
	Microconcentric	
	Electrospray	
	Electrothermal vaporization	

Data from Montaser (1998).

5.1 GASEOUS SAMPLES

Samples in the vapor state utilize the most straightforward introduction approach. In most conventional ICPs, an approximately 1 L/min Ar gas flow is injected into the base of the plasma to create its toroidal shape (see Chapter 3). Addition of sample to this gas stream provides an ideal mode of sample introduction. Simple gas plumbing systems are employed to combine samples in the vapor state with the injector gas flow (Figure 5.1).

Gaseous samples can originate from sources as simple as process gas streams in industrial settings to more sophisticated systems such as gas chromatographic effluents or hydride gas generation apparatus. An example of the direct sampling of a gas stream from an industrial process is illustrated by the analysis of impurities in silane (SiH_4) gas used in the fabrication of silica films for the manufacture of photo-voltaic cells. Arsenic and iodine are measured at the subnanogram concentration level, with detection limits of approximately 0.5 parts per billion (v/v).

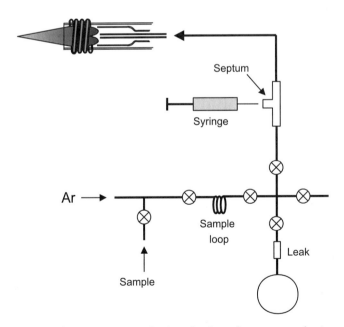

FIGURE 5.1   Plumbing arrangement for introduction of gaseous samples into the ICP, using either a fixed-volume sample loop or syringe.

5.1.1 Vapor Generation

The direct introduction of gaseous samples into the inductively coupled plasma for chemical analysis offers a variety of advantages over conventional nebulization techniques. One advantage is the ability to achieve relatively higher sensitivity than is attainable by other techniques. This higher sensitivity is primarily the result of the ability to achieve essentially 100% efficiency of transport of analytes to the plasma as compared to approximately 1%, which is generally what is obtained for pneumatically nebulized aqueous solutions.

The scope of analysis is usually very restricted. It is often limited to a single element or, at best, to a small group of elements with similar physical and chemical properties. Volatile chemical species of the analyte, which are stable at room temperature and are easily transported to the plasma in the Ar sample injection stream, are generated by chemical reactions as part of the sample introduction process. The process of chemically generating the volatile species offers the opportunity for substantial preconcentration of the analyte. In addition, this approach has the important added advantage of simultaneously removing potentially interfering components from the sample matrix. Also, it is convenient to separate the solvent vapor (originating from the original sample) from the gas stream that is transporting the analyte to the plasma, which can significantly reduce the presence of potentially interfering oxide molecular species (when H_2O is the solvent) in the resulting mass spectrum.

Hydride Generation

There is a selected group of elements, which include As, Bi, Ge, Pb, Sb, Se, Sn, and Te, that can be chemically reduced in aqueous solution to form a hydride molecular compound that is a stable vapor at room temperature. This reaction can be controlled to select a specific form of the analyte. This includes specific oxidation states of ionic analyte species. After the reduction reaction the gaseous hydride compound that is formed is separated from the aqueous solution by sparging with Ar gas, when a batch reaction vessel is used for analysis, or by directing the reaction mixture through a gas–liquid separator when a flow–injection–type system is utilized. With either system, the hydride compound that is formed by the reduction reaction is transported by the argon carrier gas directly to the plasma, where atomization and ionization occur. The most common and convenient reducing agent used for this hydride generation reaction is sodium borohydride in acidic solution. Figure 5.2 shows the plumbing arrangements for a typical hydride generation system.

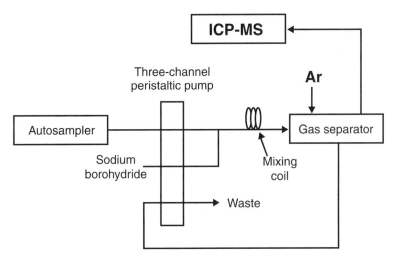

FIGURE 5.2   Plumbing arrangement for hydride generation sample introduction system.

In a flow-injection hydride generation system, the reagents, including the $NaBO_4$ and acid, are dynamically pumped (i.e., in the continuous-flow mode), usually with a multichannel peristaltic pump, to a mixing vessel (or mixing coil) with the sample, where they are efficiently mixed for a controlled period of time. When the reaction is complete, the products are transported into a gas–liquid separation device (see Figure 5.3) where the molecular hydride compound is isolated from the liquid and non-hydride-forming matrix components of the sample. The Ar carrier stream then transports the hydride compound, in the vapor state, with the injection gas directly into the plasma. A transient signal is generated by this technique, requiring a data collection system capable of acquiring and processing this type of data, either in the peak height or peak area mode.

Multiple hydride-forming elements can be determined simultaneously by this process. However, as previously mentioned, the hydride generation reaction is critically dependent on the oxidation state or form of the analyte element. For example, a prereductant, such as KI, must be added to samples containing As^{5+}, prior to the hydride reduction step, to convert it to As^{3+}, which is the active species for the hydride formation reaction. Without this prereducing step, As^{5+} will be missed in a total As determination.

When using the batch generation system, which involves producing the hydride compound from a fixed volume of sample, by the addition of the hydride-generating reagents in a reaction vessel, the analyte vapor is

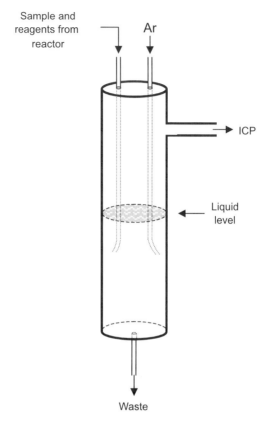

FIGURE 5.3   Gas–liquid separation device for hydride generation sample introduction system.

directly transported, after gas–liquid separation, to the plasma or is condensed in a cold trap maintained at liquid nitrogen temperatures. In the latter case, the cold trap is rapidly warmed to room temperature, and the concentrated hydride compound is volatilized and transported in the Ar gas stream to the plasma. In addition to the separation of excess hydrogen gas produced during the hydride generation reaction, a sharper more well-defined signal peak usually results in better measurement precision and higher sensitivity.

Osmium Tetroxide Vapor

Osmium, which is a particularly difficult element to analyze, is determined by converting it to the oxide form (OsO_4), which can be distilled directly into an Ar carrier gas stream and ultimately transported to the plasma for

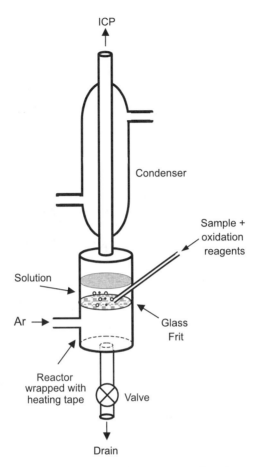

FIGURE 5.4   Batch process reactor for the determination of osmium as osmium tetroxide.

atomization and ionization. A diagram of a batch process reactor, shown in Figure 5.4, illustrates the configuration of apparatus used to quantitatively convert Os to its oxide form. This procedure separates Os from its potential interferents, particularly Re, and offers a 1 to 2 orders of magnitude pre-concentration enhancement in sensitivity. The condenser shown in the diagram is maintained at a temperature of about 5°C, removing most of the water vapor from the oxidation process. Approximately 2.5% w/v of periodic acid (HIO_4) is required for the optimum oxidation of Os to OsO_4.

Utilizing a batch process, sample is injected into the injection port on the side of the heated reactor, which contains the oxidation reagents. The OsO_4 that is formed by the oxidation is distilled through the condenser to separate water vapor, and transported to the plasma for atomization and ionization. Because of serious memory effects of Os on the walls of the apparatus, the equipment must be thoroughly cleaned between samples to maintain high accuracy. This analytical technique has been used successfully for the determination of Os isotope ratios on analyte concentrations in the low nanogram/milliliter range. Isotope ratio precisions on the order of 0.3% relative standard deviation (RSD) have been reported.

Cold–Vapor Mercury

Mercury is uniquely determined by the reduction of inorganic mercury compounds (Hg_2^+ and Hg^{2+}) to the elemental state (Hg^0) using a reducing agent, followed by the transport of the elemental Hg vapor to the plasma in the Ar carrier stream. A unique physical property of elemental Hg is its high vapor pressure at room temperature. The chemical stability of elemental Hg in the vapor form provides for an operationally simple apparatus and vapor transport system.

Stannous chloride ($SnCl_2$) in hydrochloric acid is commonly used as the reducing agent for cold-vapor Hg generation, although $NaBO_4$ has also been effectively used. Absolute detection limits in the low picogram range have been reported using this technique. Even greater sensitivity can be achieved using a gold amalgamation preconcentration step. An added advantage of ICP-MS over AAS or AFS, which are commonly used approaches in commercial instrumentation, is the ability to make isotope ratio measurements on the stable isotopes of Hg. This technique has been utilized for the accurate isotope dilution determination of Hg in environmental water samples.

5.1.2 Chromatography

Chromatographic techniques involve the separation of constituents by differences in their solubility in a stationary liquid phase or by differences in adsorption affinity on a solid substrate. By passing a mixture of the vaporized analyte constituents in a carrier gas, through a column containing the substrate, separated species can be sequentially eluted in the effluent.

Vapor Phase Chromatography

Volatile elements such as B, C, Br, Cl, Si, P, O, I, and S can be determined in the gaseous effluent from a vapor phase chromatograph. These components, or molecular entities containing these elements, are separated from each other by their differences in solubility in a stationary liquid phase in conventionally substrate-packed or capillary columns, or using a polymer, which selectively sorbs these species. The effluent chromatographic gas (usually He) is mixed with the Ar injector gas stream for the ICP. After injection into the plasma, atomization and ionization of the analyte elements occur.

For example, a high-resolution capillary column gas chromatograph (GC) can be used to separate five volatile alkyllead compounds, with temperature programming from 40 to 180°C. An interface using a heated transfer line was used to couple the GC to the ICP-MS. A typical gas chromatogram showing the ion current for lead plotted as a function of time (eluent volume) for the compounds tetramethyllead, trimethylethyllead, dimethyldiethyllead, methyltriethyllead, and tetraethyllead is shown in Figure 5.5.

The advantage of coupling a vapor phase chromatograph to an ICP-MS is the realization of the ability to perform highly sensitive measurements of specific species of the elements in question. Disadvantages include the ability to determine only elements contained in compounds that are volatile and stable at the temperatures employed to induce chromatographic separation. In addition, the determination requires sufficient time for the elution of all species of interest to the analyst, often increasing the elapsed time of the analysis.

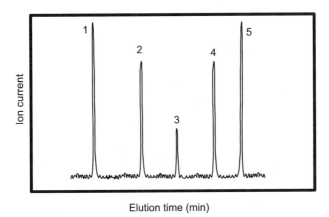

Elution time (min)

FIGURE 5.5   Gas chromatogram showing the separation of (1) tetramethyllead, (2) trimethylethyllead, (3) dimethyldiethyllead, (4) methyltriethyllead, and (5) tetraethyllead.

Supercritical Fluid Chromatography

Using supercritical fluids to extract and subsequently chromatographically separate thermally labile, nonvolatile, and high molecular weight compounds prior to introduction into the ICP for atomization and ionization, elemental chemical speciation analysis is accomplished. Determination of these types of analyte compounds is not feasible by gas chromatography, and requires impractically long elution times by liquid chromatography. Unlike liquid chromatography, no liquid nebulizer or spray chamber is required for sample injection into the plasma.

Flowing liquid carbon dioxide (CO_2) in the supercritical state (70 atm and 40°C) is used to dissolve and separate the organometallic analyte compounds in a capillary column fitted with a gas restrictor at its exit. As the compounds sequentially leave the separation column, they are transported in the vapor state, along with the now gaseous carbon dioxide, to the plasma in the heated argon carrier gas stream. The introduction of the CO_2 into the plasma creates potential interferences from the formation of carbide molecular compounds. Species such as $^{12}C^+$, $^{12}C^{16}O^+$, and $^{40}Ar^{12}C^+$ significantly contribute to the background spectrum and can seriously interfere with the determination of some elements.

An example of an application that effectively utilizes supercritical fluid chromatography technology is the determination of low-volatility organotin compounds. A chromatogram showing the separation of tetrabutyltin, tributyltin chloride, triphenyltin chloride, and tetraphenyltin by supercritical fluid chromatography is shown in Figure 5.6. The chromatogram shows the response of the mass spectrometer for the ion current of tin plotted versus the retention time in the capillary column.

5.2 LIQUID SAMPLES

Both liquid samples and solutions of dissolved analyte species are analyzed either by direct introduction into the plasma or by evaporation of the solvent and introduction of the residue by direct insertion, thermal vaporization, or laser/spark ablation. Of the two approaches, the direct introduction techniques are by far the simplest and most straightforward. Samples can be reproducibly introduced with a minimum of sample handling. However, significant advantages, for example, enhancement of sensitivity by preconcentration, can be realized using the solvent evaporation approach for some types of samples, except for the analysis of volatile analytes.

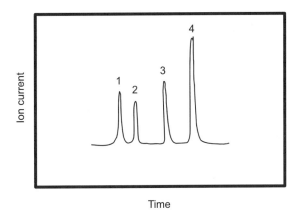

FIGURE 5.6 Supercritical fluid chromatogram showing the separation of (1) tetra-butyltin, (2) tributyltin chloride, (3) triphenyltin chloride, and (4) tetraphenyltin.

5.2.1 Nebulizers

Nebulizers perform the function of converting liquid samples to an aerosol, consisting of finely divided droplets, which are suspended in the plasma carrier gas. This aerosol, which is presumably representative of the composition of the original sample, is transported to the plasma torch, where it is injected into the central channel of the plasma for subsequent atomization and ionization.

The process of converting a sample from its initial form, prior to introduction into the ICP-MS, to the ions extracted from the plasma for measurement in the mass spectrometer, is shown schematically in Figure 5.7. Each step of the process is shown as a function of the incremental increased addition of energy from the plasma to perform the necessary conversion. From this figure it is seen that initially, as the liquid aerosol is introduced into the "thermal front" at the base of the plasma, the solvent is evaporated when the temperature of the aerosol reaches the solvent's boiling point. The resulting "dry aerosol" continues to be transported into a higher temperature region of the plasma. When the melting point, followed by the decomposition temperature of the residue is reached, simple molecules and atoms are formed. Finally, in the highest temperature region of the plasma, complete atomization occurs, and the resulting atoms are efficiently ionized prior to being sampled by the mass spectrometer interface and transported

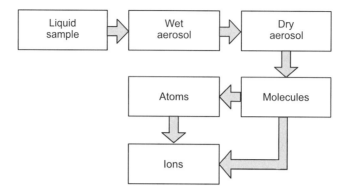

FIGURE 5.7   Block diagram for converting liquid samples to ions in ICP.

to the mass analyzer for analysis. In addition, molecules formed from the decomposition of the dry aerosol or recombination by collisions of the atoms can form ions also.

The aerosol formed by the nebulization process creates a population of droplets that have a distribution of sizes, ranging from a mean diameter of about 1 to 80 μm. The more uniform the droplet size (i.e., the narrower the size distribution) the more precise the results of the subsequent analytical determinations. Larger droplets require more energy to evaporate the solvent and subsequently more energy to vaporize and atomize the *postsolvent removal residue*, resulting in local instability in the plasma. This instability is reflected in the measured ion currents of the analyte elements.

Spray Chamber

To improve this situation, techniques are used to classify the droplets before the aerosol is transported to the plasma. This results in a narrow distribution of droplet sizes being introduced into the plasma, preferably with a small mean droplet diameter. The most effective way to accomplish this classification is to use a spray chamber, which provides an expansion chamber and a circuitous route for the droplets to travel en route to the plasma. The larger droplets, which have higher momentum, collide with the walls of the spray chamber where they are condensed. Only the smallest diameter particles survive the process and are transported through the system. The graphs in Figure 5.8 illustrate the need

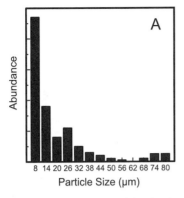 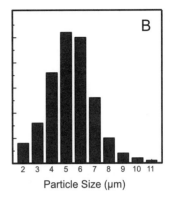

FIGURE 5.8   Comparison of droplet size distribution before and after classification by a spray chamber: (A) distribution of droplets produced by nebulizer and (B) size distribution of droplets after classification with a spray chamber.

for the use of a spray chamber to classify aerosol droplet size. Figure 5.8A shows the distribution of droplets produced by the nebulizer. Note the bimodal distribution with a measurable quantity of large-diameter droplets. After passing through a spray chamber (Figure 5.8B), the size distribution is much narrower and exhibits a smaller mean droplet diameter. Several configurations and designs of spray chambers are prevalent and are used with different types of nebulizers.

A diagram of the Scott-type spray chamber is shown in Figure 5.9. Spray chambers of this type have both an inner and outer compartment. The aerosol enters the inner compartment and is forced to change direction 180°, returning through the outer compartment, where it exits the spray chamber and is directed to the plasma. An external drain is provided to remove condensed solvent. For some specialized applications a water jacket or an electrical Peltier cooler is provided that acts as a thermostat for the spray chamber. By lowering the temperature, the vapor pressure of the solvent can be significantly reduced. This results in a substantial reduction in molecular oxide (in the case of water as the solvent) formation and potentially interfering peaks in the background spectrum of the plasma originating from solvent species. Many variations of this type of spray chamber are available.

A second popular type of spray chamber utilizes a cyclonic design. A diagram of this spray chamber is shown in Figure 5.10. Instead of using two compartments as with the Scott-type spray chambers, this device uses a

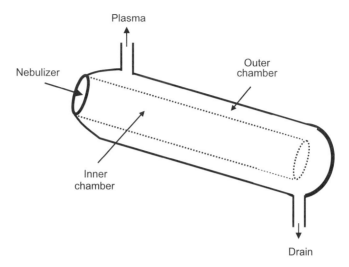

FIGURE 5.9   Diagram of Scott-type spray chamber for pneumatic nebulization.

tangential rotary flow path in a single circular compartment. The centrifugal force of the aerosol, as it ascribes a circular path in the spray chamber,
results in the larger droplets being forced to the outside where they collide
with the wall. The smaller diameter droplets are swept through the spray
chamber and exit to the plasma. Usually the internal volume of cyclonic
spray chambers is small (20–40 mL), reducing the washout time by a factor

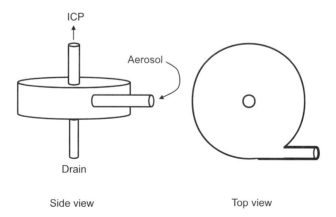

FIGURE 5.10   Diagram of cyclonic spray chamber for pneumatic nebulization.

of 2–3. The use of a cyclonic-type of spray chamber benefits the analyst from the point of view of a simple design and ease of operation.

The nature of the operation of these spray chambers is that only a small percentage (<10%) of the original sample is actually transported to the plasma for analysis. The bulk of the sample is diverted to waste. Therefore, when limited sample is available, special modifications or techniques are required for accurate and sensitive analysis.

Droplet size classification can also be accomplished by using an impinger bead. This involves directing the aerosol against a glass or PTFE bead that serves to fragment larger droplets into smaller ones, and to coalesce droplets after collision with the surface of the bead. The small-diameter droplets are directed around the impinger bead and transported on to the plasma. The larger diameter droplets are thereby removed from the aerosol stream and directed to the drain.

Pneumatic Nebulizer

The most common type of nebulizer for producing an aerosol of the sample is the pneumatic nebulizer, of which there are several different designs. All pneumatic nebulizers share a common feature: They use the force of a flowing gas, passing through an orifice or capillary tube, to create microdroplets from the liquid sample. These droplets are transported via the flowing gas stream to the plasma for decomposition, atomization, and ionization.

Cross-Flow Nebulizer

A diagram of the cross-flow nebulizer is shown in Figure 5.11. This type of nebulizer consists of two capillary tubes, usually made of glass or quartz, positioned at right angles to each other so that gas flowing through one capillary creates a pressure differential at the tip, which naturally aspirates solution from the other capillary tube at about 3 mL/min. The efficiency and performance of the aerosol generation process is highly dependent on the alignment of the two capillary tubes. The energy imparted by the flowing gas disrupts the liquid and results in the generation of an aerosol at the capillary tip. The drop-size distribution is usually very large for aerosols generated by this type of nebulizer. Better precision can be obtained by forcing sample solution to the nebulizer by use of a peristaltic pump (see Figure 5.12). Pumping minimizes variations in sample delivery rate due to variability of sample viscosity and surface tension.

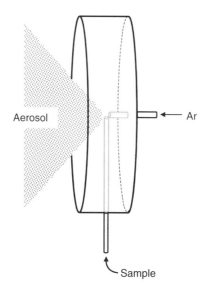

FIGURE 5.11   Diagram of a cross-flow nebulizer.

Since the sample solution must pass through one of the nebulizer capillaries, partial clogging or blockage can result from suspended particulate matter in the sample or to a lesser extent high dissolved solids concentrations in the samples that precipitate at the tip as a result of solvent evaporation. This can be a particularly insidious problem because partial blockage

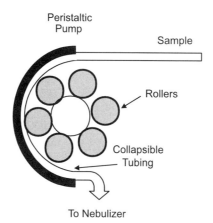

FIGURE 5.12   Use of a peristaltic pump for sample delivery to a pneumatic nebulizer.

can result in an irreproducible inhibition of aerosol formation. This decrease in aerosol production results in a variable loss of sensitivity that may not be apparent to the analyst.

A modification of the cross-flow nebulizer is the Meddings-Andersen-Kaiser (MAK) nebulizer. This nebulizer has "fixed" or nonadjustable capillaries that are constructed of heavy-walled glass. This nebulizer typically operates at a gas flow rate of about 500 mL/min. Operating precision of better than 0.5% RSD has been reported.

Concentric Nebulizer

A very popular type of pneumatic nebulizer is the concentric or Meinhard nebulizer. This nebulizer is a one-piece device, usually made of glass, that has an internal capillary tube (10–35 μm in diameter) mounted in a concentric fashion axial to an external tube (see Figure 5.13). Nebulizer gas is passed through the external tube at a flow rate of about 1 L/min, which results in sample being aspirated (or preferably pumped) through the internal capillary at a rate of about 0.5–1 mL/min, with aerosol formation occurring at the tip. As with the cross-flow nebulizer, samples with suspended matter or high dissolved solids concentration can result in partial blocking of the nebulizer, thus inhibiting aerosol formation.

The high-efficiency nebulizer (HEN) is a low-volume low-flow version of the concentric nebulizer. It operates with a sample delivery rate as low as 10 μL/min. Therefore, this is a suitable nebulizer for use when low sample volume is available. This nebulizer has good long-term stability, but because its internal capillary diameter is only 75–100 μm, it is prone to blockage from suspended particulate matter.

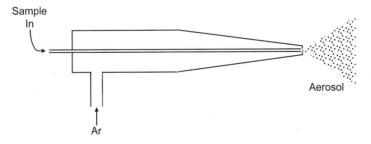

FIGURE 5.13   Concentric pneumatic nebulizer.

The microconcentric nebulizer, which is made entirely of an inert material, is highly suitable for the analysis of corrosive materials such as hydrofluoric acid. This nebulizer operates at a solution uptake rate of about 30 μL/min. Detection limits achieved with this system are comparable to those obtained with other pneumatic nebulizers.

Babington-Type-Nebulizer

Babington nebulizers include several specifically configured devices that are based on a common feature. This feature involves the process of passing the nebulizer gas through an orifice over which the sample is flooded in excess volume. This results in a nebulizer that is relatively immune to clogging or plugging due to suspended material or high dissolved concentration solids.

The USGS Babington nebulizer is shown in Figure 5.14. It consists of a glass tube with a hemispherical end and a 0.1-mm-diameter orifice in the side. This tube is mounted in a PTFE housing with a sample delivery tube positioned directly above it. Argon gas passes through the orifice at about

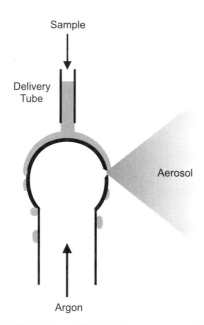

FIGURE 5.14   USGS Babington-type nebulizer.

600 mL/min. Sample is introduced through the top delivery tube as 3–5 mL/min. An aerosol is produced when a thin film of sample flows over the orifice from the top.

The primary advantage of this nebulizer is its ability to produce an aerosol from virtually any liquid material, irrespective of viscosity or suspended solids content. The performance, in terms of precision and sensitivity is essentially equivalent to the concentric or cross-flow nebulizers.

A slot or V-groove nebulizer consists of a simple slot machined in PTFE or plastic, which has an orifice positioned in the bottom of the valley of the groove. Sample is flowed down the groove, passing over the orifice, through which gas is directed. The aerosol is produced as the sample film passes over the orifice. This type of nebulizer is shown in Figure 5.15. Slot nebulizers have been successfully used with argon gas flow rates of 0.4–1.2 L/min. A specialized version of this design is called the GMK nebulizer. This nebulizer uses a glass impact bead to fragment larger droplets that are formed.

The cone-spray nebulizer is a more efficient design of the previously described slot nebulizer. A diagram of this type of nebulizer is shown in Figure 5.16. Instead of a slot, where sample approaches the orifice from only one direction, the cone spray consists of a three-dimension funnel-shaped depression made in an inert material, such as sapphire, with an orifice positioned in the bottom of the funnel. Sample is introduced into the funnel from above and is concentrated on the orifice as it flows to the bottom. A commercially available version of this nebulizer has an orifice with a 216-μm diameter and operates at about 32 psi. The long-term precision of this nebulizer has been shown to be about 1% over an 8-h period of time.

Another type of Babington nebulizer is the Hildebrand grid nebulizer. This nebulizer consists of two parallel 100-mesh platinum grids positioned about 2 mm apart. Sample, delivered by a V-groove to the first grid, is nebulized by gas passing through the grid from the back. The aerosol that is

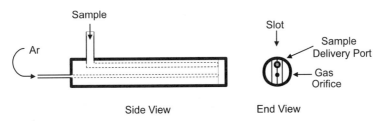

FIGURE 5.15   Slot-type nebulizer.

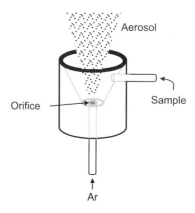

FIGURE 5.16   Cone-spray-type nebulizer.

produced is immediately impacted on the second grid where the larger droplets are fragmented. Although this is a more complicated nebulizer to operate (difficult to clean between samples), it produces a larger quantity of aerosol compared to the concentric or cross-flow nebulizers.

The final type of Babington nebulizer is called the Lichte or fritted disk nebulizer. The configuration of this nebulizer is shown in Figure 5.17. The nebulizer consists of a glass-enclosed, medium to coarse fritted disk, configured so that argon nebulization gas passes through the frit. Sample pumped

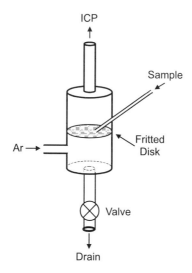

FIGURE 5.17   Fritted disk nebulizer.

through small-bore tubing is deposited on the surface of the frit. As the argon passes through the frit, each pore behaves as if it were an individual orifice. The sample is nebulized at the surface of the frit. Because of the large area for aerosol generation, this nebulizer operates with very high efficiency (>90%). The flow rate of sample pumped to the surface of the frit is approximately 50 μL/min, which makes this technique suitable for low sample volumes. Primary disadvantages of this device are the long stabilization times required for suitable analysis and the lengthy washout time (a minimum of 2 min) between samples.

Direct Injection Nebulizer

The direct injection nebulizer (DIN) is a total consumption device. The entire aerosol generated by the nebulization process is injected directly into the plasma. No spray chamber is used to classify the size of the droplets formed by nebulization, so this technique is 100% efficient. Precision and washout times were significantly improved using this approach. Use of this device also leads to reduced memory effects.

Because the DIN operates at 100% efficiency, a significant solvent load is introduced into the plasma. This can result in increased interferences from molecular compounds formed from solvent constituents. Detection limits achieved by this method are equivalent to or better than those obtained by other nebulization techniques.

An advantage of the DIN is its ability to function with micro-size samples. Sample volumes as small as a few microliters can be analyzed, making this approach an ideal solution to many special analytical applications. However, a major drawback to this device is its intolerance to suspended matter in the sample solution. Transport of sample, which is forced through a microcapillary as part of the nebulization process, can be seriously affected by partial clogging from particles or fibers. This alteration of the sample transport will have a direct impact on the accuracy of the analysis.

Ultrasonic Nebulizer

The ultrasonic nebulizer is an atomization device that is popular for use when enhanced sensitivity is required for analysis applications. The apparatus consists of a piezoelectric crystal transducer, which is driven by an ultrasonic generator operating at a frequency of 200 kHz to 10 MHz. Sample is delivered to the front surface of the transducer through tubing from a peristaltic pump at a flow rate of up to 1 mL/min. An aerosol is formed by the standing longitudinal waves at the surface of the transducer crystal. When the amplitude of the wave is large enough to disrupt the film of solution deposited on the surface of the transducer, an aerosol is generated, which is removed from the nebulizer by the argon flow. A typical ultrasonic nebulizer is shown in Figure 5.18.

Delivery of sample at a rate of 1 mL/min yields an analyte transport efficiency of about 20%. By reducing the delivery rate of sample to the transducer, high transport efficiencies can be obtained. Using a sample delivery rate of 5–20 μL/min, an aerosol transport efficiency of close to 100% can be obtained. Between samples, the surface of the transducer is flushed with large quantities of water to remove all traces of the previous sample.

Since the aerosol production rate is high and is independent of aerosol carrier gas flow rate, more sample is transported to the plasma at lower injector gas flow rates than that obtained with conventional pneumatic nebulizers. This results in enhanced sensitivity obtained using this nebulizer, with a corresponding improvement in detection limits for essentially all elements measured. Table 5.2 shows the comparison of detection limits obtained with conventional pneumatic nebulizers and a commercial ultrasonic nebulizer. In general, an improvement in detection limits by 1 to 2 orders of magnitude is realized for simple sample matrices. With the presence of high dissolved solids in aqueous samples, no significant improvement in detection limits is obtained. Usually, precision is marginally poorer for the ultrasonic nebulizer than with conventional pneumatic nebulizers. For comparable analyte concentrations, the ultrasonic generally is capable of 2–3% RSD, while conventional pneumatic nebulizers can achieve about 1% RSD.

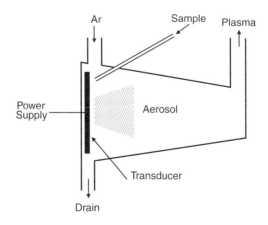

FIGURE 5.18   Ultrasonic nebulizer.

TABLE 5.2   Comparison of Detection Limits (ng/L) Obtained Using Conventional Pneumatic and Ultrasonic Nebulization with ICP-MS

Element	m/z	Pneumatic[a]	Ultrasonic[b]
Aluminum	27	6	0.5
Barium	138	1	0.3
Beryllium	9	2	0.4
Cadmium	111	2	0.1
Calcium	40	800	270
Chromium	52	3	0.2
Cobalt	59	1	0.07
Copper	63	2	0.5
Lead	208	2	0.1
Magnesium	24	2	0.2
Manganese	55	2	0.3
Molybdenum	95	2	0.3
Nickel	58	12	0.4
Silver	107	1	0.2
Sodium	23	20	3
Strontium	88	1	0.5
Thorium	232	0.4	0.03
Thallium	205	1	3
Titanium	48	4	0.4
Vanadium	51	1	0.2
Yttrium	89	1	0.1
Zinc	66	5	0.5

[a] Concentric glass nebulizer with Scott spray chamber.
[b] Ultrasonic nebulizer with desolvation.

Due to the high nebulization efficiency and the relatively small droplet size, a large quantity of solvent from the sample is transported to the plasma. This results in a deleterious impact on plasma atomization and ionization capacity. In effect, the excess solvent can extinguish or "cool" the plasma, which reduces the efficiency of converting unatomized analyte species to a form suitable for ionization. In addition, the high quantity of hydrogen and oxygen, in the case of water, or carbon, in the case of organic solvents, results in high concentrations of molecular compound formation, which can cause serious isobaric spectral interference.

The problem occurring due to excess solvent reaching the plasma can be significantly reduced by the use of a desolvation system. A desolvation system consists of passing the nebulized aerosol through a heated chamber where the solvent component of the aerosol droplets is thermally evaporated. This is accomplished by using an infrared heat lamp or by wrapping the chamber with resistance heating wire that can increase the internal temperature to above 100°C when working with aqueous samples. Immediately following this heated chamber, the gas is passed through a condenser where the solvent vapor is condensed and removed through the drain (see Figure 5.19).

In some systems, an alternative that is employed consists of the use of a microporous PTFE membrane to selectively remove solvent. Aerosol, after passing through the heated chamber to vaporize solvent and the condenser, flows through the membrane separator before being directed to the plasma. The vaporized solvent molecules diffuse through the pores into the gas stream on the exterior side of the membrane where they are swept from the system and vented. Aerosol particulates do not pass through the pores.

By either technique the removal of solvent from the aerosol can potentially cause an interference effect depending on the dissolved solids concentration. A maximum limit of 5% dissolved solids can usually be tolerated.

5.2.2 Electrothermal Vaporizers

Electrothermal vaporizers (ETVs), commonly used in atomic absorption spectrometry, have been successfully employed as discrete sample introduction devices for ICP-MS. The graphite furnace, carbon rod, and metallic filaments have all been used as ETV devices. In general, these devices provide

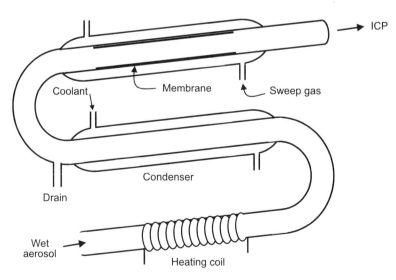

FIGURE 5.19   Desolvation apparatus.

better detection limits than continuous introduction pneumatic nebulizers. Because a discrete quantity of sample is introduced into the plasma, the response signal produced from the detector of the mass spectrometer is transient in nature. The characteristics and processing of this type of response signal are described in detail in the section on laser ablation sample introduction.

The most commonly used ETV is the graphite furnace. The configuration and operation of the graphite furnace (Figure 5.20) is essentially identical to that used in graphite-furnace atomic absorption spectrometry (GFAAS). A graphite tube is housed in electrode clamps, which are connected to a power supply. By passing an electrical current through the graphite tube, it is resistively heated, which vaporizes sample deposited on its inner wall.

A small volume sample (typically 10–100 μL) is deposited in the furnace through the hole at the top of the tube. While passing an argon carrier gas through the tube furnace, a small electrical current, sufficient to warm the tube to the boiling point of the solvent, vaporizes the solvent. This is immediately followed by an ashing step, where the electrical current is increased to the point that the sample residue is *in situ* ashed to remove organic matter or other nonanalyte sample components. Finally, the internal temperature of the tube furnace is rapidly increased by applying a high electrical current, such that the sample residue is thermally vaporized in 0.1–10 s. The vaporized analyte is transported from the graphite tube to the plasma for atomization and ionization. Performance can be enhanced by pyrolytically coating the interior surface of the graphite tube or coating it with a deposit of palladium metal. This reduces the potential interferences of the carbon from the tube reacting at high temperature to form carbide molecular compounds that are transported to the plasma. Also similar to GFAAS, a platform placed in the graphite tube furnace improves the thermal characteristics of the ETV, resulting in better stability.

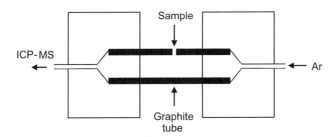

FIGURE 5.20   Graphite furnace electrothermal vaporization device.

Another commonly used ETV is the metal filament device. A metal ribbon made of tantalum, tungsten, rhenium, or platinum is mounted between two electrodes connected to a power supply (see Figure 5.21). The system operates in a similar fashion to the previously described graphite furnace. Sample is volumetrically deposited directly on the metal ribbon. As with the graphite furnace, the temperature of the metal ribbon is controlled by the magnitude of the current provided by the power supply.

When multiple analyte species are present in the sample, each has a characteristic atomization rate dependent on the vapor pressure of the material. This results in different species being vaporized at different times. Figure 5.22 shows signal responses for Cd, Ni, and Cr in a single sample deposited in an ETV. Cadmium has the lowest vaporization temperature and, as predicted, is the first of the three elements to be vaporized from the ETV. This process is followed by Ni and Cr, in the order of their vaporization temperatures. Depending on the heating rates and the physical and chemical characteristics of each analyte, various elements may vaporize from the heated surface at different times; therefore, optimization for each type of sample is required. Use of matrix modifiers, as in GFAAS, can stabilize this process.

Because no spray chamber is necessary, and the sample is present as a vapor in the argon carrier gas (compared to an aerosol produced from pneumatic nebulizers), the transport efficiency is very high ($>75\%$), resulting in at least an order-of-magnitude increase in sensitivity with a corresponding improvement in detection limits. Compared to typical results from GFAAS, detection limits by ETV for first and second row transition

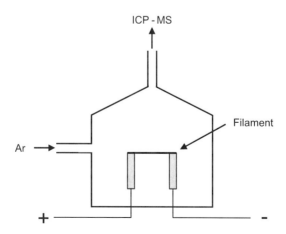

FIGURE 5.21   Metal filament electrothermal vaporization device.

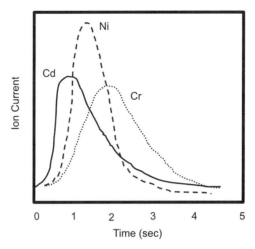

FIGURE 5.22   Graph of the vaporization rates of Cd, Ni, and Cr.

elements improved by a factor ranging from 2 to 350 times, while the rare earth elements improved by 10^4–10^6 times. The removal of the solvent during the vaporization process greatly improves the plasma atomization and ionization characteristics. In addition, this process minimizes the formation of molecular compounds originating from the components of the solvent. This technique is also ideal for solving analytical problems involving limited quantities of sample.

Limitations of this technique include the extended length of time required for analysis, the lack of automation capability, and its relatively poor repeatability compared with other techniques, all of which limit its usefulness for routine analysis. Because of sample pipetting uncertainties and variation in the sample vaporization step, the relative precision (10–15% RSD) of this technique is not as good as that achieved by solution nebulization methods. Table 5.3 shows a summary of precision and the percent recovery of various analyte elements for a 1 µg/L standard solution.

5.3 SOLID SAMPLES

In classical spectrographic analysis (i.e., direct current arc and controlled-waveform spark emission spectrography) the easiest method for sample introduction involved the direct analysis of solid samples. This approach was also used as the conventional sampling technique for trace element analysis by spark–source mass spectrometry. Solid samples were easy to prepare, usu-

TABLE 5.3   Typical Precision and Recovery for a 1 μg/L
Multielement Calibration Standard Using Electrothermal Vaporization
ICP-MS

Element	Precision (% RSD)	Percent recovery
Aluminum	10	96
Bismuth	20	92
Cadmium	16	110
Cobalt	6	82
Copper	23	98
Gallium	21	104
Indium	23	119
Lead	18	106
Lithium	25	109
Magnesium	21	91
Manganese	14	89
Nickel	8	84
Rhodium	20	94
Silver	21	109
Sodium	10	100
Strontium	8	60
Thallium	19	96
Zinc	18	100

ally requiring only grinding, mixing and subsampling, or polishing prior to sample introduction. Trace analysis was performed directly, minimizing problems of sample contamination and analyte loss due to volatilization during chemical dissolution.

Because of the general ease of operation, liquid sample introduction has traditionally been used for ICP-MS methods. However, in the last decade, the requirement for direct analysis of solid samples has grown in popularity. Techniques have been developed for solid sample analysis and have evolved from highly specialized research procedures to more routine applications. The obvious advantage of the minimal handling of samples, thereby maintaining the integrity of the sample trace element composition, makes the direct analysis of solids highly desirable. In addition, the absence of solvent potentially reduces the spectral interferences that arise from the formation of molecular species in the plasma. Coupled with the ability to perform specialized analysis techniques, such as surface analysis, depth profiling, spatial discrimination, the ability to determine highly exotic and refractory elements that are difficult to dissolve, and speciation of analytes that would be destroyed by chemical dissolution, the general ease of minimal sample preparation has driven the research and development in this area of mass spectrometric analysis.

5.3.1 Ablation Methods

Ablation methods involve the application of external energy to the surface of a solid specimen, which liberates a representative portion of the sample by vaporization or ejection of solid particulates, directly forming a solid aerosol. Techniques utilizing the ablation process are probably the most versatile means of solid sampling for ICP-MS applications. By varying the conditions and form of the energy applied to the specimen, the characteristics of the sampling of the solid material can be made highly specific for the solution of chemical analysis problems involving solid samples.

Laser Ablation

Energy in the form of a focused laser beam is directed to the surface of the sample specimen. Sample material is ablated from the surface of the specimen and partially vaporized. This material is then transported to the plasma for atomization and ionization. A typical laser ablation setup is shown in Figure 5.23. The sample specimen is placed on a movable stage within the cell, which allows the specimen to be precisely positioned relative to the fixed laser beam.

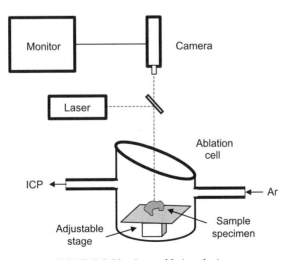

FIGURE 5.23   Laser ablation device.

Aerosol carrier gas (Ar) is passed through the cell at a constant flow rate and directed through the plumbing system to the injector tube in the plasma torch. The laser beam is focused on the surface of the sample specimen through a window in the top of the cell, which is transparent to the frequency of the laser. A video camera or microscope is used to spatially align the position of the laser beam on the sample. Microsample analysis and spatial studies require the ability to focus the size of the laser beam to a spot approximately 10 μm in diameter or less. A pulse of energy from the laser strikes a specific region of the sample specimen, removing a fixed quantity of material depending on the duration and energy of the pulse. The volume of the cell is kept as small as possible to minimize the dispersion of the plume of sample. Small dead volume coupled with the rapid transport of the sample aerosol from the cell reduces deposition of the sample on the walls of the cell. The plume of the ejected particulate and vapor is transported as a discrete slug of sample to the plasma. As this slug of material enters the plasma, it is atomized and ionized, and subsequently extracted with the sampler cone into the ICP-MS interface.

A transient signal is measured at the detector for the ion current of the specific analyte species of interest. The nature of this transient signal is shown in Figure 5.24. As the analyte enters the plasma and ions are formed, the magnitude of the signal increases to a maximum and then gradually tails off to background as the analyte is depleted. Either the peak height of this profile can be measured or, as shown in the figure, the area under the curve can be integrated. Either form of signal measurement is proportional to concentration of the analyte in the original sample.

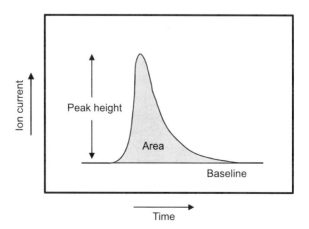

FIGURE 5.24   Transient signal.

Many types of lasers are used for this analysis technique, including Nd-YAG, ruby, CO_2, N_2, and excimer. For satisfactory operation, lasers must be capable of sufficient power density to effectively ablate and vaporize refractory sample matrices. Both the Nd–YAG and ruby lasers can be operated in two modes: "normal" (free-running) mode and the fixed Q-switched mode. In the normal mode, ablation craters are created with uniform and reproducible diameters, the depth of which are determined by the number of pulses used. The Q-switched mode is used for samples that are particularly difficult to energy couple because of their transparent nature. Although the Q-switched mode provides more energy and hence a larger amount of material is ablated, performance in terms of efficiency and reproducibility are usually not a good as operation in the normal mode.

Depending on the energy of the laser beam, the power applied, laser repetition rate or duration of the pulse, and the melting point and volatility of the sample material, the depth and configuration of the laser sampling crater vary. A photograph of a polished agate geological specimen is shown in Figure 5.25. A series of ablation craters is shown (each created with equal energy characteristics). The craters in the agate silicate mineral matrix show a relatively well-defined configuration. The crater that is formed in the galena (PbS) mineral inclusion, which is a relatively soft mineral and has a lower melting point than the agate, is considerably larger and shows a nonuniform configuration with significant splattering around the edges. This difference in crater configuration translates to a significant difference in analyte ablation efficiency.

An attractive aspect of laser ablation inductively coupled plasma mass spectrometry (LA-ICP-MS) is its versatility. It is capable of performing analysis in a variety of conducting and nonconducting solid samples, including metals and alloys, geological materials, environmental materials, semiconductors, biological materials (tissues), graphite, air particulates, and many other substances. Both bulk and spatially discriminated (horizontal and vertical) microanalyses are commonly performed. Spatial resolution of the sample under analysis depends on the wavelength of the laser and optics used to focus it on the specimen. Typically, a spot size of 5–60 μm can be achieved with Nd:YAG lasers, which are operated in the frequency-quadrupled mode at a wavelength of 266 nm in the far ultraviolet. Larger spot sizes (~100 μm) are produced with a standard Nd:YAG laser operated in the near-infrared region of the spectrum at its fundamental frequency with a corresponding wavelength of 1064 nm. In addition, high-energy mixed-gas excimer lasers can also be used to obtain small spot sizes. An argon fluoride laser operating in the far ultraviolet at 193 nm can produce a spot size as small as 1 μm in diameter.

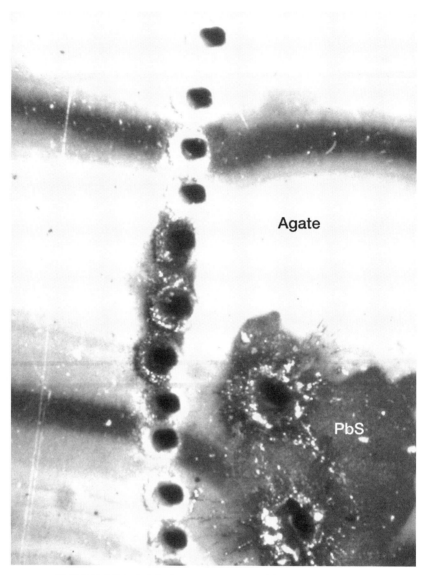

FIGURE 5.25   Polished agate specimen with laser ablation craters.

Detection limits are directly related to the amount of material ablated from the sample. The larger the crater diameter, the more material is ablated; hence, higher sensitivity is achieved, resulting in better detection limits. Typical detection limits obtained for sample ablated from a 25-μm-diameter crater are in the 0.1–10 μg/g range. Detection limits can be improved by optimizing the flow rates of the carrier gas and the energy characteristics of the laser for various sample matrices. For bulk analysis, detection limits can also be improved significantly by rapidly rastering the laser beam across a large area of the sample specimen. This produces a pseudo-continuous signal, which greatly increases the amount of sample introduced into the plasma per unit time.

Calibration of the system for quantitative analysis is a problematic situation. Most applications depend on solid calibration standards, with varying concentration of the analyte elements. By ablating these matrix-matched standard materials under identical operating conditions as the samples, a calibration curve can be obtained for quantitation of the analyte. Although this is the ideal approach to quantitative calibration, it is often difficult to find solid standard materials that simulate the sample matrix and have all of the analyte elements of interest.

An alternate approach is to use a fusion technique. The samples are dissolved in a molten flux such as $LiBO_4$, Na_2CO_3, or other appropriate salts. Spike additions of the analyte elements can be added to the molten sample and flux in varying concentrations. The fused melt is solidified by pouring the molten flux onto a flat surface (aluminum or stainless steel plate), where it cools, forming a uniform glassy specimen suitable for direct laser ablation analysis. Ablation of the fluxed material, after solidification, is used to establish calibration curves for quantitation or for standard addition analysis.

Another unique approach is to produce a dry aerosol from a standard aqueous solution with a pneumatic nebulizer and desolvator. The aerosol produced by this apparatus is combined with the aerosol from the ablation cell using a dual gas-flow sample introduction system and a gas-mixing cell. By varying the concentration of the analyte in the standard aqueous solution, calibration curves can be created by this procedure. The composition of the standard solution can be adjusted to match the matrix composition of the bulk material of the solid sample, thereby simulating the behavior of the ablated solid in the plasma. Secondary standards can be created by this technique from selected samples. Future calibration curves can be prepared from ablating these secondary standards.

Spark Ablation

Another technique used for the generation of an aerosol from a solid sample involves the application of an alternating current (approximately 100 Hz) spark discharge between a counter electrode (anode) and the electrically conducting sample (cathode). The sample is mounted in an isolation chamber, through which the argon carrier gas flows to the plasma. When the spark interacts with the surface of the sample specimen, its energy physically ablates the surface creating a dry sample aerosol. As this aerosol is generated, it is continuously transported to the plasma for atomization and ionization, resulting in steady-state ion currents for each of the analyte species.

The anode is commonly constructed of a high-purity refractory conducting material, such as tungsten metal. The use of a refractory material minimizes the rate of vaporization of the anode during operation, which can modify the spark ablation efficiency, resulting in calibration drift. In addition, the material from which the anode is constructed also contributes as a contaminant to the sample aerosol. It is therefore desirable to use a material consisting of a single element to reduce the magnitude of potential interferences. Usually, an uncommon exotic element, not normally of interest as an analyte, is selected for the anode material.

The sample (cathode) consists of an electrically conducting specimen with at least one flat surface, against which the spark can impact a fairly large surface region. The surface is usually created by grinding and/or polishing the specimen until it has a flat side suitable for spark ablation analysis. Powdered samples are often compacted under pressure into a pellet or briquette, which has a flat side. If the sample is a powder, which is nonconducting, an electrically conducting binder, such as graphite or silver powder, is used as part of the pelletizing process, to create a specimen suitable for analysis.

This technique can be used either for bulk chemical analysis of the sample or, more specifically, for the analysis of the surface material of an unpowdered sample. This approach can be carried a step further, providing a crude method for depth profiling of the composition of the sample specimen, by varying the sampling time of the ablation process, or performing consecutive analysis of the same region of the sample. However, unlike laser ablation, this technique does not provide for specific spatial isolation of the sample specimen, or accurate control over the amount of material removed, thus limiting its utility for depth profiling.

The primary application of this approach to the compositional analysis of solid materials is the determination of major and trace components of metals and alloys. The advantage that is offered by this technique is the relative ease and speed of analysis. No lengthy sample dissolution is required, which often is very difficult for the preparation of sample material such as high-duty nickel alloys, which are used in applications such as jet engine turbine blades.

5.3.2 Slurry Nebulization

Solid sample in the form of a finely ground powder can be analyzed direct-ly by ICP-MS without chemical dissolution. Small-diameter particulate matter is transported to a nebulizer as a suspension in a solvent, which is called a slurry. This slurry is nebulized with a conventional Babington-type nebulizer and transported to the ICP as an aerosol containing suspended particulates in the plasma carrier gas. The total mass of particulate matter injected as a function of time must be kept to a minimum, so as not to overwhelm the plasma and reduce its atomization and ionization efficiency. Also, to ensure 100% vaporization and decomposition of the crystal struc-ture and matrix of the sample material in the plasma, the particle size must be reduced to less than 8 μm in diameter. Particles with a diameter larger than this require more energy to vaporize and ionize than is readily avail-able in the plasma under normal operating conditions. Therefore, the parti-cles will be passed through the plasma without being decomposed. Because no chemical decomposition process is used, the only possibility for contam-ination in this method originates from the sample grinding and pulverizing process used to reduce the particle size to that suitable for analysis.

A high-solids Babington-type nebulizer, such as the Cone-spray or V-groove type, used with a nonbaffled spray chamber permits the nebuliza-tion and transport of a slurry with a solids concentration up to about 1% m/v. A Babington-type nebulizer must be used to prevent clogging, to maintain stable aerosol generation. A dispersing agent such as Triton 100-X or Aerosol TO is often used to stabilize the slurry and to prevent floccula-tion of the particulates. Sodium pyrophosphate $(Na_2P_2O_7)$ has also been effectively used to maintain the suspension of slurries.

If the mean diameter of the particles is very small (1–2 μm), standard cali-bration curves can be prepared using an aqueous standard, direct calibration technique (see Chapter 7). One or more suitable internal standard elements are required to compensate for the differences in sensitivities of analytes that occur between the aqueous standards and the solid sample matrix. The use of internal standards ensures maximum accuracy of the calibration process.

Alternatively, matrix-matching procedures can be used, which in-volve approximating the matrix of the sample by the addition of a simi-lar quantity of the major component elements in the solids to the cali-bration standard solutions. This approach can potentially reduce the magnitude of interelement matrix interference effects. The technique can be problematic in that it may be difficult to add a sufficient quanti-ty of salts of the major components to approximate the solid matrix composition.

For larger particulate slurries (5–8 μm), the use of "similar matrix" calibration standard materials is required for attaining suitable accuracy. This consists of preparing solid calibration standards of an identical or similar matrix material as the samples, by adding known quantities of the analyte elements to the matrix material. This is essentially a *solid phase standard additions* process (see Chapter 7). Analytes can be added as powdered salts, followed by thorough blending and mixing, or as aqueous standard solutions, followed by immediate solvent evaporation. An alternate approach consists of preparing calibration curves using National Institute of Standards and Technology Certified Standard Reference Materials, U.S. Geological Survey Rock Standards, or other similar commercial standard reference materials. Finally, secondary standards can be used for direct calibration. This is accomplished by the preparation of calibration curves using solid samples that have been previously analyzed by a conventional chemical dissolution process.

Table 5.4 shows the comparison of results of the analysis of selected elements in NIST SRM 1623b and 1635 coals by slurry nebulization ICP-MS. For the analysis of these standards, aqueous standard calibration curves were used. From this data it is observed that high-concentration elements such as Al and Fe are simultaneously determined on the same sample as the low-concentration elements Mo and U.

5.3.3 Direct Insertion

The analysis of solid materials by their direct insertion into an ICP has been utilized as a method for compositional determination without the need for sample dissolution. The technique involves the packing of 10–30 mg of finely pulverized and homogenized (to ensure satisfactory subsampling statistics) sample into a depression or cavity in the end of a high-purity graphite rod. Often conventional direct current arc emission spectrographic electrodes can be used for this purpose.

A rather complicated mechanical apparatus is used to reproducibly insert this graphite rod axially through a custom configured plasma torch into the base of the ICP. This is usually accomplished by modifying the plasma torch to permit the rod to be inserted through the space where the aerosol injector tube is normally located (see Figure 5.26). Because of the thermal gradient in the plasma, the precise distance to which the sample is inserted into the plasma and the rate used to thrust it in are critical for reproducible and hence accurate sample analysis.

TABLE 5.4   Typical Analysis of Selected Elements (μg/g) in National Institute of Standards 1623b and 1635 Standard Reference Materials Using Slurry Nebulization Quantitation

Element	NIST 1623b		NIST 1635	
	Certified	Measured	Certified	Measured
Aluminum	3460	8550	1950	(3200)
Arsenic	3.8	3.7	0.6	0.42
Barium	57	68	75	—
Cadmium	3.7	0.06	75	0.03
Chromium	13	(11)	6.7	2.5
Cobalt	2.5	2.3	1.3	(0.65)
Copper	5.8	6.3	3.6	3.6
Iron	5980	7570	2210	2390
Lead	3.5	3.7	2.1	1.9
Lithium	12	(10)	2.6	—
Magnesium	368	383	818	—
Manganese	13	12	21	21
Molybdenum	0.9	(0.9)	0.5	—
Nickel	12	6.1	4.9	1.7
Strontium	87	(102)	110	—
Thorium	2.8	2.8	3.3	0.6
Titanium	312	454	205	(200)
Uranium	1.0	0.43	1.4	0.24
Vanadium	9.1	(14)	3.8	5.2
Zinc	13	12	8.8	4.7

Numbers in parentheses represent recommended or uncertified values.
Data from Ebdon et al. (1998).

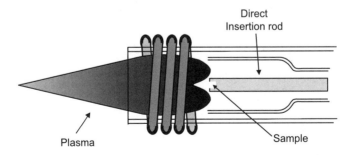

FIGURE 5.26   Direct insertion device.

When the solid sample comes in contact with the base of the plasma, a combination of thermal energy and impact from high-energy plasma gas ions vaporizes the analyte elements from the solid material. Very fine diameter particulate matter can also be sputtered from the sample. It is entrained in the plasma gas and transported into a high-energy region of the plasma. At this point, the particulates are thermally vaporized. The vaporized sample material is atomized in the high-energy region of the plasma and, finally, ionized prior to extraction into the mass spectrometer through the interface. The sample is sometimes mixed with graphite powder before it is packed into the electrode cavity to improve its volatilization characteristics.

As the sample is heated, analyte elements are vaporized from the sample in the order of the boiling points of the predominant compound of the element, prevalent in the sample (i.e., the most volatile compound is vaporized first followed by each of the other compounds in the increasing order of their respective volatilities). Because a finite quantity of each of these analytes is present in the sample, as they are volatilized, transported to the plasma, atomized, and ionized, a transient signal is produced from the mass spectrometer detector. Multiple analyte elements, which have different volatility rates, produce nonsuperimposed transient signals similar to those observed for electrothermal vaporization (see Figure 5.22). This process significantly complicates the suitability of this technique for multielement determinations.

This approach demonstrates a significant advantage over solution sample introduction techniques, in not requiring lengthy and contamination-prone sample dissolution prior to analysis. In addition, the absence of water as a solvent minimizes molecular interferences, especially in the low-mass region of the mass spectrum. However, if graphite is used as a binder, carbide molecules are expected to be more prevalent in the spectrum, which can contribute significant interferences.

CHAPTER 6

Special Techniques

Numerous special techniques have been devised to enhance the unique properties and capabilities of ICP-MS. For the most part, these techniques have centered around highly specialized sample introduction methods. Most of these techniques such as flow injection sample introduction have been borrowed from other analytical chemistry technology, but have been modified to take full advantage of the high-sensitivity and multielement detection capabilities of ICP-MS.

6.1 FLOW INJECTION

As previously described in the direct analysis mode of operation, sample is continuously introduced into the plasma, resulting in a constant production of ions, which are continuously sampled by the mass spectrometer interface. This results in a steady-state ion current signal produced by the detector. This ion current is integrated for a predetermined period of time. The resulting count rate is directly related to the population of ions in the plasma and hence the concentration of the analytes in the original sample. Blanks, standards, and samples are all measured under identical conditions providing a calibration strategy for quantitative chemical analysis.

An alternate approach uses a flow injection (FIA) sample introduction strategy. Discreet volumes of sample are repetitively injected into a continuously flowing carrier stream. This carrier stream can be either gas or liquid, but with conventional FIA systems it is usually the latter. These discreet volumes are either separated by air bubbles (segmented flow) or are not physically isolated in the flowing stream (unsegmented flow). The advantage of the segmented flow approach is to minimize dispersion of the sample in the carrier solvent. Although dispersion does occur with the unsegmented flow system, which increases as a function of the distance that the sample is transported, it offers a much simpler plumbing arrangement. Unsegmented flow systems provide more reliable and trouble-free operation.

As the sample reaches the nebulizer, aerosol is generated that contains discreet quantities of sample on a repetitive basis. When these parcels of sample aerosol reach the plasma and ions are produced, a series of transient signals is measured at the detector of the mass spectrometer corresponding to the sampling of ions by the interface and isolation by the mass filter. Figure 6.1 shows the transient detector responses of repetitive sequential flow injections. This figure shows 7 successive injections of 50 μg/L concentration copper (measured at m/z 65) present in a 3% m/v nickel solution over a 12-min time period. This data demonstrates a precision of 0.8% RSD.

A schematic diagram of a typical flow injection apparatus is shown in Figure 6.2. This apparatus consists of a rotary injection valve, which under computer control reproducibly alternates the transport of carrier solvent or

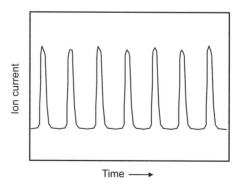

FIGURE 6.1   Repetitive flow injection responses for a Cu solution.

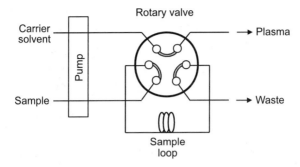

FIGURE 6.2   Flow injection apparatus.

a fixed volume of injected sample. The quantity of sample is precisely reproduced by the use of a fixed volume of the sample loop, which is alternately filled by pumping sample solution through it for a sufficient length of time to ensure that the loop is completely filled with undiluted samples. While filling the sample loop with sample, the carrier solution is directed to the nebulizer. After the sample loop is completely filled with sample, the valve is switched to allow the carrier solution to flow through the sample loop, thereby rapidly and quantitatively flushing the sample into the tube to transport it to the nebulizer. A precision pump is required to maintain a constant flow rate of the carrier and/or sample.

Using a system such as this, small volumes of sample can be analyzed without the requirement of lengthy stabilization times, which are normally required for the continuous-flow approach. Time and sample are not wasted while the transport plumbing/nebulizer/spray chamber assemblies are stabilized with sample to ensure an equilibrated system that will produce reproducible results. In addition, preallocated time is not required between samples for washout to minimize cross-contamination problems. Short washout times result in sample processing rates of about 100 samples per hour.

This system is particularly beneficial for the analysis of samples with high dissolved solid matrices. Usually, with direct sampling techniques, dissolved solid concentrations must be less than 0.2% m/v. Because only small discreet quantities of sample are injected into the plasma, immediately followed by a region of solvent only, plasma matrix effects are minimized, without loss of sensitivity by preanalysis dilution. The plasma and mass spectrometer are not exposed continuously to high dissolved solids from the sample, but only intermittently, as the discreet quantities of injected sample sequentially pass through the nebulizer. Problems normally encountered when analyzing high dissolved solid samples, such as nebulizer clogging due to salt buildup, and deposition of sample residue on the extraction cones of the mass spectrometer interface, both of which result in unstable ion current measurement, are not significant with flow injection sample introduction.

When samples with high analyte concentrations are analyzed, the flow injection system can be modified to incorporate an on-line dilution process. This on-line dilution can be achieved by adjusting the dispersion of the injected sample by increasing the length of the delivery tube between the injection valve and the nebulizer (dispersion of the segment of the sample is directly related to the length of the transport tubing, the total dead volume of the apparatus, and the delivery flow rate) and increasing the time between injections. This approach with other required adjustments to the timing process of the flow injection can result in a suitable fixed percentage dilution.

Chemical modifications can be achieved by the addition of chemical reagents to the solvent in the transport line before the sample has been injected. The analyte elements undergo reactions with the reagents during the time that they are transported to the nebulizer, the transport tube acting as the chemical reaction vessel. The length of the transport line and the flow rate of the carrier liquid control the time of the reaction. After the reaction is completed and before the sample is introduced into the plasma, separations can be carried out on the analyte species. Two examples of this process include the separation of volatile species formed by a hydride generation reaction by a gas–liquid separator; and the separation of complexed analyte species by a continuous liquid–liquid solvent extraction. This procedure has been effectively used for speciation measurements to isolate a specific form or oxidation state of the analyte prior to elemental analysis.

Another specialized application of the flow injection system includes an on-line matrix separation procedure. This approach can be used to either simply remove matrix components prior to analysis or to serve as a preconcentration step for individual analyte elements. The technique uses cartridges packed with cation or anion ion-exchange resins to sequester the matrix components, allowing the trace analyte species to pass through unsorbed to the nebulizer. An alternate approach uses chelating resins, such as Chelex 100, 8-hydroxyquinoline, or polydithiocarbamate to sequester the analyte elements, separating them from the sample matrix. The analytes are then removed from the cartridge using an elution reagent. The latter approach can also serve as a preconcentration technique.

6.2 CHROMATOGRAPHY

Chromatographic techniques are used to separate specific analyte compounds or forms of the analyte that are present in the sample. For medium to high molecular weight organometallic compounds, liquid chromatographic techniques are usually employed. For specific oxidation state ionic species or ionic complexes, ion-exchange chromatography is the method of choice. As specific species are eluted from the chromatographic column, they are transported to a nebulizer, nebulized, and introduced directly into the ICP for atomization, followed immediately by ionization. Vapor phase chromatographic and supercritical fluid chromatographic methods were discussed in Chapter 5.

The ability to measure the concentration of the exact form of the analyte present in the sample yields important information regarding the chemical and biological reactivity of the material. Because of the significant differences in toxicity between various forms of trace elements in the sample, occurrence standards are often established by regulatory agencies specific to a particular form of the analyte. This speciation analysis approach provides the information required to study and monitor these toxic compounds. In addition, to determine the fate of specific compounds in chemically reactive situations, quantitative information about the specific forms of the materials present is required to understand the process chemistry and kinetics that are occurring.

Speciation applications require analytical technology that can measure the presence and concentration of forms of the analyte elements that are present in the original specimen at the time of sampling. Extensive sample preparation usually results in modification of the form of the species being determined, thereby precluding accurate speciation measurements. To avoid this problem, on-line chromatographic techniques are utilized to measure the species present in the sample with a minimum of sample pretreatment.

6.2.1 Liquid Chromatography

Liquid chromatography is the most effective technique for separating polar, nonpolar, neutral, and complex compounds of intermediate to high molecular weight, low volatility, and low thermal lability. Reversed-phase liquid chromatography using combinations of aqueous and organic solvents as mobile phases has commonly been used. As compounds containing analyte

elements are separated by the chromatographic process, they are sequentially eluted from the column and directed to an appropriate nebulizer to form
an aerosol, which is transported to the plasma.

Organic solvents, ion-pair reagents, and salts in buffered solutions can all
be used as mobile phases. When organic solvents are utilized, the increased
carbon content in the ICP results in the formation of molecular carbide
species, which can produce isobaric interferences with some analytes. Also,
high concentrations of carbon in the plasma can increase the probability of
its deposition on the walls of the plasma torch and on the surface of the
sampler cone. By passing the organic-rich aerosol through a water-cooled
spray chamber, the vapor pressure of the solvent is decreased, which reduces
the quantity of carbon reaching the plasma. This approach, coupled with an
increase in the applied RF power used to generate the ICP, assists in stabi-
lizing the plasma. Also, the addition of 1–3% v/v of oxygen to the argon
plasma support gas can reduce the carbon buildup in the torch and on the
sampler cone. However, this process does result in the increase of oxide
molecular species in the spectrum, especially carbon-oxygen compounds.

The direct injection nebulizer (see Chapter 5) can be effectively used to
couple the liquid chromatograph to an ICP-MS. This approach was used
to obtain the chromatograms shown in Figure 6.3. Organolead and

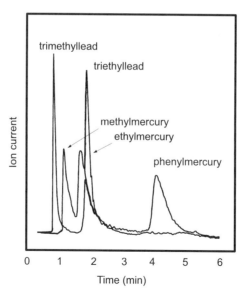

FIGURE 6.3   Chromatograms of organolead and organomercurial compounds.

organomercurial compounds were separated using reversed phase chromatography with ion-pair reagents. Ammonium pentane sulfonate, ammonium dodecanesulfonate, and ammonium heptanesulfonate were used with acetonitrile to perform this separation. Signal traces were obtained by measuring the ion currents from $^{208}Pb^+$ and $^{202}Hg^+$.

6.2.2 Ion-Exchange Chromatography

Ion-exchange chromatography inductively coupled plasma–mass spectrometry (IC-ICP-MS) uses anion or cation ion-exchange resins or other solid phases with ion-exchange properties to separate ionic species prior to their sequential introduction into the ICP. Ions are separated according to their ionic affinity to the resin, rather than their differences in solubility, which is the principle used in liquid chromatography. Separations are also highly dependent on the properties of the mobile phase. The mobile phase is usually composed of an aqueous-salt solution, which competes for the weakly electrostatically bonded analyte species.

As previously described in an earlier section, a sample containing a mixture of analyte species is injected into the mobile phase and sorbed onto the analytical column. Specific individual analyte species are sequentially eluted from the analytical column into a nebulizer where an aerosol is formed, which is transported into the plasma for atomization and ionization (see Figure 6.4.). Measurement of the individual species is accomplished by monitoring the transient signal for the ion that is specific to the analyte species that is being eluted from the column. This capability offers an ideal technique for use in speciation analysis.

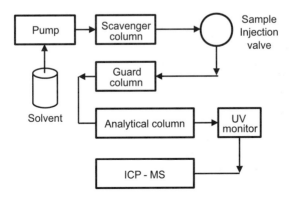

FIGURE 6.4   Ion-exchange chromatographic sample introduction device.

One application involves the separation and measurement of rare earth species with IC-ICP-MS. A silica-gel-based sulfonated cation-exchange column is used with a mixture of citric-hydroxyisobutyric acid as the mobile phase. Ion currents for each of the rare earth analytes are measured as a function of elution time to record a family of ion-specific chromatograms. This technique provides a means to eliminate several spectral isobaric interferences from rare earth oxides, hydroxides, and molecular hydrides on the analyte rare earth elements. For example, the isobaric interferences from the isotopes of Gd oxide and hydroxide on the primary isotopes of Yb and Lu were completely eliminated. In addition, the interference from the $^{139}LaH^+$ molecule, which directly interferes with the measurement of $^{140}Ce^+$, is totally removed by this approach.

6.3 FIELD FLOW FRACTIONATION

A technique utilizing sedimentation field flow fractionation (SdFFF) to separate suspended particulate matter prior to introduction into an ICP-MS provides a method for their elemental chemical characterization according to the mean particle size. The technique, invented by Taylor in 1990 (for further information, see Taylor *et al.*, 1992), is a useful tool for studying industrial processes involving sub-micron-size particulates, such as in the production of paint pigments. It has also been extensively used for studying and monitoring environmental processes involving the transport or interaction of suspended colloidal particulate matter in surface or ground water. The primary advantage of this approach is that the coupling of a high-resolution particle size separation technique, SdFFF, with the high sensitivity and element specificity of the ICP-MS allows rapid comprehensive analysis of particulates for major, minor, and trace chemical constituents.

An important characteristic of SdFFF is that particles in the 0.05- to 2-μm-diameter range can be separated from each other at high resolution. A schematic diagram of the SdFFF-ICP-MS instrument is shown in Figure 6.5. The technique is similar to a chromatographic procedure in that a discrete quantity of sample is loaded onto a channel (analogous to a chromatographic column). By passing a carrier solution (consisting of a surfactant, such as Triton 100-X or $Na_2P_2O_7$) through the channel at about 2 mL/min, various size particles are sequentially eluted from the exit port of the channel. The carrier solution, with the eluted particles, is passed through a particle light-scattering detector operating at 254 nm. As the

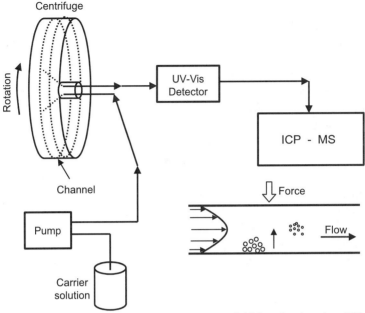

FIGURE 6.5   Schematic diagram of a sedimentation field flow fractionation-ICP-MS.

ultraviolet (UV) light is scattered in the detector, a fractogram is obtained that is analogous to a chromatogram. The carrier solution is then routed to a fraction collector, to collect samples of the fractogram for discrete analysis; or directly to a robust pneumatic nebulizer (insensitive to particle clogging) for direct introduction into the ICP-MS. In the latter case, which is most convenient for obtaining element-specific fractograms, the suspended particles are decomposed, atomized, and ionized in the plasma similar to the process used in slurry nebulization (see previous chapter on sample introduction). A specific m/z ion current, characteristic of each of the analyte elements, is monitored in the multielement analysis mode. Processing of this data yields element-specific fractograms for the composition of the separated particles.

The channel to induce particle separation is mounted on the circumference of a centrifuge. The rotational movement of the centrifuge creates a centrifugal force, that drives the particles to the outer accumulation wall of the channel. A diffusion process in opposition to the centrifugal force moves specific small-size particles off of the accumulation wall where they interact with the laminar flow of the carrier stream. As the particles move down the channel, this process separates them according to size. A typical rotation rate is about 1000 revolutions per minute (rpm). However, analogous to temperature programming in chromatographic methods, by using an exponential decay of the rotation rate, down to about 10 rpm, great improvements in resolution can be achieved.

Early work demonstrating the utility of this technique relied strictly on the collection of fractions at specific volumes eluted from the channel. Performing complete chemical analysis of each of these fractions, followed by plotting of the ion currents of interest as a function of the elution volume, provides a time-based fractogram that can be converted to a size-based fractogram by mathematically transforming the data using the UV scattering response. The fraction collection approach is a laborious and time-consuming process. More recent developments avoid the use of a fraction collector and involve the direct interface of the SdFFF with the ICP-MS, greatly enhancing the efficiency of the process. As particles are eluted from the channel, ion current output from the ICP-MS is continuous. Fractogram signals are produced directly, transformed for size analysis, and plotted in real time. Figure 6.6 shows the generic form of a typical fractogram. This fractogram demonstrates the size discrimination of the chemical composition of some types of particulates.

An advantage of using the wide dynamic concentration range capability of ICP-MS is the ability to perform the determination of major composition constituents simultaneously with the ultra-low-concentration trace elements. The major component analysis can be used to characterize the

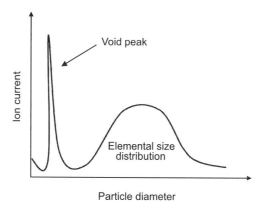

FIGURE 6.6   Typical generic fractogram.

nominal composition of the suspended matter, for example, the type and form of mineralization of the particles at different size fractions. The ability to accurately determine the trace composition of the particles provides a convenient way to study processes, such as trace element contaminant adsorption.

CHAPTER 7

Quantitation Techniques

Inductively coupled plasma–mass spectrometry is versatile enough to provide many approaches for performing a variety of types of analyses. The technique's high sensitivity, coupled with multielement detection capability, allows a wide scope of qualitative, semiquantitative, and quantitative determinations to be achieved over a large dynamic concentration range (minor to ultratrace concentration levels). By optimizing these quantitation methods, solutions to specific analysis problems can be arrived at readily.

A unique capability of ICP-MS is its ability to measure the concentration of specific analyte isotopes. This characteristic of performing isotopic ratio measurements provides a powerful alternative method called *isotope dilution analysis,* comparable to gravimetric determinations, which do not require a direct relationship to a calibration standard. The ICP-MS technique allows both single and multielement isotope dilution measurements to be performed on a single sample aliquot.

7.1 QUALITATIVE ANALYSIS

Qualitative analysis is the process of the determination of the presence (or absence) of a particular element or group of elements in a sample. The ability to perform a comprehensive qualitative analysis is directly related to the sensitivity of the analysis method and hence the detection capability. Ideally, it is desirable to determine major, minor, trace, and ultratrace concentration level elements simultaneously on the same sample aliquot, which requires an instrument and technique that exhibits a wide dynamic range of measurement of the ion currents for the various element isotopes. However, in practice, it is often difficult to determine the high intensity of major elements on the same sample dilution as that required for ultratrace concentration element determination (usually measured on undiluted sample). This problem is often accommodated by the use of very low abundance isotopes of the major concentration analyte, reducing the analysis sensitivity. Where no low abundance isotope is available, instrumentation with dual detectors (electron multiplier for low concentration analytes and Faraday analog detectors for high concentration elements) can be used effectively.

The simplest mass spectrum for qualitative analysis is normally obtained by using a spectral scanning procedure. The spectrum is continuously scanned from an m/z value of 1 to 240, with an ion current measurement being made at each m/z value. The spectrum then consists of a file with a single data point at each mass. A more comprehensive spectrum can be obtained by collecting additional data points for each nominal mass unit. For instance, 10 data points can be obtained every m/z, resulting in a mass spectrum with more peak structure. When a limited number of elements are to be sought, the peak hopping approach can also be used.

Because of the comprehensive nature of the ICP-MS spectrum (an uninterfered response for essentially all elements in the periodic table at roughly similar sensitivity, after correction for isotopic abundance), a very rapid determination of the presence or absence of a given analyte is established by observation for each of the m/z values for isotopes of the elements. This is illustrated by the comprehensive (10 points per mass unit) multielement spectrum from m/z 43 to m/z 70 shown in Figure 7.1. No signal observed at a specified m/z value confirms the absence of the element to a concentration level commensurate with the detection limit achievable under the operating conditions selected for the determination. Although the appearance of a signal at the specified m/z value does not immediately indicate the presence of a specific element with a high degree

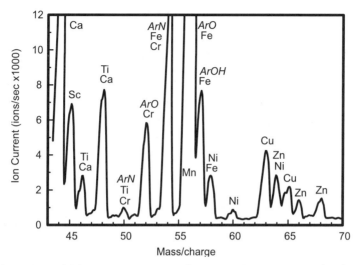

FIGURE 7.1   Multielement mass spectrum from m/z 43 to m/z 70. Molecular ions are shown in italics.

of certainty, it establishes a candidate for a potential confirmation. This confirmation is obtained by searching for another known isotope of the element in question. If a signal is observed at the expected m/z, a higher degree of probability of the element's presence is achieved. A final confirmation is established by measuring the intensity ratios of the specified m/z signals. If this ratio is within about 20% of the calculated ratio of the isotopic abundances (see Appendix 2), then confirmation of the presence of the element in the sample is complete. If possible, isotopes should be selected that have no obvious isobaric interferences, because this could seriously impact the measured ratio and results in erroneous identification of the analyte.

Qualitative analysis is normally used for an initial characterization of a totally unknown sample. After developing some general information about the composition of the material, a more sophisticated analysis scheme can be devised for further analysis. Although qualitative analysis is the simplest form of elemental characterization, with a marginal amount of additional effort, a semiquantitative assessment of the sample can be made, which provides the analyst with a significant amount of additional information, beyond the identification of the elements that are present.

7.2 SEMIQUANTITATIVE ANALYSIS

The qualitative analysis procedure can be expanded to provide semiquantitative data by the addition of a simple standardization step. A multielement semiquantitative analysis consists of the measurement of the concentration of multiple elements in a single sample to an accuracy of about $\pm 50\%$ or better, of the amount of each element present. To achieve this level of accuracy, it is necessary to use a calibration standard for each potential element to be measured.

A single multielement calibration standard is used to establish a relative sensitivity factor (R_i) for each analyte (i) to be determined in the multielement analysis. For solution analysis, this multielement standard is usually prepared from high–purity metal salts dissolved in deionized water, with sufficient nitric acid added to stabilize their concentrations (pH 2 or less). Because this is only a semiquantitative analysis, matrix matching of the calibration standard to the matrix of the sample is not required. When using solid analysis techniques (i.e., slurry nebulization, laser ablation, etc.), an appropriate solid phase multielement calibration standard is most desirable; however, novel approaches for the use of a solution standard have been used with some methods.

The magnitude (intensity) of the ion current peak in the mass spectrum is directly proportional to the concentration of the analyte. The relative sensitivity factor for each element is calculated by dividing the known concentration of the analyte in the standard (C_i) by the measured intensity (I_i) of the ion current for the appropriate m/z ion. To achieve the desired accuracy, only m/z values that are known to be interference free are used; otherwise a significant systematic bias will be introduced in all subsequent determinations using this R_i. The formula used to calculate the R_i follows:

$$R_i = \frac{C_i}{I_i}. \tag{7.1}$$

The magnitude of the relative sensitivity factors will vary depending on the abundance of the specific analyte isotope measured and the operating conditions selected for the analysis.

Concentrations of analytes in the unknown samples are obtained by measuring the ion currents for the specified m/z values under the same operating conditions that the R_i values were established. The measured ion currents multiplied by the predetermined R_i will yield a concentration in the same units as the calibration standard. These concentrations are usually accurate to within a factor of approximately 2 of the amount present. Because relative sensitivity factors are highly dependent on the specific operating conditions selected, they must be reestablished whenever adjustments or modifications are made to the instrument operating parameters.

More accurate semiquantitative determinations can be made by the use of an internal standard technique. Accuracies on the order of $\pm 50\%$ of the amount present can be achieved by this approach. A known quantity of an element that is not indigenously present in the sample being measured is added to the calibration standards before the relative sensitivity factors are determined. An internal standard element normalized relative sensitivity factor $(R_i^\star)$ is calculated as follows:

$$R_i^\star = \frac{R_i}{I_{IS}}, \tag{7.2}$$

where I_{IS} is the intensity of the internal standard element isotope ion current. When the exact same quantity of internal standard element is added to the sample, prior to analysis, its intensity is measured along with the appropriate analyte ion currents. The concentrations of the analyte elements in the sample are then computed from the following equation:

$$C_{analyte} = I_{analyte} \cdot I_{IS(sample)} \cdot R_{analyte}^\star. \tag{7.3}$$

Indium is often used as an internal standard element for semiquantitative analysis, because it is usually present at low concentration (or totally absent) in most types of samples; it can be obtained at modest cost in a highly purified form; it has two isotopes available for measurement at significantly different abundances (4.3% for m/z 113 and 95.7% for m/z 115) that are relatively interference free in most sample matrices; and these isotopes fall in the middle of the mass spectrum, so they can be used for both low and high mass analyte elements. Typical semiquantitative analyses, using both calibration methods, are illustrated in Table 7.1. Other elements can also be used for internal standards. Internal standardization is described in more detail in the following section.

This technique provides a useful method for the rapid multielement survey analysis of many types of materials. It is particularly valuable when there is a need for the rapid assessment of large numbers of samples, and the limited accuracy of semiquantitative determinations is acceptable for the application.

The semiquantitative analysis method also provides an excellent way to rapidly measure the approximate concentration of analyte elements in

TABLE 7.1   Typical Semiquantitative Analysis (μg/L) of Selected
Elements in National Institute of Standards and Technology 1643a
Trace Elements in Water Standard Reference Material, Using
Relative Sensitivity Factor and Internal Standard Calibration

Element	RSF[a]	IS[b]	Certified
Arsenic	90	78	76
Barium	30	37	41
Beryllium	20	20	19
Cadmium	5	8	10
Chromium	20	19	17
Cobalt	20	18	19
Copper	10	16	18
Iron	80	90	88

[a] RSF, relative sensitivity factor calibration.
[b] IS, internal standard calibration.

unknown samples. This approach can be used effectively to acquire preliminary data prior to designing a more detailed high accuracy and precision
quantitative analysis. Information about potential interelement interferences and approximate concentration levels of analytes is invaluable in optimizing such methods as standard addition or isotope dilution quantitation
techniques.

7.3 QUANTITATIVE ANALYSIS

Quantitative analyses are used to measure highly accurate and precise concentration data of constituents in sample specimens. With techniques such
as ICP-MS, the capability to perform multielement determinations at very
high sensitivity provides a unique approach to acquiring high-quality data.
In addition, the unique ability of ICP-MS to perform stable isotopic measurements offers a highly accurate approach to quantitation without the
need for high-quality standard reference materials. To obtain high accuracy,
potential interferences must either be absent from the analysis procedure or
a suitable means of correction must be available to the analyst. Therefore,
quantitative analysis procedures must take advantage of methods to remove
interferences or provide techniques for their correction.

7.3.1 Direct Calibration

Calibration Curves

The most fundamental approach for quantitative analysis involves the use of a conventional analytical calibration curve. This curve establishes a functional relationship between the ion current of the analyte and known concentrations of carefully prepared calibration standards. For dilute solution analysis, these standards usually consist of gravimetrically prepared aqueous dilutions of high-purity metal salts. These compounds should be either nitrate or oxide salts. Chloride salts should be avoided whenever possible to prevent the possibility of molecular ion interferences (see chapter on interferences). An alternate approach is to use commercially available calibration standard solutions that have been obtained from a trusted source. After dilution, the standards are preserved with high-purity nitric acid to prevent precipitation or adsorption of the trace analyte species onto the interior walls of the sample holding container.

Stock standard solutions are usually prepared at the 1–10 mg/L concentration level. These solutions are sufficiently stable, if properly preserved, that they have a shelf life of approximately 1 year. Intermediate serial dilutions (μg/L concentrations) are prepared on a monthly basis and final low-concentration working standards (ng/L concentrations), made from the intermediate dilutions, are freshly prepared daily or weekly. All standards should be stored in tightly sealed PTFE bottles at room temperature in the dark to minimize the possibility of photochemical reactions.

For the analysis of more complex materials, such as acid digests, fusions, or other samples that have high concentrations of components from the original material, or components introduced by the sample preparation procedure, a matrix-matching technique must be employed to achieve accurate results. Matrix matching involves preparing the calibration standard solutions with approximately the same composition and concentration of elements that are present in the original sample. This will normalize any effects from these major components on the relative sensitivity of the analytes, thereby minimizing matrix interference effects. If the concentration of acids (either mineral or organic) exceeds about 1% (by weight), they also should be incorporated in the matrix-matching process.

For the best accuracy, the calibration curve should be constructed so that there is at least one standard with a concentration above and one below the concentration of the unknown being measured. This effectively brackets the unknown concentration so that extrapolation is not required. An example of a typical calibration curve is shown in Figure 7.2. Although some ICP-MS calibration curves may have a curvilinear functional relationship, the best accuracy is achieved when the calibration curves are linear. Nonlinear curves usually indicate that matrix suppression effects or background effects might be influencing the integrity of the calibration. It is recommended that the degree of linearity of the calibration curve be used as a measure of the quality of the calibration process.

It is common practice to perform a statistical fit of the calibration standards to produce a calibration equation for each element. A statistical least squares regression fit of the data is recommended. The "goodness of fit" of the regression is used as a measure of the reliability of the calibration. The correlation coefficient (R^2) should be computed at 0.999 or better. If it is less than 0.999, the calibration standards should be rerun or reprepared to achieve the 0.999 level. For the least squares regression analysis to be used effectively, the calibration standards must be uniformly spaced throughout the calibration range. Once established for a given run of samples, the concentration of each element is directly computed from the approximate regression equation.

Scanning techniques can be employed for the measurement of the analyte ion currents at the appropriate m/z values. When scanning is used, several replicate scans for each sample analysis should be employed to minimize the measurement variance of each standard and sample. If scanning is used for quantitation, either peak heights or peak areas can be used to establish the intensity of the ion current.

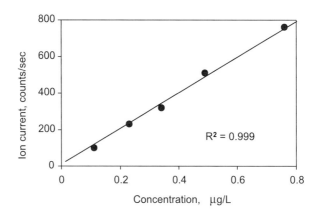

FIGURE 7.2   Graph showing typical calibration curve.

For the best quantitation process, peak hopping with signal integration is the preferred data acquisition technique. The simplest procedure is to collect a single data point per m/z peak being measured, integrated for a controlled period of time. This takes the shortest total elapsed time for the determination. To obtain better precision, multiple data points per peak are used to better define the top of the mass spectral peak. Usually, three points per peak is adequate for most quantitative analysis methods. A typical dwell time of 150 ms is used with three to five replicates. The average of these points is used to establish the signal intensity. For high-resolution (magnetic-sector) instruments operated in the low-resolution mode, flat top peaks are observed, which produce significantly higher precision than the quadrupole instrument, with even only one point per peak.

Because this technique offers a combination of high sensitivity (detection limits in the sub-part-per-billion range) and a wide linear dynamic range, calibration curves are often designed to cover several orders of magnitude (8–11) of concentration. This is particularly advantageous when operating in the multielement mode, because widely varying concentrations of various elements can be determined simultaneously with a single calibration curve for each element. Calibration curves are sometimes plotted on a log-log scale to cover the wide dynamic working range. It is recommended that a minimum of five calibration standards, equally spaced throughout the concentration range, be used to establish the calibration curve.

When comprehensive multielement determinations are to be performed, mixed-element calibration standards are prepared. Elements in each mix must be carefully selected to ensure compatibility and hence stability of the standards. An example of the combination of elements in each mix is shown in Table 7.2. A set of these mixed-element calibration stan-

TABLE 7.2   Recommended Composition of Mixed-Element Calibration Standards for Direct Calibration Curve Quantitation

---

Calibration mix A (HNO$_3$)
  Al, As, Be, Bi, Cd, Ce, Cs, Co, Cu, Eu, Fe, La, Pb, Li, Lu, Mg, K, Pr, Sm, Sc, Se, Sr, Tl, Th, U, V, Yb, Y, Zn

Calibration mix B (HCl)
  Sb, Ba, Ga, Ge, Mo, Nd, Ni, Pd, Na, Ta, Te, Ti, W, Zr

Calibration mix C (HNO$_3$)
  B, Ca, Cr, Dy, Er, Gd, Ho, Mn, Re, Rb, Si, Tb, Tm,

---

dards is prepared for each concentration level to be used for the creation of the calibration curves.

The results of a typical direct calibration curve multielement quantitative analysis are shown in Figure 7.3. This is a correlation plot of the measured concentration of selected elements in National Institute of Standards and Technology (NIST) 1643a and 1643b, Trace Elements in Water, Standard Reference Materials; and the concentration of selected elements in U.S. Geological Survey (USGS) Standard Reference Water Sample, SRWS 53, versus their certified values. To cover a wide concentration range, this is shown as a log-log plot. The plot demonstrates that essentially all elements measured fall on the "line of perfect agreement" of the correlation plot demonstrating the merit of this quantitation technique.

Internal Standardization

To obtain the highest quality concentration data using the direct calibration curve approach, the use of internal standardization is required. Internal standards are elements that are added to the calibration standards and samples at a fixed known concentration, sufficiently high to allow ion current measurements to be made without limitation by counting statistics. Several criteria must be met to achieve satisfactory internal standardization: (1) The internal standard element(s) must be absent or at an insignificantly low concentration in the samples, so as to not have the indigenous levels miti-

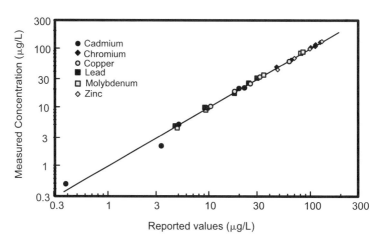

FIGURE 7.3   Multielement correlation plot of direct calibration curve quantitation for NIST 1643a and 1643b trace elements in water and USGS SRWS 53.

gate the process; (2) the internal standard elements must be available in a high-purity form so that there is no contamination of analyte elements; (3) there must be at least one uninterfered isotope available for measurement; and (4) it must not interact with indigenous matrix or analyte elements from the sample.

The purpose of the use of internal standards is to compensate for intermediate to long-term drift in the measurement components of the instrument during an analysis. By using appropriate internal standards, the stability of the measurement can be improved by an order of magnitude, which is reflected in the overall accuracy of the quantitation. Figure 7.4 shows a plot illustrating a typical internal standard correction. Several elements that have been successfully used as internal standards are listed in Table 7.3 along with the isotopes used for measurement.

The most appropriate internal standard element is one that is closely located to the analyte in the mass spectrum. When the multielement analysis covers a wide mass range, several separate internal standard elements are used simultaneously to achieve optimal results. For optimal results, analytes being determined using low m/z values require an internal standard with a low m/z value. Similarly, analytes being measured at intermediate or high m/z values require internal standard elements located in these mass ranges.

The procedure for the use of internal standardization requires that the same concentration of the selected internal standard element be added to all of the calibration standards and the unknown samples and blanks. After acquiring ion current data from the mass spectrum, the ratio of the analyte

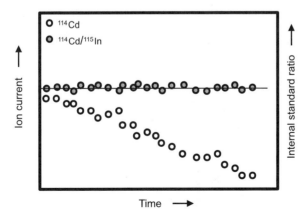

FIGURE 7.4   Plot of mass spectral ion current as a function of time showing effect of drift correction by internal standardization.

TABLE 7.3   Elements That Have Been
Successfully Used for Internal Standards

Element	m/z
Germanium	72
Germanium	74
Indium	113
Indium	115
Lithium	6 (enriched)
Rhodium	103
Scandium	45
Terbium	159
Thallium	169
Thorium	232
Yttrium	89

ion current (I_i) to the internal standard ion current (I_{IS}) is plotted versus the known concentration of the analyte in the calibration standard. A typical plot illustrating this is shown in Figure 7.5. As before, a least squares regression equation is obtained for these data. Concentrations of the analytes in the unknown samples are computed from these equations after measurement of the ratio of analyte to internal standard ion currents in the sample.

Internal standard elements can be added to the calibration standards and the samples by a simple volumetric addition to the appropriate solutions. Although this provides an accurate means of adding internal stan-

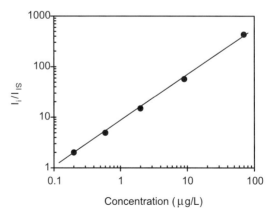

FIGURE 7.5   Calibration curve using internal standard drift correction.

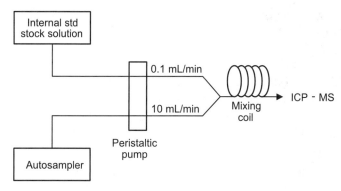

FIGURE 7.6   Automated internal standard addition setup.

dard elements, it is tedious and time consuming, especially when many samples are to be analyzed. An alternate approach is the use of an on-line continuous internal standard addition process. A schematic diagram of this device is shown in Figure 7.6. This system is inserted between the autosampler and the nebulization apparatus. A constant ratio (usually 1:10) of internal standard stock solution and sample or calibration standard solution is continuously mixed as the sample is being transported to the nebulizer. This same apparatus can be used to make an on-line dilution of the sample, greatly reducing the manual manipulation of the sample during volumetric dilutions.

Some workers have suggested that internal standardization can be effectively used to correct for matrix interference effects. Studies have shown that under a given set of plasma operating conditions (i.e., forward RF power, nebulizer gas flow rate, sampling position, etc.), groups of analyte elements tend to behave similarly with respect to matrix enhancement or suppression. By selecting one of these elements as an internal standard, some correction for variations in these effects can be made. Generally, matrix-matching or dilution protocols are more effective in dealing with these problems.

7.3.2 Standard Addition

The standard addition quantitation technique is often employed when verification of results is required. The advantage of the method is that a calibration curve is prepared with the exact matrix of the unknown sample.

Preliminary analysis of the sample to determine the approximate composition for matrix matching is not required and inconsistencies in this process are avoided.

The procedure involves making a minimum of three known quantity spike additions of the analyte to equal aliquots of the sample. For optimum results the first addition should be of a magnitude that is approximately equal to the estimated quantity of the analyte in the sample. The second addition should be twice the quantity of the first addition, and the third addition should be three times the quantity of the first addition. These three spiked addition samples together with an unspiked aliquot of the sample are all measured under identical analysis conditions. The ion current for the analyte in each sample is plotted versus the quantity of the analyte added in the spike. An example of a standard additions plot is shown in Figure 7.7. The fit of these three points must result in a linear function with a minimum of variation in any of the three data points. If this is not essentially a perfect fit, the process must be repeated.

The point at which the extrapolated curve intersects the abscissa axis represents the quantity of the analyte in the original sample. In addition to the requirement of curve linearity, any interference or background signal must be negligible, because it will not be differentiated from the analyte ion current and will result in a biased concentration determination.

Although this technique can yield excellent results, as illustrated in Table 7.4, it is a time-consuming process and for this reason is only used for limited analytes in small numbers of samples. It is advisable that a preliminary semiquantitative analysis be performed so that the magnitude of the spike concentration can be chosen to provide optimal results. In general, this approach is not suitable for multielement analysis.

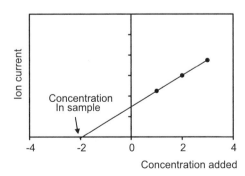

FIGURE 7.7   Graph showing a standard addition plot.

TABLE 7.4　Typical Analysis (μg/g) of Selected Elements in National Research Council of Canada CASS-1 Standard Reference Material, Using a Standard Additional Quantitation Technique

Element	Certified	Measured SA[a]
Cadmium	0.026 ± 0.005	0.027 ± 0.001
Copper	0.29　± 0.03	0.263 ± 0.007
Lead	0.25　± 0.03	0.22　± 0.02
Nickel	0.29　± 0.03	0.302 ± 0.005
Zinc	0.98　± 0.1	1.05　± 0.004

[a] SA, standard addition.

7.3.3 Isotope Dilution

The ability to independently measure the ion current of individual stable isotopes of an element provides the analyst with an opportunity to perform a definitive quantitation technique called *stable isotope dilution analysis*. The sensitivity and measurement precision of ICP-MS permits stable isotope dilution quantitation to be performed at trace concentration levels, without preconcentration. In addition, with a carefully designed analysis strategy, simultaneous multielement determinations can be accomplished (Garbarino and Taylor, 1987).

As discussed in Chapter 2, many elements have multiple stable isotopes of varying abundance. The abundances for stable isotopes of the elements are listed in Appendix 2. The ratio of two stable isotopes of a given element can be determined by independently measuring their ion currents. The isotope ratio is then computed by dividing one m/z isotope ion current by the other m/z isotope ion current. If no isobaric mass spectral interferences are present, and the sample has a natural isotopic abundance for the element in question (i.e., no fractionation or radiogenic processes have affected the abundances), the calculated ratio should be close to the theoretical value computed from the Appendix 2 table entries. This ability to perform isotope ratio measurements provides an opportunity to carry out isotope dilution analysis.

The principle of isotope dilution quantitation involves the precise addition of an accurately known quantity of an enriched stable isotope of the analyte element to the unknown sample. After thorough equilibration with the unknown sample, the modified analyte isotope ratio is measured and the concentration of the analyte in the original sample is calculated by the isotope dilution equation. This equation is as follows:

$$C_{analyte} = \frac{M_e K (A_e - B_e R)}{W(BR - A)},$$
(7.4)

where $C_{analyte}$ is the concentration of the analyte in the original sample, M_e is the mass of the enriched isotope spike, W is the weight of the sample, K is the ratio of the natural atomic weight to the atomic weight of the enriched material, A is the natural abundance of the reference isotope, B is the natural abundance of the enriched isotope in the sample prior to spiking, A_e is the abundance of the reference isotope in the enriched spike solution, B_e is the abundance of the enriched isotope in the spike, and R is the final measured isotope ratio of the reference isotope to the enriched isotope (after equilibration). For maximum accuracy, the measured abundances of the two isotopes in the original sample are usually used rather than assuming natural abundances.

Representative mass spectral diagrams for copper, illustrating this procedure, are shown in Figures 7.8, 7.9, and 7.10. Two uninterfered stable isotopes for copper are available for isotope dilution analysis, ^{63}Cu and ^{65}Cu. The natural abundances for these isotopes show ^{63}Cu being approximately twice as abundant as ^{65}Cu (see Figure 7.8). An enriched isotope of ^{65}Cu can be commercially purchased such that $^{65}Cu/^{63}CU = 322$. A mass spectrum of the enriched isotope is shown in Figure 7.9. After a known quantity of

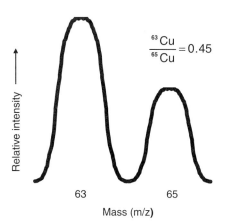

FIGURE 7.8   Mass spectrum showing the natural abundance of ^{63}Cu and ^{65}Cu.

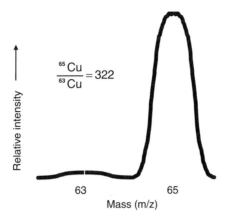

FIGURE 7.9 Diagram of mass spectrum showing the abundance of enriched ^{65}Cu.

the enriched ^{65}Cu spike is equilibrated with the original sample, a modified isotope ratio is measured. The mass spectrum of this equilibrated mixture is shown in Figure 7.10. For maximum precision of the analysis, the concentration of the enriched isotope is adjusted so that its ion current is approximately equal to that of the reference isotope, or that the isotope ratio is close to a value of 1.

For selected elements, a summary of isotope pairs suitable for isotope dilution analysis is shown in Table 7.5. In addition, this table lists the enrichment values that are commercially available for these elements. For some

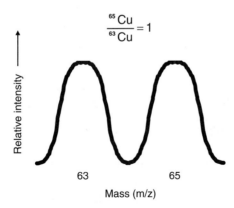

FIGURE 7.10 Diagram of mass spectrum showing equilibrated isotope dilution of ^{63}Cu and ^{65}Cu.

TABLE 7.5   Isotope Pairs for Selected Elements Suitable for Isotope
Dilution Quantitation

		Abundance ratio	
Element	Ratio	Natural	Enriched[a]
Barium	135/138	0.092	3.40
Cadmium	116/114	0.27	45.1
Copper	65/63	0.45	322
Lead	206/208	0.48	3325
Nickel	61/60	0.048	14.5
Thallium	203/205	0.42	23.0

[a] Commercially available.

elements the preparation of enriched isotopes is a very difficult process. This difficulty is reflected in the degree of enrichment, the availability, and the cost of the enriched isotope.

A typical procedure for performing a multielement isotope dilution analysis consists of the following sequential steps:

1. Perform a suitable semiquantitative analysis of the sample (see previous section) to estimate the concentrations of the analyte element or elements (in the case of multielement analysis) present.
2. Knowing the estimated concentrations, spike the sample with the quantities of enriched isotopes that will result in approximately equal ion currents between the reference isotopes and the enriched isotopes.
3. Equilibrate the sample to ensure intimate mixing of the natural isotopes and the enriched spikes.
4. Measure the altered isotope ratios of the reference isotopes and the enriched spikes
5. Calculate the original concentration or concentrations of the analytes from the isotope dilution equation, Eq. (7.4).

An enriched isotope spike stock solution for each analyte should be gravimetrically prepared at a known concentration. All samples and spike additions should be weighed rather than volumetrically transferred to provide maximum accuracy for the analysis.

Unlike other isotope dilution analytical methods (i.e., thermal ionization mass spectrometry), separation and isolation of the analyte species are not required, greatly simplifying the determination procedure.

Using this approach, highly precise analyte isotope ratios can be measured. This high precision directly translates to high analysis accuracy, commensurate with the accuracy with which the sample and enriched isotope is weighed. Of all techniques suitable for use with ICP-MS, the isotope dilution method universally provides the highest accuracy. In fact, this procedure is widely used for the certification of natural matrix secondary reference standards. An example of the accuracy obtainable by this technique is shown in Table 7.6. This table shows the results of the determination of selected elements in NIST 1643a and 1643b, Trace Elements in Water, Standard Reference Materials, by multielement isotope dilution analysis. In addition to the mean values, error terms representing the 95% confidence level are presented.

The isotope dilution analysis method provides a definitive technique that does not depend on the comparison to one or more external calibration standards for quantitation. The method is similar to the standard addition technique in that a known quantity of analyte material is added to the sample to be analyzed, thereby performing the quantitation *in situ*. This approach eliminates matrix interference problems. The method is also similar to the internal standardization procedure, with the exception that a form of the analyte itself serves as the internal standard. All of these factors combine to provide analysis characteristics comparable to the most basic gravimetric methods.

To utilize this methodology, one or more uninterfered stable isotopes must be available for measurement. Monoisotopic elements cannot be

TABLE 7.6   Typical Analysis (μg/g) of Selected Elements in National Institute of Standards and Technology 1643a and 1643b Trace Elements in Water Standard Reference Materials by Isotope Dilution Quantitation

Element	1643a		1643b	
	Certified	Measured	Certified	Measured
Barium	46.8 ± 2	45.6 ± 1	45 ± 2	45.6 ± 1
Cadmium	10.2 ± 1	11.9 ± 0.1	20.3 ± 1	21.9 ± 0.3
Copper	18.3 ± 2	20.1 ± 0.8	22.3 ± 0.4	24.3 ± 1
Nickel	55.9 ± 3	58.1 ± 3	49.8 ± 3	55.1 ± 1
Lead	27.4 ± 1	27.8 ± 0.3	24.1 ± 0.7	21.4 ± 0.2
Strontium	243 ± 5	243 ± 2	230 ± 6	234 ± 2
Thallium	—	—	8.1 ± 2	8.0 ± 0.02

TABLE 7.7   Elements That Are Not Suitable (Monoisotopic) or Potentially Not Suitable for Isotope Dilution Quantitation

Monoisotopic
Al, As, Au, Be, Co, Cs,
Ho, I, Mn, Na, Nb, P, Pr,
Rh, Sc, Tb, Th, Tm, Y
Potential isobaric interference on at least one isotope
Cd, Ce, In, La, Lu, Pd,
Rb, Re, Sb, Ta, V

determined. In principle, radioisotopes with a long half-life may be used in extreme cases. Table 7.7 lists all of the elements that are not suitable for analysis by ICP-MS. Because of the several steps in the analytical procedure, including the initial semiquantitative analysis to estimate the concentrations of the analytes prior to the isotope dilution, rapid analysis is not feasible. Therefore, this method is usually not employed for large-scale analytical work, but more often is used as a referee or quality assurance technique.

7.3.4 Quantitation Strategy

To achieve the best accuracy and precision for quantitative analysis, samples must be analyzed in—at least—triplicate. This accomplishes two objectives: (1) The agreement between the three replicates provides a basis for editing the data to reject outliers and (2) a measure of the determination of variance can be calculated from the three replicates. Generally, by making a minimum of three replicate determinations, reruns, dictated by quality assurance protocols (see Chapter 10) will not be required or will be necessary only in unusual situations. Reporting results as a mean and standard deviation ($\pm$ error term) provides data that have more interpretation value.

Blank samples must always be analyzed with the unknown samples to evaluate the potential for contamination and to correct for trace concentration levels of the analytes in the reagents used for preparation or preservation of the samples. Great care must be employed when using data obtained from blank determinations. If the blank consists of a representative quantity of the reagents or other components of the sample, such as deionized water, which may have been added for dilution prior to analysis, the analyst is justified in subtracting the analyte concentrations in the blank from that measured in the sample. This results in a net, blank subtracted analysis, which theoretically represents the concentration of the analytes in the original sample. When there is any question about the validity of the blank determinations, they must be reanalyzed before corrections are applied.

Any blank specimens that represent the handling process to which the sample was exposed, either from the original collection of the sample or from the handling of the sample up to the time of analysis, cannot be subtracted from the observed concentrations in the sample. The utility of these blanks is strictly quality control, as positive observations call into question the validity of the analyses.

CHAPTER 8

Interferences

To obtain the most accurate quantitative analysis, potential interferences that could affect the measurement of ion currents must be eliminated or minimized. This task is particularly difficult when attempting to perform multielement determinations for elements with m/z values ranging from the low mass end of the mass spectrum (i.e., Li at m/z 7) to the high mass end (i.e., U at m/z 238). In addition, when simultaneous analysis is attempted for various concentrations of different elements, ranging from high levels to ultra-trace levels, the potential for interelement interference effects is increased.

Two basic categories of interference are encountered: spectroscopic and nonspectroscopic. Each of these types has characteristics extending the interference from minor to extremely severe. Interference effects of minor magnitude can usually be ignored, qualifying the reported data to accommodate uncertainties in accuracy. Also, when interference effects are of low magnitude, various types of corrections can be applied to minimize their importance or compensate for the error.

When interference effects are large in magnitude, external sample treatment is usually the only means of alleviating the problem. This is often accomplished by chemical separation of the analyte species from the sample matrix prior to analysis. If this approach is not possible, then the interference probably precludes the determination of the desired analyte element under the prevailing conditions.

8.1 SPECTROMETRIC EFFECTS

Effects that impact the measurement of specific isotope ion currents in the mass spectrum are identified as spectrometric interferences. In general, these types of interferences result in a positive error on the analyte ion current measurement. There are four basic types of mass spectrometric interferences: isobaric spectral overlap, polyatomic molecular ion overlap, multiple charged species (usually doubly charged ions), and background contribution to the measurement of the ion current. Each of these types of interferences is described in this section.

8.1.1 Isobaric Spectral Overlap

As described in Chapter 2, each element has at least one isotope that appears at a specific m/z value in the mass spectrum. Most elements in the periodic table have multiple isotopes. Each isotope, for a given analyte element, has a specific and unique m/z value. It is fortunate that nature has provided a distribution of isotopes (see Appendix 2) for the elements such that most of them have at least one isotope that does not nominally overlap with isotopes from one or more adjacent elements. This is why satisfactory quantitative analysis can usually be accomplished for most elements with unit-mass resolution quadrupole mass spectrometers. This process of isotope overlap is called *isobaric spectrometric interference*. In general, most isotopes with odd m/z values are free from isobaric interference, while many with even m/z values suffer from this interference. Below m/z 36, there are no isotopes exhibiting isobaric interferences. Even if isotopes do nominally overlap, by using mass spectrometers with sufficiently high resolving power (double-focusing magnetic sector), uninterfered ion currents can be measured.

An additional factor in isobaric overlap concerns each isotope having a specific abundance ranging from a fraction of a percent to a maximum of

100%. When elements that are present in the sample at high concentration have even minor isotopes that overlap another element present at low concentration, an isobaric interference exists that can be significant. However, when the converse is true, the interference will probably be insignificant. The most abundant isotope for a given element is always selected to provide the most sensitive analysis and, hence, the best detection limits. If other elements are present in the sample at measurable concentrations that have isotopes that overlap with this m/z value, they can have a detrimental effect on analysis accuracy. For example, the most abundant isotope of zinc is ^{64}Zn, which would be the obvious selection for ultra-low-concentration level determinations. But a substantial quantity of nickel is present in the sample, the isotope ^{64}Ni, even though it is only 0.92% abundant, will interfere with the trace analysis of zinc. This situation must be evaluated on a case-by-case basis to determine if an operational isobaric interference is present that will affect accuracy.

In addition to isobaric overlaps from other species in the sample matrix, a common source of interference is from the plasma support gas. For example, Ar, by far the most common support gas for *atmospheric pressure* ICPs, has a major isotope at m/z 40 (99.6% abundant). This isotope has a direct isobaric interference with ^{40}K$^+$ and ^{40}Ca$^+$. Because the ion current for ^{40}Ar$^+$ is always very large, a measurement of ^{40}K$^+$ or ^{40}Ca$^+$ is not possible, even with a high-resolution mass spectrometer. Also, Kr is often a detectable impurity in Ar gas. One of its isotopes, ^{86}Kr$^+$ has an isobaric interference with ^{86}Sr$^+$. Therefore, if an attempt is made to measure the isotope ratio ^{86}Sr/^{88}Sr, which is commonly performed in geochemistry applications, the presence of Kr in the Ar can result in a distorted ratio.

Isobaric interferences can be corrected by a simple mathematical approach. By subtracting the calculated contribution of the interfering isotope ion current from the total ion current measured at the interfered m/z value, a net ion current for the analyte isotope is obtained. The contribution to the ion current by the interfering isotope is computed from the known ratio of its abundance to that of an uninterfered alternate isotope of the same element. This calculation is shown in the following equation:

$$I_{net} = I_{total} - \left(I_{unint} \cdot \frac{A_{int}}{A_{unint}} \right), \qquad (8.1)$$

where I_{net} is the net ion current of the analyte isotope, I_{total} is the total ion current measured at the interfered m/z value, I_{unint} is the ion current measured at an uninterfered alternate isotope of the interfering element, A_{int} is the abundance of the interfering isotope, and A_{unint} is the abundance of the uninterfered isotope of the interfering element. As with other computations of this type, this correction is only accurate if the correction term is small relative to the total ion current measured at the interfered m/z value. Otherwise, subtracting two comparable magnitude values from each other results in a relatively inaccurate net ion current.

The previous example of nickel interfering with the determination of zinc is used to demonstrate this correction. Zinc is determined by measuring the ^{64}Zn isotope. Because ^{64}Ni (0.925% abundant) interferes with the ^{64}Zn, measurement of the ion current of the ^{60}Ni (26.2% abundant) is used to compute a correction for the ion current measurement at m/z 64. This calculation is shown below:

$$I_{^{64}Zn} = I_{^{64}total} - \left(I_{^{60}Ni} \cdot \frac{26.2}{0.92}\right). \tag{8.2}$$

Isobaric interferences are usually more prevalent when very complicated matrix samples are analyzed. The more elements present in the sample at nominally high concentrations, the greater the probability of isobaric overlap occurring at desirable m/z values, used for analyte element determinations. If no uninterfered isotopes of the analytes are available for use, or if those that are available have such a low abundance that an analysis with adequate sensitivity is not possible, then a chemical separation to remove interferents is required to permit an accurate analysis. The only other alternative is to perform the determinations on a high-resolution mass spectrometer, which is capable of making uninterfered ion current measurements.

8.1.2 Polyatomic Molecular Spectral Overlap

Another prevalent feature of the mass spectrum, which is often more of a problem than elemental isobaric overlap, is the occurrence of peaks attributed to ionized molecular species. The formation of molecular ion species is discussed in Chapter 2. As discussed there, usually only diatomic homogeneous molecular species, such as Ar_2^+, and diatomic heterogeneous molecular species, such as ArO^+, are observed in the mass spectrum from the ICP, although some triatomic species, such as $ArOH^+$, are observed when the components that form these polyatomic molecules are present at high concentrations. When molecules are present, only singly charged ions are observed. The principal m/z value observed in the mass spectrum for the molecular species is determined by the sum of the most abundant isotope for each of the atoms for which the molecule is comprised. Other peaks are observed that correspond to combinations of the minor isotopes of the atoms. As with isobaric interferences, the mass deficit, or exact mass, of the molecule can often be separated from the analyte isotope with a high-resolution mass spectrometer. Appendix 3 lists the most common of these polyatomic molecules.

Polyatomic Molecules from Plasma Gas Components

A number of molecular ions originate from the components of plasma support gas. For atmospheric pressure ICPs, which uses argon as the support gas, important interferences arise from Ar-containing molecules. The second most abundant species in the argon ICP is the $^{40}Ar_2^+$ dimer molecule at m/z 80, second only to the $^{40}Ar^+$ ion at m/z 40. This molecule interferes with the measurement of the ion current of $^{80}Se^+$. A second Ar dimer, composed of $^{40}Ar^{36}Ar^+$, directly interferes with $^{76}Se^+$ and $^{76}Ge^+$. Finally, there is a third dimer of argon composed of $^{40}Ar^{38}Ar^+$ that, although it is of much lower intensity, directly interferes with $^{78}Se^+$. Even though it is an inert gas, Ar forms molecules with other abundant species in the plasma. These species are discussed in detail in subsequent sections.

Molecular Oxides

Molecular oxide species are formed from excess oxygen present in the plasma. This excess oxygen can originate from several sources depending on the type of sample introduction being utilized. For example, with laser ablation techniques, the primary sources of oxygen are impurities in the Ar support gas, air leaking into the gas plumbing system, and oxide compounds in the original sample. When using solution nebulization sample introduction, the most predominant source of oxides is from the aqueous solvent that is used to transport the sample to the plasma. This water vapor is easily dissociated in the high-energy plasma to provide a large quantity of oxygen atoms, which can combine with other atoms to form molecular species.

Oxide molecular species can result in either a positive or negative interference on analyte analysis. The positive interference is the result of a mass spectral overlap, analogous to an isobaric-type interference. This type of interference can be avoided by increasing the resolution so a spectral separation can be achieved or by decreasing the magnitude of the formation of the molecular oxide molecule so that its magnitude is insignificant relative to the analyte ion current. The magnitude of the population of oxide molecules can often be reduced by optimization of the operating conditions to minimize their formation. A negative interference is observed from the formation of an oxide molecular species with the analyte isotope being measured. This results in a decrease in the magnitude of the analyte ion current, because of a smaller quantity of analyte ion being available for detection at the designated m/z value in the mass spectrum. This negative interference effect is usually compensated by the calibration process; however, if the composition of the sample is significantly different than that of the calibration standards, it may result in decreased analytical accuracy.

Other molecules can be formed from components introduced into the plasma originating from water vapor including hydride (H) and hydroxide (OH) species. In addition to MO^+, it is common to detect MH^+ and MOH^+ polyatomics in the mass spectrum. The most abundant species detected are ArO^+, ArH^+, and $ArOH^+$, which are observed in the background spectrum shown in Figure 8.1. Each combination of the isotopes of the various atomic components making up the polyatomic molecules is present in the mass spectrum with ion currents corresponding to their combined isotopic abundances. For other than the principal isotopes, the relative ion current intensities of these species are generally rather low.

Reduction of oxide and the other water-related molecular species in the plasma can be attained by reducing the partial pressure of H_2O in the Ar gas

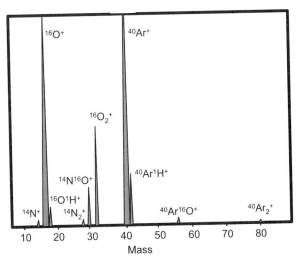

FIGURE 8.1   Typical background spectrum.

stream. This is accomplished by using the desolvation apparatus described in Chapter 5. In addition, the use of a cooled spray chamber with pneumatic nebulizers can significantly lower the introduction of H_2O vapor into the plasma.

Similar to the oxide species, which are formed when aqueous solvents are used, carbon and hydrogen containing molecular ions are observed in the mass spectrum when organic solvents are used as a sample carrier. The prevalence of these molecules in the spectrum introduces problems similar to those encountered with the oxide molecules. As before, the use of a cooled spray chamber can have substantial success in reducing the magnitude of the interferences.

Sample Matrix / Acid Components

Molecular ions are formed from the presence of any major chemical component in the plasma. These can originate from the major constituents present in the sample matrix. For example, for the analysis of samples of gadolinium (Gd), significant interferences in the mass spectrum can be expected from the following ions: $^{155}Gd^{16}O^+$, $^{156}Gd^{16}O^+$, $^{157}Gd^{16}O^+$, $^{158}Gd^{16}O^+$, and $^{160}Gd^{16}O^+$. In addition, lower intensity interfering ions are observed for the ^{18}O, ^{16}OH, and ^{18}OH molecules for each of the Gd isotopes. Also the presence of $^{155}Gd^{16}O^{16}O^+$, $^{155}Gd^{18}O^{16}O^+$, $^{155}Gd^{18}O^{18}O^+$, $^{156}Gd^{16}O^{16}O^+$, and so on, is also probable. When multiple major constituents are included in the sample, interfering molecules for all of the isotopes of each of the components are expected.

The impact of molecular oxides formed with the light rare earth elements can cause serious interference of the heavier rare earth elements. The lowest mass rare earth is ^{139}La. Therefore, any element above m/z 155 can potentially be affected by one or more of the rare earth oxide species. Mathematical corrections of the type previously described can be utilized to assist in alleviating this problem. However, it can be expected that the accuracy of the heavier rare earth elements can be seriously compromised by these interferences.

Another source of sample matrix components, other than the sample itself, is any reagents that have been added to the sample for preservation or dissolution. The most common type of added reagent encountered is a high concentration of one or more mineral acids used to convert a solid sample to a solution. The most common types of acids encountered include HCl, HNO_3, H_2SO_4, HF, and $HClO_4$. When these acids have been added to the sample in modest to high quantities, polyatomic molecules containing atoms of Cl, F, N, O, and S are prevalent in the mass spectrum. These inter-ferences can be so insidious that accurate analyses of some analytes are essentially impossible to achieve.

Chloride presents a particularly difficult problem to accommodate. Not only does Cl have two high abundance isotopes (^{35}Cl at 75.8% and ^{37}Cl at 24.2%), it forms stable polyatomic molecules with both metals and non-metals. One of the most severe interferences is the direct overlap of the mol-ecule $^{40}Ar^{35}Cl^+$ with the monoisotopic analyte element $^{75}As^+$. Because no alternate isotopes are available for As, either Cl must be absent or eliminated from the sample prior to analysis to ensure the most accurate determination. For modest concentrations of Cl in the sample, a mathematical correction can be made that results in a relatively accurate As determination.

One method of correction is analogous to that used with isobaric inter-ferences. It involves subtracting the contribution from $^{40}Ar^{35}Cl^+$ to the total ion current at m/z 75 to yield the net ion current, which represents the concentration of $^{75}As^+$. This is shown as follows:

$$I_{^{75}As} = I_{^{75}total} - I_{^{40}Ar^{35}Cl}, \tag{8.3}$$

where $I_{^{75}As}$ is the ion current for $^{75}As^+$, $I_{^{75}total}$ is the total ion current meas-ured at m/z 75, and $I_{^{40}Ar^{35}Cl}$ is the ion current for $^{40}Ar^{35}Cl^+$. Since the ion current for $^{40}Ar^{35}Cl^+$ cannot be measured directly, the ion current at m/z 77, which represents the ion $^{40}Ar^{37}Cl^+$, must be measured and an appropri-ate correction applied. This correction is simply the ratio of the theoretical isotopic abundances of ^{35}Cl and ^{37}Cl. The isotopic abundance of ^{35}Cl is 75.77% and the abundance of ^{37}Cl is 24.23%. By substitution, the final equation for calculating the corrected ion current for ^{75}As is as follows:

$$I_{^{75}As} = I_{^{75}total} - \left(\frac{75.77}{24.23} \cdot I_{^{40}Ar^{37}Cl} \right). \tag{8.4}$$

The corrected net ion current for $^{75}As^+$ is then calibrated in the usual man-ner for the calculation as the As concentration.

Care must be exercised to ensure that the alternate m/z that is measured to make the interference correction must itself not be interfered. If the cor-rection isotope is interfered, a second correction can be made. For example, in the case of As, it is possible that if a moderate amount of Se is present in the sample, a secondary correction for an interference on m/z from $^{77}Se^+$ can be made. Another isotope for Se is measured to compute the contribu-tion of $^{77}Se^+$ on m/z 77. The net ion current at m/z 77 is due to $^{40}Ar^{37}Cl^+$, which is then used to compute the ion current from $^{40}Ar^{35}Cl^+$, on m/z 75, which is then subtracted from the total ion current at m/z 75, to compute the ion current of $^{75}As^+$. When implementing this *cascading correction process*, precision of the analysis is reduce for each step employed.

Another correction approach that is somewhat more straightforward is to compensate for polyatomic molecular spectral overlap interferences on trace element analyses is by making a correction in the concentration domain. This correction is accomplished by measuring the *equivalent analyte concentration* of the interference in the absence of the analyte (i.e., $C_{analyte} = 0$). This measurement is made for a known concentration in the sample of the interfering component of the molecular ion. Typically, a standard solution (known concentration) of the interferent, with no analyte present, is measured under the same conditions as the calibration and analysis of the analyte element. The *apparent analyte concentration* for this interfering molecular ion is computed from the analyte calibration curve (or regression function). An *interferent correction constant* (K) is calculated by the following equation:

$$K = \frac{C_{equiv}}{C_{int}},\qquad(8.5)$$

where C_{int} is the concentration of the interferent and C_{equiv} is the equivalent analyte concentration. Therefore, for any unknown sample that contains both the interferent and the analyte, a specific correction can be established by knowing (or measuring) the concentration of the interferent, and multiplying it by the interferent correction constant. This constant needs to be reestablished during each calibration process. Similar to isobaric interference corrections, this technique is only suitable for use when the interference correction is small compared to the analyte concentration:

$$C_{analyte} = C_{total} - K \cdot C_{int}.\qquad(8.6)$$

This technique can only be employed for molecular interferences that have a component that can be measured. This situation generally means that it is not possible to make corrections unless the interfering molecular ion contains an atom whose concentration can be measured in the sample or is previously known from an alternate determination.

For example, to perform a trace determination of As, the total ion current is measured at m/z 75. This is representative of any As present in the sample, as well as an unknown quantity of $^{40}Ar^{35}Cl^+$. The amount of $^{40}Ar^{35}Cl^+$ is computed from the quantity of Cl measured in the sample, either by an alternate analytical method or from a calibration curve obtained from standardization of the peak occurring at m/z 35 or

m/z 37, multiplied by the previously determined value of the constant K. The computed equivalent As concentration for $^{40}Ar^{35}Cl^+$ is subtracted from the total measured As concentration to obtain the "true" As value.

In general, the most serious polyatomic interferences result from the most abundant isotopes of the elements Ar, C, Cl, H, N, O, and S. The less abundant isotopes of these elements can also create molecular ions that can cause relevant interferences, but usually they are not as severe as those formed from the major isotopes. By far the best solution to the problem of molecular ion interference is to avoid the use of acids or other reagents that have multiple isotope elements at relatively high mass, such as Cl and S. A general rule of thumb for the accurate analysis of the greatest number of elements is to avoid the use of any reagents that contain Cl or S at moderate to high concentration.

Reaction/Collision Cells

A recent development in instrumentation that can potentially remove or minimize the prevalence of molecular ions in the mass spectrum is the *reaction cell* (sometimes called a *collision cells*). The reaction cell is a device that is mounted in the mass spectrometer, positioned between the ion lenses and the mass analyzer. A diagram of a typical reaction cell is shown in Figure 8.2. The sample ion beam produced by the ICP, which is collimated by the ion lens, is directed into the cell through an aperture. The beam, which is composed of analyte ions, matrix component ions, and polyatomic molecules, is transmitted through the cell by the action of a quadrupole (reaction cell) or hexapole (collision cell) transmission optic element. An externally supplied fill gas in the cell selectively reacts with the polyatomic molecular

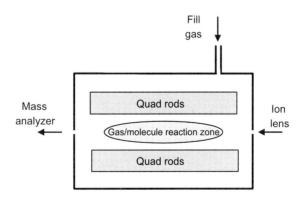

FIGURE 8.2   Typical reaction cell.

ions in the ion beam. These reactions effectively remove the molecular ions from the beam prior to its entrance into the mass analyzer for ion separation, with no appreciable loss of sensitivity for the analyte species. It has been stated by instrument manufacturers that this process is so efficient that the magnitude of the polyatomic molecules can be reduced by as much as a factor of 10^6.

The three main types of ion–molecule reactions are charge transfer, proton transfer, and hydrogen ion transfer. Typical reaction cell gases include ammonia and helium. A common reaction using ammonia as the cell gas is the removal of the $^{38}ArH^+$ ion, which interferes with the analysis of $^{39}K^+$. This reaction is shown below:

$$NH_3 + {}^{38}ArH^+ \rightarrow NH_4^+ + {}^{38}Ar^0$$

The $^{38}ArH^+$ is effectively removed from the ion beam by converting it to a neutral Ar atom, which is not transported into the mass analyzer.

Other manufacturers have reported the use of He in the collision cell at a flow rate as low as 7 mL/min. An example of the reaction process is shown in Figure 8.3. In this figure an ArO^+ molecule is shown colliding with a He atom. The energy transfer results in the molecule breaking up into Ar and O^+, removing the interference at m/z 56 in the mass spectrum. The process also results in the He obtaining kinetic energy from the collision.

Secondary interferences from reaction products can be a problem under some conditions. However, by careful selection of operating conditions and by the choice of the hardware geometry, the problem of secondary ions can be controlled. In addition, by carefully optimizing the operating parameters of the quadrupole/hexapole, and synchronizing with the mass analyzer, interferences by secondary ions at the analyte m/z can be rejected.

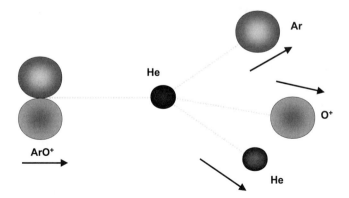

FIGURE 8.3   Collision of He with ArO^+ in collision cell.

Although this technology has demonstrated real merit in solving the polyatomic molecular ion interference problems for some selected applications, there is still much research to be accomplished before its universal applicability is understood. This is particularly true for multielement analyses because the operating conditions and choice of parameters appear to be specific for the solution of each analytical problem.

8.1.3 Doubly Charged Ions

As discussed in Chapter 2, some elements, which have sufficiently low ionization potentials, form doubly charged ions. These elements are primarily the alkaline earths, some rare earths, and a limited number of transition metals. Even for those ions that favor the formation of doubly charged ions, under normal operating conditions their production is normally very small (<1%).

The formation of doubly charged ions for the analyte element can have a negative interference on its determination. This is because the pool of ions, which contribute to the ion current at the m/z value for the isotope selected for analysis, is depleted by one for every doubly charged ion formed. Therefore, the larger the population of doubly charged ions, the lower the sensitivity for the analysis using the singly charged species.

The formation of doubly charged ions also can contribute a positive interference for the analysis of other elements. Because of the mass-to-charge (m/z) relation ship, doubly charged ions will appear in the mass spectrum at one-half the m/z value as observed for the nominal m/z value of the single charged isotope. If there is an overlap with another element's principal m/z value for analysis, there will be an isobaric interference. As with other isobaric overlaps, a high-resolution mass spectrometer may be able to resolve this interference. Clearly, this will only be a problem with even mass isotopes of the element forming doubly charged ions, because doubly charged species of the odd mass isotopes will appear in the mass spectrum at a half-mass location, which will not interfere with other elements.

8.1.4 Background

The background of the mass spectrum is the count rate in the absence of any specific species at the m/z value where an analyte ion is to be measured for quantitative analysis. The background should not be confused with the blank spectrum, which is often also called the *background spectrum*. True background count rates generally result from extraneous ions, which impinge on the cathode of the electron multiplier, unrelated to the operation of the mass analyzer (usually very low probability of occurrence with modern mass spectrometers); photons, which result from stray light inside the mass spectrometer; and random electronic noise generated in the detector (usually as a function of temperature).

Typical background count rates with modern mass spectrometers equipped with well-designed electron multiplier detectors and stable electronic circuitry are 10 counts per second (Hz) or less. Usually the lower the magnitude of the background signal, the better the signal-to-background ratio and hence the better the sensitivity of the analysis. However, with better designed mass spectrometers, higher analyte ion current count rates (10 −20 million counts per second per part per million for a monoisotopic element) are achieved without a significant increase in the background count rate, which also results in an increase in the signal-to-background ratio.

Because of the very high sensitivity of ICP-MS measurements, and the relatively constant background count rate, background correction procedures normally used for other plasma quantitation techniques (i.e., ICP-AES) are not required in ICP-MS measurements for accurate and sensitive analysis. However, background measurements are often used in ICP-MS as a diagnostic tool to ascertain whether the instrumentation is performing property or requires maintenance to function reliably.

8.2 NONSPECTROMETRIC EFFECTS

Various chemical and physical interference effects can seriously impact the accuracy of ICP-MS analyses. In general, these interference effects are independent of the spectral overlap or isobaric interferences discussed in the previous section. These interferences can manifest themselves in either suppression or enhancement of the ion currents that are measured for quantitation. Some of these interferences can also have a deleterious effect on stability of signals and analysis precision. Because these effects can originate from multiple sources and often are very complex, they can be difficult to ascertain and mediate. In addition, the magnitude of these interferences can be hardware or apparatus dependent.

8.2.1 Matrix Effects

High concentrations of a matrix constituent (element) can cause a suppression of the ion current of analyte species. This interference is not necessarily specific in nature and is usually not limited to a single interferent element. However, in general, more serious effects are observed with higher

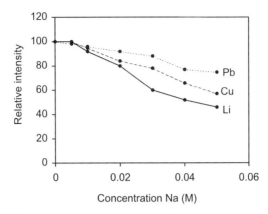

FIGURE 8.4 Interference effects of Na on analyte elements.

mass matrix elements on lighter mass analytes. The extent of the matrix effect is dependent on the absolute amount of the matrix element present rather than the relative proportion of the matrix element to the analyte element. Therefore, by reducing the absolute concentration of the matrix components (by dilution), suppression effects can be reduced to an insignificant level.

Figure 8.4 and 8.5 show, respectively the effects of a light element, Na, and a heavy element, Cd, on several selected analytes. As seen from these figures, the analyte elements are all depressed at different concomitant concentration levels of the interferent. The heavier element, Cd, on average

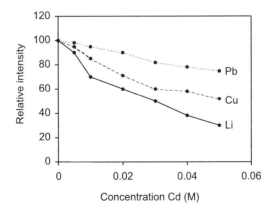

FIGURE 8.5 Interference effects of Cd on analyte elements.

begins to have an effect at lower concentrations than the lighter Na. Under some experimental conditions, interelement enhancements are also observed.

The reason for these matrix effects is not well understood; however, it is thought that they are influenced by a combination of ionization properties and space charge effects from the ion lenses. The magnitude of these interference effects can be reduced by adjusting instrument parameters such as ion lens voltage, RF power, and nebulizer gas flow rates. Matrix separation techniques can be used to reduce the amount of matrix interferent in the sample. In addition to being time consuming, some loss or contamination of very low concentration analyte elements can occur during this process.

Flow injection sample introduction has been successfully used to overcome matrix problems (see Chapter 6). By using this technique, determination can be performed without the need to do external dilutions or chemical separations. The method has the added benefit of rapid analysis of multiple samples. An example of this approach is the successful direct analysis of seawater.

8.2.2 Physical Effects

One type of physical interference effect is that associated with a high dissolve solids content in the sample. As the concentration of the total solids in the sample increases, the possibility of drift in analyte ion current signals becomes more problematic. An example of this drift is shown in Figure 8.6.

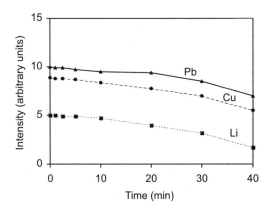

FIGURE 8.6   Interference effect of 10mg/L Ca on selected analytes as a function of time.

This figure shows the magnitude of selected analyte ion signals with a constant 10 mg/L calcium concentration, plotted as function of time. As seen from this figure, severe signal degradation is observed after a few minutes.

It is understood that this problem is directly related to the buildup of salts on the orifice of the interface sampler cone. Typical sampler orifice diameters of 1–1.5 mm slowly constrict as a coating of sample matrix salts accumulates on the sampler cone. As the diameter of the sampler orifice decreases, the transmission of ions into the mass spectrometer decreases, which results in a decreases in analyte ion currents.

Normally when dissolved solids concentrations exceed about 2 mg/L, depending on the specific matrix of the sample, it can be anticipated that potential stability problems will be encountered. Several solutions to solving this problem can be employed. Perhaps the simplest is to dilute the samples so that matrix concentrations are below levels that encourage the buildup of salts on the sampler cone. This approach can be used effectively because of the high sensitivity enjoyed by the ICP-MS methodology, thereby maintaining suitable detectability of analyte species.

Internal standards can be used to correct for the depression of analyte signal caused by this buildup on the sampler cone. This approach does not require sample dilution for correction of the drift. However, it can be use in conjunction with sample dilution to achieve optimal correction.

Although more tedious to perform, matrix matching of the calibration standards to the sample composition will improve this situation. However, using this approach, frequent restandardization is required to maintain analysis accuracy.

Other types of physical interference effects include those affecting the nebulization/sample introduction process. These sample transport effects, which result from differences in viscosity, surface tension, and volatility, can be minimized by dilution. The use of a delivery pump to transport the sample solution to the nebulizer will normalize these effects to a limited extent. Whenever mineral acids are used in sample decomposition or preservation, equivalent quantities added to the calibration standards compensate for differences in solution properties and hence minimize sample transport effects.

Finally, a serious problem can result from contamination of the analytical system (i.e., sample transport, nebulizer, spray chamber, plasma torch, and attendant tubing and fittings) by high concentrations of analytes originating from sample solutions. If care is not exercised when analyzing multiple samples, analyte carryover can mitigate the quality of the analytical determinations. This is a situation not unlike blank problems originating from unpurified reagents or poorly cleaned apparatus.

Analyte carryover is related to the chemical properties of various analyte species. Some analytes, such as boron, adsorb strongly to the surfaces of the apparatus, which come into contact with the sample aerosol, while others have no particular affinity. Nebulizer spray chambers can be a particularly obstinate component of the apparatus to suffer analyte adsorption. Because of its relatively large volume, a correspondingly large quantity of sample aerosol is needed to reach steady-state transport of analyte to the plasma. After the ion current measurement is made (i.e., signal integration is performed), the sample aerosol must be flushed from the system before the next sample can be processed. This is usually accomplished by nebulizing deionized water.

After the aerosol is flushed from the apparatus, additional washout is required to remove traces of the previous sample that is physically attached to the walls of the nebulizer, spray chamber, and tubing. Any analyte that is chemically adsorbed onto the walls of the apparatus will require more extensive flushing for removal. The material (glass, plastic, Ryton, or PTFE) from which the spray chamber or other components of the sample introduction system are constructed can play a significant role in the difficulty of analyte carryover. More extensive flushing can be accomplished by passing voluminous quantities of solvent (water) through the system, relying on dilution to remove the analyte. An alternate approach, especially for the removal of strongly bound constituents, is the augmentation of the solvent with chemical components that can reverse the adsorption process. these components are commonly dilute mineral acids or metal complexing agents that can thoroughly clean the system. Most of the material removed by this process is not transported to the plasma, but is flushed through the drainage outlet of the sample introduction system.

An example of this process is shown in Figure 8.7. The diagram is a plot of ion current as a function of time. The steady-state ion current signal represents the introduction of a constant quantity of sample containing the analyte to the plasma. At 120 s, the sample is removed from the introduction system, and is replaced by a constant flow of deionized water (the same flow rate as the sample). The removal of the analyte from the sample introduction system is shown by the exponential decay of the analyte ion current. The time required to achieve an acceptable background, sufficiently low in magnitude to permit the introduction of the next sample (about 1% of the magnitude of the steady state signal), is reproducible, but dependent on the characteristics of each individual analyte species.

This characteristic washout time represents the minimum required time between sample analysis, and it must be specifically determined during a methods development study for each analyte to be measured in a

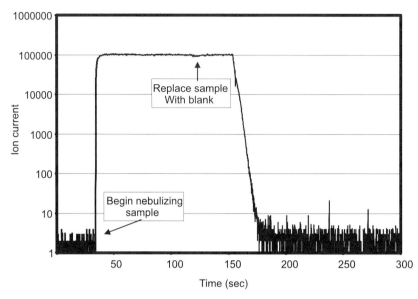

FIGURE 8.7   Plot of ion current as a function of time demonstrating washout times in a pneumatic nebulization system.

multielement determination. This time will also depend, to some extent, on the nature of the matrix of the sample being analyzed. Typical washout times to reach the 1% carryover criteria range from about 60 to 180 s under normal conditions. As previously mentioned, techniques such as increasing flush rates during the washout period or flushing the apparatus with chemically modified washout solution can, in some cases, drastically reduce washout times. This can drastically reduce multiple sample throughput rates for the determination.

The more complicated the sample introduction system, the longer the washout times required. This is particularly significant when using ultrasonic nebulization where the nebulizer itself has a relatively large internal surface area. This factor, coupled with a desolvation system and the higher relative sensitivity characteristics of this type of nebulizer, can result is substantially longer washout times.

CHAPTER 9

Optimization

To obtain maximum sensitivity and precision and to minimize certain interference effects in ICP-MS, careful optimization of instrument operating parameters is required. A depiction of this optimization process is illustrated in Figure 9.1. The plot of ion current as a function of a variable that affects the signal usually passes through a maximum that represents optimal sensitivity of the analysis. In addition, the region where maximum signal is observed is often associated with minimal variation as a function of the optimizing variable. In the figure, ΔI represents the change in ion current as a function of ΔV, which is the change in the magnitude of the optimizing variable.

As the maximum of the ion current is approached (ΔV_1), the variation in ion current ΔI_1, is minimized, thereby obtaining the lowest variability in the ion current measurement. Compared to a steeper region of the curve, ΔV_2, the change in ion current, ΔI_2, is significantly larger, resulting in less precise measurements.

This process is particularly important for multielement determinations, where optimal performance for one analyte may not be optimal for another, requiring a compromise of conditions. The reduction or elimination of one interference effect may exacerbate the occurrence of another. The magnitude and sacrifices of this compromise must be understood by the analyst before the choice of analytical conditions is selected.

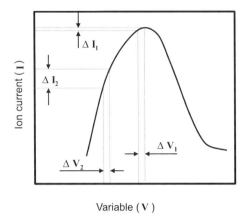

Variable (**V**)

FIGURE 9.1   The parameter optimization process, showing the effect of measurements near the maximum ion current.

These parameters include such variables as applied RF power, type of nebulizer and nebulizer gas flow rate, ion sampling position in the plasma, and applied voltages on the ion lenses. Additional parameters such as plasma support gas flow rates and intermediate gas flow rates have significantly less effect and are not considered critical.

9.1 PARAMETER OPTIMIZATION

By far the best diagnostic tool for evaluating the effects of operating parameters is the use of nebulizer gas (injector gas) flow rates plotted as a function of analyte ion current intensities (Horlick *et al.*, 1985). The relationship between nebulizer gas flow rate and analyte ion current at various applied RF power settings and a constant sampling distance from the top of the load coil is shown in Figure 9.2, which is generally characteristic of essentially all analyte elements. In this diagram a family of curves is observed where the variables, applied RF power and nebulizer gas flow rate, are shown to be interdependent, with the optimal signal continuously increasing with increasing nebulizer flow rate.

Figure 9.3 shows three panels, each with plots similar to Figure 9.2. Each panel represents a different sampling distance above the load coil. This demonstrates that at a constant RF power, as the ions are sampled from the ICP at a distance further from the load coil, the nebulizer flow rate must be

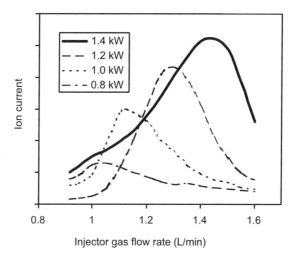

FIGURE 9.2   Plot of ion current at various applied RF power settings as a function of nebulizer gas flow rate.

increased to maintain the magnitude of the ion current. These observations are consistent with the description and behavior of the normal analytical zone (NAZ) described in Chapter 3. The NAZ, which is the discrete region of the ICP where maximum analyte ionization occurs, moves in a relative manner toward the load coil, or away from the sampler orifice, as the RF power increases. Conversely, as the nebulizer gas flow rate is increased, the NAZ moves back toward the sampler orifice. In general, at a specific sampling depth, a range of combinations of RF power and nebulizer gas flow rates will provide similar analyte ion current signals. A different set of conditions will be prevalent when the sampling depth is changed. To select a set of operating conditions for multielement analysis, care must be exercised to ensure that the best optimal conditions are chosen for maximum signal, which will result in the highest analysis sensitivity.

Optimization of ion lens settings is described in considerable detail in Chapter 4. The processes affecting the ion transmission of analyte species are very complex in nature and are difficult to predict based on the operating conditions of the analysis. In addition, the transmission has a dependence on the history of the recent operation of the plasma and mass spectrometer, which may not be reproducible. In summary, the space charge effects, which defocus the ion beam, can severely alter the ion trajectories in the ion lens system. This defocusing can be highly dependent on the matrix of the sample. Some mass spectrometers have dynamically pro-

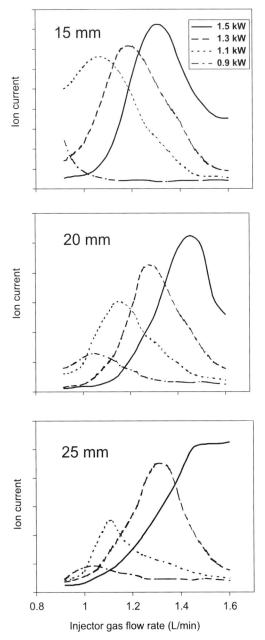

FIGURE 9.3   Plots of ion currents as a function of nebulizer gas flow rates for a variety applied RF power for sampling distances of 15, 20, and 25 mm from the load coil.

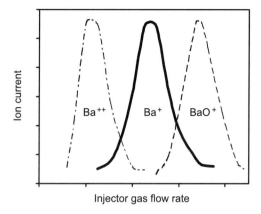

FIGURE 9.4   Profiles of plots of ion currents for Ba^{2+}, Ba^+ and BaO^+ a function of nebulizer gas flow rate.

grammed ion lens voltages to reduce the mass dependence of the analyte on the space charge effect, greatly reducing the potential matrix effects.

The magnitude of the formation of molecular oxide species depends on several instrument-operating parameters. These parameters include nebulizer flow rate, applied RF power, sampler orifice size, sampler cone positioning in the ICP, oxygen content of the plasma, and efficiency of aqueous solvent removal. Each of these variables has an effect on the prevalence and magnitude of the oxide molecular formation process. The dependence on oxide formation of Ba in the ICP as a function of nebulizer gas flow rate is shown in Figure 9.4. The magnitude of this effect either increases or decreases as a function of applied RF power with different instrument designs. High nebulizer gas flow rates typically tend to increase the amount of MO^+ or MOH^+ relative to M^+ ions. It is very difficult, and nearly impossible, to achieve both minimum molecular oxide formation and maximum sensitivity for the M^+ measurement simultaneously. Therefore, a compromise of conditions must be achieved to attain acceptable oxide interference levels and satisfactory measurement sensitivity.

9.2 "COLD PLASMA" OPERATING CONDITIONS

With the appropriate hardware interface configuration (to control plasma potential) and modification of the operating parameters of the ICP, conditions can be achieved that will reduce or minimize certain polyatomic spectral interferences. Reducing the applied RF power and increasing the nebulizer gas flow rate, the background spectrum (blank solution), consisting of high-intensity Ar species, can be reduced significantly. For example, by reducing the applied RF power from 1.2 to 0.6 kW and increasing the nebulizer gas flow rate from about 0.8 to 1.1 L/min, the magnitude of the ion current from $^{40}Ar^+$ can be reduced so that the isotope ratios of $^{39}K^+$, $^{40}K^+$, and $^{41}K^+$ are measured with acceptable accuracy. This is called operating under *cold plasma* conditions.

When operating in this cold plasma mode, the dominant species in the background spectrum are NO^+, O_2^+, and H_3O^+, replacing the normally observed Ar species. The magnitude of the $^{40}Ar^{16}O^+$ molecular species at m/z 56 can also be reduced to levels suitable for the determination of $^{56}Fe^+$. For example, if the ion current at m/z 56 under normal operating conditions is about 10^6 Hz, it is typically reduced to about 20 Hz under cold plasma conditions.

To maintain stability, most plasmas cannot be operated at applied RF power settings substantially lower than about 0.6 kW. Therefore, when operating in the cold plasma mode, careful optimization of parameters must be made to accomplish satisfactory operation of the ICP while introducing sample. The cold plasma has significantly lower energy available for atomization and ionization than plasmas operating at normal conditions, limiting its use to relatively clean samples (i.e., low dissolved solid samples). In addition, its usefulness is also limited when performing multielement determinations including elements that have relatively high ionization potentials.

CHAPTER 10

Figures of Merit

Analytical figures of merit are used to evaluate the performance of instrumentation and methodology for the determination of analyte elements. The analytical characteristics of a specific manufacturer's instrumentation, types of instrumentation, options for sample introduction, and techniques used to perform quantitative analysis can be compared by their characteristic figures of merit. In addition, the repetitive determination of figures of merit during routine analysis can be used by the analyst as a diagnostic tool to evaluate the performance of their instrumentation and dictate the need for maintenance, recalibration, or troubleshooting.

The primary figures of merit that are used to characterize ICP-MS instruments and methods include selectivity, stability, robustness (susceptibility to interferences), sensitivity, detection limits, accuracy, and precision. Selectivity, stability, and robustness have all been discussed in some detail in previous chapters. The focus in this chapter will be on sensitivity, detection limits, accuracy, and precision.

The use of figures of merit can provide a convenient method for assisting with quality assurance monitoring. When the sensitivity, for example, changes markedly during an analysis session, accuracy and/or precision may be compromised. Levels of both accuracy and precision can be used to assist in the interpretation of analytical data to achieve the stated goals of the analysis.

10.1 SENSITIVITY

Sensitivity in the context of quantitative analysis is the ability to measure small differences in concentration of an analyte species. Sensitivity is usually defined as the slope of the analytical calibration curve (plot of ion current as a function of concentration) divided by the precision (expressed as standard deviation of the ion current) of an intermediate concentration calibration standard (see Figure 7.2). The equation to compute sensitivity is as follows:

$$\gamma = \frac{dI/dC}{s} = \frac{m}{s}, \tag{10.1}$$

where, γ is the sensitivity, I ion current, C is concentration of the analyte, m is the slope of the calibration curve, and s is the experimentally measured standard deviation of the ion current of an intermediate concentration standard. Sensitivity is expressed as units of inverse concentration.

The sensitivity of various elements is roughly proportional to the abundance of the isotope that is used for ion current measurement. Monoisotopic elements are normally the most sensitive. In addition, the degree of ionization has an impact on the measurement sensitivity of an analyte. Because most elements are ionized greater than 90% in the ICP, in principle, they demonstrate similar sensitivities. Elements that are ionized to a lesser degree, such as Hg or Se, will have lower sensitivities.

Detectability of an analyte is closely related to sensitivity. When there is a requirement to determine analytes at very low concentrations, conditions providing the highest sensitivity are selected. However, better results (superior accuracy and precision) for the determination of intermediate to high concentration elements is achieved if lower sensitivity isotopes are measured. This approach can often alleviate the need to rerun samples with a higher dilution.

Because sensitivity is related to the transmission of the mass spectrometer (i.e., the total number of ions reaching the detector of the spectrometer per unit concentration), it is also related to the resolution setting of the instrument. In general, the higher the resolving power, the lower the transmission and hence the lower the sensitivity. For this reason, resolution settings are usually selected to provide the minimum resolving power necessary to eliminate analyte interferences.

There also is a small dependence of sensitivity on the mass of the analyte ion being measured. The relationship is a complex combination originating from many sources. For this reason it is universally determined empirically. This effect is of particular importance in isotope ratio measurements. Typically, when highly precise isotope ratio measurements are required, a known certified isotope standard is used to establish the mass bias, which is then used to correct the sample ratio. Usually this technique is only appropriate over a small mass range.

10.2 DETECTION LIMITS

A specialized case of sensitivity is used to determine the smallest concentration of an analyte that can be detected. The controlling principle is the ability to measure a signal, or ion current, representative of the concentration of the analyte comparable in magnitude to the blank (absence of analyte) signal. This ability to measure the ion current representative of the analyte concentration is proportional to the slope of the analytical calibration curve and the ability to make the measurement near the ion current of the blank at the same m/z value. This results in the sage quotation attributed to Morrison and Skogerboe (1965), "given adequate precision, the greater the sensitivity, the better the detectability." The detection limit is defined as the smallest concentration of a given analyte for which a reliable measurement can be made.

The detection limit (lower limit of detection of the analyte) can be expressed either in absolute terms or relative terms. The absolute limit is usually reported in weight (i.e., nanograms or picograms). The relative limit is reported in concentration units in the total sample (i.e., nanograms/gram or micrograms/liter). Both approaches are essentially equivalent and either is appropriate, depending on the application and the details of the methodology used for a particular analysis.

Because the detection limit has become a favorite figure of merit for the selection and evaluation of instrumentation and analytical methodology, a technique for computation that is universally applicable is desirable. The following approach fulfills this need. The lowest detectable concentration of the analyte is computed by the following equation:

$$C_{DL} = \frac{t_{(1-a,n-1)} \cdot s_b}{m},\qquad(10.2)$$

where C_{DL} is the detection limit, $t_{(1-a,n-1)}$ is Student's t statistic at a specified confidence level and n degrees of freedom, m is the slope of the analytical calibration curve, and s_b is the standard deviation of the background (computed from at least 10 replicate measurements). Student's t statistic is obtained from published tables for various levels of statistical certainty (see Appendix 4). Typical values of the t statistic for $n = 9$ (10 replicates) are 1.83, 2.26, and 3.25 for the 90, 95, and 99% symmetric, two-tailed confidence limits, respectively.

An example of the use of this equation for the calculation of the detection limit for Cd at the 95% confidence level would be:

$$C_{DL} = \frac{(2.262)(3\ \mathrm{cps})}{1400\ \mathrm{cps}\over \mu g/L} = 0.005\ \mu g/L,$$

where the standard deviation of the background is 3 cps, based on 10 independent measurements of the background ion current, and the slope of the calibration curve is 1400 cps/μg/L. The detection limit is justifiably reported only to a single significant figure.

Note that in spite of the mathematical definitions cited, detection limits are rather nebulous quantities. Because they depend on many variables, a factor of 2–3 times uncertainty in the values can be anticipated. They can vary significantly between various manufacturers' instrumentation and are especially sensitive to different modes of sample introduction. They can also be modified by the optimization for the determination of specific elements. When performing multielement analyses, a compromise of optimization must be tolerated. This compromise usually results in the achievement of optimal detection limits for only a few elements, with the remainder often being a factor of 2–3 times their optimized values. Also, because detection limits are so dependent on operating parameters, it is prudent to frequently (i.e., with each batch of samples analyzed) compute detection limits to reliably report ultra-trace concentration levels. Care must be taken to not report too many significant figures when stating detection limits, so as to be consistent with the probability level selected in the computation. Typical published detection limits for various types of instrumentation are tabulated in Table 10.1.

The International Union of Pure and Applied Chemistry (IUPAC, 1976) strongly recommends a value of 3 be used for the proportionality

TABLE 10.1   A Summary of Published Instrumental Detection Limits for a Variety of Types of ICP-MS Instruments

Element	m/z	Sciex Elan 6000[a]	Sciex Elan 6100[b]	VGE PQ3[c]	Varian Ultramass[d]	Micromass Plasmatrace2[e]
Ag	107	2		1	1	0.03
Al	27	7	0.07	6	10	0.06[f]
As	75	60		2	10	10[g]
B	11		1	35	170	
Ba	137	8		1		
Ba	138				6	0.004
Be	9	8	0.6	2	2	0.01
Bi	209	1		1	1	0.001
Br	79			100	50	
Ca	40		1	800		
Cd	111	2		2		
Cd	114				1	0.09
Ce	140	0.2		1		
Co	59	1	0.07	1	3	0.08
Cr	52	50	0.25	3	30	0.09
Cs	133			1	1	0.009
Cu	63	5		2	3	0.08
Dy	163	0.4		1	1	0.002
Er	167	0.5		1	0.7	
Eu	151	0.5		1	0.4	0.003
Fe	57				3	2
Ga	71			2	70	0.09
Gd	158	0.5		1	1	
Ge	74			7	7	0.2
Ho	165	0.1			0.3	0.0005
I	127			14	5	
In	115		0.01		1	0.003
K	39		1	2000	5	
La	139	0.3		1	0.3	0.001
Li	7		0.08	1	1	0.02
Lu	175	0.2		1	0.2	0.009
Mg	25					0.2
Mn	55	2		2	8	0.1
Mo	95			2	3	0.03
Mo	98	3				
Na	23		0.3	20	15	
Nd	146	1		2	1	0.006
Ni	60	4			20	0.2
P	31			600	6000	18[f]
Pb	208	4	0.03	2	7	0.005

(continues)

TABLE 10.1 *(continued)*

Element	m/z	Sciex Elan 6000[a]	Sciex Elan 6100[b]	VGE PQ3[c]	Varian Ultramass[d]	Micromass Plasmatrace2[e]
Pr	141	0.5		1	0.2	
Rb	85			1	2	0.02
Re	185			1	1	0.004
S	34			30000	16000	94[f]
Sb	121	2	0.06	2	2	0.03
Sc	45			9	40	0.02
Se	77			80	280	
Se	80		5			
Se	82	150				
Si	28			4000	8000	360[f]
Sm	147	2		1	2	0.005
Sn	118			2	4	0.07
Sr	88			1	1	0.007
Ta	181			1	1	0.0009
Tb	159			1	0.2	0.0008
Th	232	1		0.4	1	0.001
Ti	49				10	1
Tl	205	0.4		1	2	0.0009
Tm	169	0.2		1	0.2	
U	238	0.3	0.01	0.5	1	0.007
V	51	20		1	1	0.03
W	182			3	4	0.004
Y	89			1	1	0.007
Yb	174	0.4			0.8	
Yb	172			1	1	0.002
Zn	66	20		5	20	0.06
Zr	90			5	2	

Note: These detection limits (ng/L) are calculated using the IUPAC specified (3 sigma) definition.
[a]Quadrupole with pneumatic nebulization.
[b]Quadrupole with dynamic reaction cell (Denoyer *et al.,* 1999).
[c]Quadrupole with pneumatic nebulization (Brenner *et al.,* 1999).
[d]Quadrupole with pneumatic nebulization (Nham, 1996).
[e]Double-focusing sector with ultrasonic nebulization (Becker and Dietze, 1998).
[f]Resolution—1000–1500 (Becker and Dietze, 1998).
[g]Resolution—>7500 (Becker and Dietze, 1998).

constant (equivalent to Student's t statistic) used in the detection limit computation. They claim that this corresponds to a 99.6% confidence level, which applies to *one-sided* Gaussian distributions. However, they specify that at low concentration levels, non-Gaussian distributions are likely, therefore resulting in a confidence level closer to about 90%.

10.3 PRECISION

Precision, as defined by Morrison and Skogerboe (1965), is the measure of the reproducibility of replicate analyses. Because reproducibility determined in this manner is considered to be random in nature and hence a normal distribution, the standard deviation is used as a measure of the magnitude of precision. To normalize the standard deviation to make comparisons between different magnitudes of measurements, the relative standard deviation is used. The relative standard deviation is expressed on a percentage basis and is called the percent relative standard deviation (% RSD). It is computed as follows:

$$\% \, RSD = \frac{s}{\bar{x}} \cdot 100, \tag{10.3}$$

where s is the measured standard deviation of replicates and $\bar{x}$ is the mean of the replicates.

Two types of precision are normally encountered in ICP-MS analyses. The first, which is *short-term precision*, is the reproducibility of the measurement of consecutive determinations. *Short-term precision* is generally regarded as the best performance capable by the analysis system. It is often used as a parameter to evaluate the optimization of the instrument and methodology.

The second type of precision is known as *long-term precision*. Long-term precision, is the reproducibility measured over a period of several hours, but within the domain of a single analysis session. Because of the nature of the analysis, long-term precision includes the instability from instrument drift, and is always poorer than the corresponding short-term precision.

Measurement precision incorporates a combination of sources of instability. In typical instrumentation, the magnitude of the precision is dependent on several factors including the stability of the detector, stability of the control electronics, the thermal stability of the operating environment, the stability of the ICP (including the regulation of gas flow rates), and the stability of the sample introduction system. By far the most important source of instability in the system is variability in the plasma ionization process. This variability is directly related to the quality of the regulation of gas flow rates, especially the nebulizer gas flow. For this reason, mass flow controllers are normally used to regulate this variable. The quality of the aerosol generation process (usually directly related to the performance of the aerosol producing nebulizer) is the limiting factor in optimizing short-term precision.

Figure 10.1 shows a plot of the % RSD as a function of concentration for Cd. The nature of this plot is typically characteristic of most elements. It is clearly seen from this plot that as lower concentrations are approached, the % RSD increases approximately exponentially. When approaching the detection limit, the % RSD becomes close to ±100, which is consistent with its definition. At nominal concentrations, most elements can be determined on a short-term basis with a % RSD of approximately 1%. As an estimate, for long-term precision, the values are about a factor of 3 greater.

When reporting the results of quantitative determinations, an error term assists the end user in interpreting the data. An example of the way the data should be reported is shown as follows:

$$1250 \pm 10 \; \mu g/g,$$

where 1250 is the mean concentration of the analyte in the sample in units of $\mu g/g$, and 10 is the standard deviation also in the same concentration units.

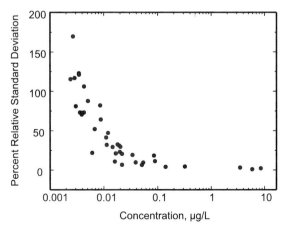

FIGURE 10.1   Plot of percent relative standard deviation as a function of concentration for Cd.

To compute this error term, replicate analyses are performed. These can consist of replicate integrations of the ion current, which is indicative of the best performance of the specific mass spectrometer used to collect the data. More realistic precision data is obtained from replicate analysis of multiple aliquots of the sample. This provides data that are representative of the overall capabilities of the analysis method. Finally, replicate samples collected in the field or at the source of the samples incorporate imprecision from all sources of variance including sampling statistics.

It is recommended that a minimum or three replicate measurements be made in order to compute a standard deviation, which is used as the error term associated with a mean of the replicate concentration results. If time and expense allow, measurement of additional replicates usually improves the accuracy of the mean value and lowers the magnitude of the standard deviation, which is the error term reported with the mean. Ten replicates are usually the practical upper limit on the number of replicates that can be effectively used. Beyond 10 replicates, improvement in accuracy or precision of the reported concentration values is normally limited.

10.4 ACCURACY

Accuracy is defined by Morrison and Skogerboe (1965) as the degree to which one can estimate the true value of the concentration of a specific analyte. It is determined by the magnitude of any systematic and random errors that affect the concentration measurement. Good accuracy is predicated on the ability to resolve peaks of analyte ion currents from interfering species or other extraneous signals originating from blank measurements or the background of the mass spectrum. Confidence, in the absence of systematic errors, can only be achieved by extensive methods development or by empirical techniques. These empirical methods usually consist of the analysis of standard reference materials of a similar composition as the samples, both for the matrix and for the concentration level of the analyte. The degree of agreement between the known concentration of the analyte in the standard reference material and the analyzed concentration is a measure of the accuracy of analysis.

Errors in the determination of the concentration of analyte species in the standard reference material that exceed the random errors associated with measurement precision are identified as *systematic errors*. These systematic errors can originate from a variety of sources. The most common source is an isobaric interference (see Chapter 8 on interferences). Other commonly observed systematic errors can result from matrix interferences, inadequate blank correction, physical interferences, or several other more subtle effects.

Sometimes these systematic errors can result from inadequate performance of the instrumentation or poor selection of operating conditions, such as resolution. When this is the situation, accuracy can normally be improved by simple adjustment of parameters or conditions. If it is not possible to alleviate the problems with these techniques, more aggressive approaches, such as chemical separation of the analyte or selection of more sophisticated instrumentation, are needed to achieve satisfactory accuracy.

Standard reference materials for the evaluation of analysis accuracy are usually obtained from government agencies, such as the United States National Institute of Standards and Technology (NIST) or the National Research Council of Canada (NRC), which are charged with the responsibility of creating standards that have certified analyte concentrations. These can be considered definitive standards suitable for use with many types of chemical analysis techniques. Appendix 5 lists some examples of standard reference materials that are available from NIST and NRC. Similar organizations in other countries also produce standard reference materials. Most analyte concentrations in standard reference materials are reported with a certified value and confidence interval statistically established by the certification process. Often concentration values for other elements that are not certified are reported as "for informational purposes only." These values can be used for both methods development and quality assurance purposes, but are generally not reliable enough for definitive accuracy evaluation or for referee purposes. The U.S. Geological Survey has produced numerous standards, evaluated by round-robin testing procedures. Data for rock standards are reported by Flanagan (1974) and for water standards by Peart *et al.* (1998).

Other organizations produce standard materials that are used for quality control purposes. These standards are usually established by so called "round-robin testing" procedures. Although these are generally not equivalent to certified materials, they have substantial utility for evaluating the accuracy of analyses. Concentration values reported by this approach are called *most probable values* and usually have error terms established by the round-robin testing procedure. Although these standards are not as good as the certified standard reference materials, they are useful when certified materials are not available at the analyte concentration levels needed.

When standard reference materials are not available, the analyst can sometimes create them by collecting an appropriate matrix sample and conducting a round-robin testing protocol with associate laboratories. This is a time-consuming process and can still produce standard materials with significant bias for selected analytes. However, if multiple analytical methods are utilized in the round robin, eliminating method-specific systematic errors, a higher degree of confidence in the results can be obtained.

The use of a definitive analytical method can also be used to establish standard reference materials. Definitive methods are ones that can produce exacting quantitative data without the need to compare measurements to a calibration standard. The gravimetric analysis method is a definitive technique. Isotope-dilution mass spectrometry, which is extensively used by NIST and other agencies producing certified standard reference materials, is also considered to be a definitive method of analysis. As discussed in Chapter 7, isotope dilution quantitation can be effectively used with ICP-MS. Therefore, a laboratory with ICP-MS instrumentation can produce reference materials in specific sample matrices for selected elements by using the isotope dilution technique. These standard reference materials still must be considered secondary standards, because they are usually not traceable to existing certified standards.

When standard reference materials are not available at the appropriate analyte concentration, the spike addition technique can be used effectively. The spike addition procedure consists of selecting representative samples that require analysis and adding a known concentration of the analyte element. This spiked sample is analyzed in the same manner as the unspiked sample. By computing the difference between the spiked and unspiked analyte determination, the recovery of the known spike can be calculated. Depending on the concentration of the analyte, and measurement parameters of the determination, it can be anticipated that recoveries between 90 and 110% can be attained under normal operating conditions. Recoveries that are determined outside this range usually are indicative of an unknown interference or a poorly designed analysis technique for the specific analyte at the concentration range indigenous in the sample.

When analyzing multiple samples, the design strategy of analytical determination can have a significant impact on the accuracy of the determination. Often when a set of samples has widely varying concentrations of the analytes, it is possible for carryover from the analysis of one sample to impact the accuracy of the determination of subsequent samples. This often happens if sample analysis rates are set too high or if flushing times between samples are too short to completely remove the analyte from the analysis apparatus before the next sample is introduced for analysis. These timing considerations are established by thorough methods development studies. However, when samples with unexpectedly high analyte concentrations are analyzed, problems described above can impact the accuracy of analysis. Specific analytes can have a severe carryover problem. For example, boron and silver demonstrate particularly extreme carryover problems.

An analysis strategy technique that can assist in mitigating this problem involves an extension of the quality control procedures utilizing replicate determinations. This strategy employs a protocol for the arrangement of samples in an analysis sequence. Initially, the first replicates of samples are analyzed in a specific sequence. The second replicates are analyzed in a sequence of reverse order. This process eliminates the systematic errors associated with the carryover of high concentration analytes in specific samples. Finally, the third replicates are analyzed in a random fashion. By carefully reviewing the individual replicate data, carryover errors can be identified and eliminated. This process also assists in the removal of drift effects during an analysis run not compensated by internal standardization.

References

Aston, F. W. (1920). The mass spectra of chemical elements, *Philos. Mag.* **40**, 628–634.

Becker, J. S., and Dietze, H. (1998). Inorganic analysis by mass spectrometry. *Spectochimica Acta Part B,* **53B**, 1475–1506.

Boumans, P. W. J. M. (1987). Inductively Coupled Plasma Emission Spectroscopy," Parts I and II, Wiley, New York.

Brenner, I. B., Liezers, M., Godfrey, J., Nelms, S., and Cantle, J. (1999). Analytical characteristics of a high efficiency ion transmission interface (S mode) inductively coupled plasma mass spectrometer for trace element determinations in geological and environmental materials. *Spectochimica Acta,* **54B**, 991–1013.

Denoyer, E. R. Tanner, S. D., and Voellkopf, U. (1999). A new dynamic reaction cell for reducing ICP-MS interferences using chemical resolution. *Spectroscopy,* **14**, 43–54.

Ebdon, L., Foulkes, M. E., Parry, H. G. M., and Tye, C. T. (1988). Direct atomic spectrometric analysis by slurry nebulization VII. Analysis of coal using inductively coupled plasma-mass spectrometry. *J. Anal. Atomic Spectrometry* **3**, 753–761.

Flanagan, F. J. (1974). Reference samples for the earth scientists. *Geochim. Cosmochim. Acta,* **34**, 1731.

Garbarino, J. R., and Taylor, H. E. (1987). Stable isotope dilution analysis of hydrologic samples by inductively coupled plasma mass spectrometry, *Anal. Chem.* **59**, 1568–1575.

Greenfield, S., Jones, I. L., and Berry, C. T. (1964). High pressure plasma emission spectroscopic emission sources. *Analyst* **89**, 713–720.

Herzberg, G. (1944). "Atomic Spectra and Atomic Structure," Dover Publications, New York.

Horlick, G., Tan, S. H., Vaughn, M. A., and Rose, C. A. (1985). The effect of plasma operating parameters on analyte signals in inductively coupled plasma–mass spectrometry. *Spectrochim. Acta* **40B**, 1555–1572.

Houk, R. S. (1986), Mass spectrometry of ICPs, *Anal. Chem.* **58**, 97A–105A.

Houk, R. S., Fassel, V. A., Flesch, G. D., Svec, H. J., Gray, A. L., and Taylor, C. E. (1980). Inductively coupled argon plasma as an ion source for mass spectrometric determination of trace elements, *Anal. Chem.* **52**, 2283–2289.

IUPAC (1976). Nomenclature, symbols and units and their usage in spectrochemical analysis, Part II, Data interpretation, *Pure Appl. Chem.* **45**, 99–103.

Koirtyohann, S. R., Jones, J. S., Jester, C. P., and Yates, D. A. (1981). Use of spatial emission profiles and a nomenclature system as aids in interpreting matrix effects in the low-power argon inductively coupled plasma *Spectrochim. Acta* **36B**, 49–59.

Montaser, A., and Golightly, D. W. (eds.). (1992). "Inductively Coupled Plasmas in Analytical Atomic Spectroscopy," 2nd ed. VCH, New York.

Montaser, A. (ed.), (1998). "Inductively Coupled Plasma Mass Spectrometry," Wiley–VCH, New York.

Morrison, G. H., and Skogerboe, R. K. (1965). General aspects of trace analysis, In "Trace Analysis: Physical Methods" (G. H. Morrison, ed.). Wiley (Interscience), New York.

Nham, T. T. (1996). Typical detection limits for Ultramass ICP-MS. *Varian Bulletin http://www.varian com/inst/osi/icpms/atwork/icpms8.htm.*

Peart, D. B., Antweiler, R. C., Taylor, H. E., Roth, D. A., and Brinton, T. I. (1998). Reevaluation and extension of the scope of elements in U.S. Geological Survey Standard Reference Water Samples. *Analyst* **123,** 455–476.

Taylor, H. E., Garbarino, J. R., Murphy, D. M., and Beckett, R. (1992). Inductively coupled plasma mass spectrometry as an element-specific detector for field-flow-fractionation particle separation. *Anal. Chem.* **64,** 2036–2041.

Wendt, R. H., and Fassel, V. A. (1965). Induction-coupled plasma spectrometric excitation source. *Anal. Chem.* **37,** 920–922.

Table of Elements with Atomic Number, Weights, and First and Second Ionization Potentials

Atomic number	Element symbol	Atomic weight	Ionization potentials (eV) First	Second
1	H	1.007	13.598	—
2	He	4.003	24.587	54.405
3	Li	6.941	5.392	75.622
4	Be	9.012	9.322	18.207
5	B	10.811	8.298	25.119
6	C	12.011	11.26	24.377
7	N	14.006	14.534	29.606
8	O	15.999	13.618	35.082
9	F	18.998	17.422	34.979
10	Ne	20.179	21.564	40.958
11	Na	22.989	5.139	47.292
12	Mg	24.305	7.646	15.032
13	Al	26.981	5.986	18.824
14	Si	28.085	8.151	16.339
15	P	30.973	10.486	19.653
16	S	32.066	10.36	23.405
17	Cl	35.452	12.967	23.799
18	Ar	39.948	15.759	27.619
19	K	39.098	4.341	31.811
20	Ca	40.078	6.113	11.868
21	Sc	44.955	6.54	12.9
22	Ti	47.88	6.82	13.6
23	V	50.941	6.74	14.2
24	Cr	51.996	6.766	16.7
25	Mn	54.938	7.435	15.636
26	Fe	55.847	7.87	16.240
27	Co	58.933	7.86	17.4
28	Ni	58.693	7.635	18.2
29	Cu	63.546	7.726	20.283

(continues)

APPENDIX 1 *(continued)*

Atomic number	Element symbol	Atomic weight	Ionization potentials (eV) First	Second
30	Zn	65.39	9.394	17.960
31	Ga	69.723	5.999	20.509
32	Ge	72.61	7.899	15.93
33	As	74.921	9.81	20.2
34	Se	78.96	9.752	21.691
35	Br	79.904	11.814	19.2
36	Kr	83.8	13.999	26.5
37	Rb	85.467	4.177	27.499
38	Sr	87.62	5.695	11.026
39	Y	88.905	6.38	12.4
40	Zr	91.224	6.84	14.03
41	Nb	92.9064	6.88	—
42	Mo	95.94	7.099	—
43	Tc	98	7.28	—
44	Ru	101.07	7.37	
45	Rh	102.905	7.46	—
46	Pd	106.42	8.34	19.9
47	Ag	107.868	7.576	21.960
48	Cd	112.41	8.993	16.904
49	In	114.82	5.786	18.867
50	Sn	118.71	7.344	14.629
51	Sb	121.757	8.641	18.6
52	Te	127.6	9.009	21.543
53	I	126.904	10.451	19.010
54	Xe	131.29	12.13	21.204
55	Cs	132.905	3.894	32.458
56	Ba	137.33	5.212	10.001
57	La	138.905		11.43
58	Ce	140.12	5.47	—
59	Pr	140.907	5.42	
60	Nd	144.24	5.49	
61	Pm	145	5.55	
62	Sm	150.36	5.63	11.4
63	Eu	151.965	5.67	11.24
64	Gd	157.25	6.15	—
65	Tb	158.925	5.86	
66	Dy	162.5	5.93	
67	Ho	164.930	6.02	
68	Er	167.26	6.101	
69	Tm	168.934	6.184	
70	Yb	173.04	6.254	12.11

(continues)

APPENDIX 1 *(continued)*

Atomic number	Element symbol	Atomic weight	Ionization potentials (eV) First	Second
71	Lu	174.967	5.58	
72	Hf	178.49	6.65	
73	Ta	180.947	7.89	
74	W	183.85	7.98	—
75	Re	186.207	7.88	
76	Os	190.2	8.7	—
77	Ir	192.22	9.1	—
78	Pt	195.08	9	19.3
79	Au	196.966	9.225	20.1
80	Hg	200.59	10.437	18.752
81	Tl	204.383	6.108	20.423
82	Pb	207.2	7.416	15.04
83	Bi	208.980	7.289	16.7
84	Po	209	8.42	
85	At	210		
86	Rn	222 7	10.748	—
87	Fr	223	0	
88	Ra	226.025	5.279	10.145
89	Ac	277	5.43	
90	Th	232.038	6.08	—
91	Pa	231.035	5.88	
92	U	238.029	6.05	
93	Np	237.048	6.19	
94	Pu	244	6.06	
95	Am	243	6	
96	Cm	247	6.02	
97	Bk	247	6.23	
98	Cf	251	6.3	
99	Es	252	6.42	
100	Fm	257	6.5	
101	Md	258	6.58	
102	No	259	6.65	
103	Lr	260		

Isotopic Composition of the Elements

Atomic number	Symbol	Mass	Abundance (at. %)
1	H	1	99.9850 (70)
		2	0.0115 (70)
2	He	3	0.000137 (3)
		4	99.999863 (3)
3	Li	6	7.59(4)
		7	92.41(4)
4	Be	9	100
5	B	10	19.9 (7)
		11	80.1 (7)
6	C	12	98.93 (8)
		13	1.07 (8)
7	N	14	99.632 (7)
		15	0.368 (7)
8	O	16	99.757 (16)
		17	0.038 (1)
		18	0.205 (14)
9	F	19	100
10	Ne	20	90.48 (3)
		21	0.27 (1)
		22	9.25 (3)
11	Na	23	100
12	Mg	24	78.99 (4)
		25	10.00 (1)
		26	11.01 (3)
13	Al	27	100
14	Si	28	92.2297 (7)
		29	4.6832 (5)
		30	3.0872 (5)
15	P	31	100
16	S	32	94.93 (31)

(continues)

APPENDIX 2 *(continued)*

Atomic number	Symbol	Mass	Abundance (at. %)
		33	0.76 (2)
		34	4.29 (28)
		36	0.02 (1)
17	Cl	35	75.78 (4)
		37	24.22 (4)
18	Ar	36	0.3365 (30)
		38	0.0632 (5)
		40	99.6003 (30)
19	K	39	93.2581 (44)
		40	0.0117 (1)
		41	6.7302 (44)
20	Ca	40	96.941 (156)
		42	0.647 (23)
		43	0.135 (10)
		44	2.086 (110)
		46	0.004 (3)
		48	0.187 (21)
21	Sc	45	100
22	Ti	46	8.25 (3)
		47	7.44 (2)
		48	73.72 (3)
		49	5.41 (2)
		50	5.18 (2)
23	V	50	0.250 (4)
		51	99.750 (4)
24	Cr	50	4.345 (13)
		52	83.789 (18)
		53	9.501 (17)
		54	2.365 (7)
25	Mn	55	100
26	Fe	54	5.845 (35)
		56	91.754 (36)
		57	2.119 (10)
		58	0.282 (4)
27	Co	59	100
28	Ni	58	68.0769 (89)
		60	26.2231 (77)
		61	1.1399 (6)
		62	3.6345 (17)
		64	0.9256 (9)
29	Cu	63	69.17 (3)
		65	30.83 (3)
30	Zn	64	48.63 (60)
		66	27.90 (27)

(continues)

APPENDIX 2 *(continued)*

Atomic number	Symbol	Mass	Abundance (at. %)
		67	4.10 (13)
		68	18.75 (51)
		70	0.62 (3)
31	Ga	69	60.108 (9)
		71	39.892 (9)
32	Ge	70	20.84 (87)
		72	27.54 (34)
		73	7.73 (5)
		74	36.28 (73)
		76	7.61 (38)
33	As	75	
34	Se	74	0.89 (4)
		76	9.37 (29)
		77	7.63 (16)
		78	23.77 (28)
		80	49.61 (41)
		82	8.73 (22)
35	Br	79	50.69 (7)
		81	49.31 (7)
36	Kr	78	0.35 (1)
		80	2.28 (6)
		82	11.58 (14)
		83	11.49 (6)
		84	57.00 (4)
		86	17.30 (22)
37	Rb	85	72.17 (2)
		87	27.83 (2)
38	Sr	84	0.56 (1)
		86	9.86 (1)
		87	7.00 (1)
		88	82.58 (1)
39	Y	89	100
40	Zr	90	51.45 (40)
		91	11.22 (5)
		92	17.15 (8)
		94	17.38 (28)
		96	2.80 (9)
41	Nb	93	100
42	Mo	92	14.84 (35)
		94	9.25 (12)
		95	15.92 (13)
		96	16.68 (2)
		97	9.55 (8)
		98	24.13 (31)

(continues)

APPENDIX 2 *(continued)*

Atomic number	Symbol	Mass	Abundance (at. %)
		100	9.63 (23)
44	Ru	96	5.54 (14)
		98	1.87 (3)
		99	12.76 (14)
		100	12.60 (7)
		101	17.06 (2)
		102	31.55 (14)
		104	18.62 (27)
45	Rh	103	100
46	Pd	102	1.02 (1)
		104	11.14 (8)
		105	22.33 (8)
		106	27.33 (3)
		108	26.46 (9)
		110	11.72 (9)
47	Ag	107	51.839 (8)
		109	48.161 (8)
48	Cd	106	1.25 (6)
		108	0.89 (3)
		110	12.49 (18)
		111	12.80 (12)
		112	24.13 (21)
		113	12.22 (12)
		114	28.73 (42)
		116	7.49 (18)
49	In	113	4.29 (5)
		115	95.71 (5)
50	Sn	112	0.97 (1)
		114	0.66 (1)
		115	0.34 (1)
		116	14.54 (9)
		117	7.68 (7)
		118	24.22 (9)
		119	8.59 (4)
		120	32.58 (9)
		122	4.63 (3)
		124	5.79 (5)
51	Sb	121	57.21 (5)
		123	42.79 (5)
52	Te	120	0.09 (1)
		122	2.55 (12)
		123	0.89 (3)
		124	4.74 (14)
		125	7.07 (15)

(continues)

APPENDIX 2 *(continued)*

Atomic number	Symbol	Mass	Abundance (at. %)
		126	18.84 (25)
		128	31.74 (8)
		130	34.08 (62)
53	I	127	100
54	Xe	124	0.09 (1)
		126	0.09 (1)
		128	1.92 (3)
		129	26.44 (24)
		130	4.08 (2)
		131	21.8 (3)
		132	26.89 (6)
		134	10.44 (10)
		136	8.87 (16)
55	Cs	133	100
56	Ba	130	0.106 (1)
		132	0.101 (1)
		134	2.417 (18)
		135	6.592 (12)
		136	7.854 (24)
		137	11.232 (24)
		138	71.698 (42)
57	La	138	0.090 (1)
		139	99.910 (1)
58	Ce	136	0.185 (2)
		138	0.251 (2)
		140	88.450 (18)
		142	11.114 (17)
59	Pr	141	100
60	Nd	142	27.2 (5)
		143	12.2 (2)
		144	23.8 (3)
		145	8.3 (1)
		146	17.2 (3)
		148	5.7 (1)
		150	5.6 (2)
62	Sm	144	3.07 (7)
		147	14.99 (18)
		148	11.24 (10)
		149	13.82 (7)
		150	7.38 (1)
		152	26.75 (16)
		154	22.75 (29)
63	Eu	151	47.81 (3)
		153	52.19 (3)

(continues)

APPENDIX 2 *(continued)*

Atomic number	Symbol	Mass	Abundance (at. %)
64	Gd	152	0.20 (1)
		154	2.18 (3)
		155	14.80 (12)
		156	20.47 (9)
		157	15.65 (2)
		158	24.84 (7)
		160	21.86 (19)
65	Tb	159	100
66	Dy	156	0.06 (1)
		158	0.10 (1)
		160	2.34 (8)
		161	18.91 (24)
		162	25.51 (26)
		163	24.90 (16)
		164	28.18 (37)
67	Ho	165	100
68	Er	162	0.14 (1)
		164	1.61 (3)
		166	33.61 (35)
		167	22.93 (17)
		168	26.78 (26)
		170	14.93 (27)
69	Tm	169	100
70	Yb	168	0.13 (1)
		170	3.04 (15)
		171	14.28 (57)
		172	21.83 (67)
		173	16.13 (27)
		174	31.83 (92)
		176	12.76 (41)
71	Lu	175	97.41 (2)
		176	2.59 (2)
72	Hf	174	0.16 (1)
		176	5.26 (7)
		177	18.60 (9)
		178	27.28 (7)
		179	13.62 (2)
		180	35.08 (16)
73	Ta	180	0.012 (2)
74	W	181	0.12 (1)
		182	26.50 (16)
		183	14.31 (4)
		184	30.64 (2)

(continues)

APPENDIX 2 *(continued)*

Atomic number	Symbol	Mass	Abundance (at. %)
		186	28.43 (19)
75	Re	185	37.40 (2)
		187	62.60 (2)
76	Os	184	0.02 (1)
		186	1.59 (3)
		187	1.96 (2)
		188	13.24 (8)
		189	16.15 (5)
		190	26.26 (2)
		192	40.78 (19)
77	Ir	191	37.3 (2)
		193	62.7 (2)
78	Pt	190	0.014 (1)
		192	0.782 (7)
		194	32.967 (99)
		195	33.832 (10)
		196	25.242 (41)
		198	7.163 (55)
79	Au	197	100
80	Hg	196	0.15 (1)
		198	9.97 (20)
		199	16.87 (22)
		200	23.10 (19)
		201	13.18 (9)
		202	29.86 (26)
		204	6.87 (15)
81	Tl	203	29.524 (14)
		205	70.476 (14)
82	Pb	204	1.4 (1)
		206	24.1 (1)
		207	22.1 (1)
		208	52.4 (1)
83	Bi	209	100
90	Th	232	100
91	Pa	231	100
92	U	234	0.0055(2)
		235	0.7200(51)
		238	099.2745(106)

Note: Numbers in parentheses indicate the uncertainty of the reported value. Data taken from *Isotopic Compositions of the Elements 1997,* International Union of Pure and Applied Chemistry.

Prominent Polyatomic Interferences Applicable for ICP-MS Determinations

Aluminum

^{27}Al $\quad\quad$ $^{12}C^{15}N$, $^{13}C^{14}N$, $^{1}H^{12}C^{14}N$

Antimony

^{123}Sb $\quad\quad$ $^{94}Zr^{16}O_2$

Arsenic

^{75}As $\quad\quad$ $^{40}Ar^{35}Cl$, $^{59}Co^{16}O$, $^{36}Ar^{38}Ar^{1}H$, $^{38}Ar^{37}Cl$, $^{36}Ar^{39}K$

Bromine

^{79}Br $\quad\quad$ $^{40}Ar^{39}K$, $^{38}Ar^{40}Ar^{1}H$

^{81}Br $\quad\quad$ $^{40}Ar^{40}Ar^{1}H$

Cadmium

^{111}Cd $\quad\quad$ $^{95}Mo^{16}O$

^{113}Cd $\quad\quad$ $^{96}Zr^{16}O^{1}H$

^{114}Cd $\quad\quad$ $^{98}Mo^{16}O$

Calcium

^{42}Ca $\quad\quad$ $^{40}Ar^{1}H_2$

^{43}Ca $\quad\quad$ $^{27}Al^{16}O$

^{44}Ca $\quad\quad$ $^{12}C^{16}O_2$, $^{14}N_2^{16}O$, $^{28}Si^{16}O$

^{46}Ca $\quad\quad$ $^{14}N^{16}O_2$, $^{32}S^{14}N$

^{48}Ca $\quad\quad$ $^{33}S^{15}N$, $^{34}S^{14}N$, $^{32}S^{16}O$

Chlorine

^{35}Cl $\quad\quad$ $^{16}O^{18}O^{1}H$, $^{34}S^{1}H$

^{37}Cl $\quad\quad$ $^{36}Ar^{1}H$, $^{36}S^{1}H$

Chromium

^{50}Cr $\quad\quad$ $^{34}S^{16}O$, $^{36}Ar^{14}N$, $^{35}Cl^{15}N$, $^{35}S^{14}N$, $^{32}S^{18}O$, $^{33}S^{17}O$

^{53}Cr $\quad\quad$ $^{35}Cl^{16}O^{1}H$, $^{40}Ar^{12}C$, $^{36}Ar^{16}O$, $^{37}Cl^{15}N$, $^{34}S^{18}O$,

$\quad\quad\quad\quad$ $^{36}S^{16}O$, $^{38}Ar^{14}N$, $^{36}Ar^{15}N^{1}H$, $^{35}Cl^{17}O$

^{54}Cr $\quad\quad$ $^{37}Cl^{16}O^{1}H$, $^{40}Ar^{14}N$, $^{38}Ar^{15}N^{1}H$, $^{36}Ar^{18}O$, ^{38}Ar

$\quad\quad\quad\quad$ ^{16}O, $^{36}Ar^{17}O^{1}H$, $^{37}Cl^{17}O$

Cobalt

^{59}Co $\quad\quad$ $^{43}Ca^{16}O$, $^{42}Ca^{16}O^{1}H$, $^{24}Mg^{35}Cl$, $^{36}Ar^{23}Na$, $^{40}Ar^{18}O^{1}H$

(continues)

APPENDIX 3 *(continued)*

Copper

^{63}Cu $^{40}Ar^{23}Na$, $^{47}Ti^{16}O$, $^{23}Na^{40}Ca$, $^{46}Ca^{16}O^1H$, $^{14}N^{12}C$ ^{37}Cl, $^{16}O^{12}C^{35}Cl$

^{65}Cu $^{49}Ti^{16}O$, $^{40}Ar^{25}Mg$, $^{40}Ca^{16}O^1H$, $^{32}S^{33}S$, $^{32}S^{10}S^{17}O$, ^{33}S $^{16}O_2$, $^{12}C^{16}O^{37}Cl$, $^{12}C^{18}O^{35}Cl$

Dysprosium

^{163}Dy $^{147}Sm^{16}O$

Erbium

^{166}Er $^{160}Nd^{16}O$, $^{150}Sm^{16}O$

^{167}Er $^{151}Eu^{16}O$

Europium

^{151}Eu $^{135}Ba^{16}O$

^{153}Eu $^{137}Ba^{16}O$

Gadolinium

^{155}Gd $^{139}La^{16}O$

^{157}Gd $^{141}Pr^{16}O$

Gallium

^{69}Ga $^{35}Cl^{16}O^{18}O$, $^{35}Cl^{17}O_2$, $^{37}Cl^{16}O_2$, $^{36}Ar^{33}S$, $^{33}S^{18}O_2$, ^{34}S $^{17}O^{18}O$, $^{36}S^{16}O^{17}O$, $^{33}S^{36}S$

^{71}Ga $^{35}Cl^{18}O_2$, $^{37}Cl^{16}O^{18}O$, $^{37}Cl^{17}O_2{}^{36}Ar^{35}Cl$, $^{36}S^{17}O^{18}O$, $^{38}Ar^{33}S$

Germanium

^{70}Ge $^{40}Ar^{14}N^{16}O$, $^{37}Cl^{17}O^{18}O$, $^{37}Cl^{16}O^{17}O$, $^{34}S^{18}O_2$, $^{36}S^{16}O$ ^{18}O, $^{36}S^{17}O_2$, $^{34}S^{36}S$, $^{36}Ar^{34}S$, $^{38}Ar^{32}S$, $^{35}Cl_2$

^{72}Ge $^{36}Ar_2$, $^{37}Cl^{17}O^{18}O$, $^{35}Cl^{37}Cl$, $^{36}S^{18}O_2$, $^{36}S_2$, $^{36}Ar^{36}S$, $^{56}Fe^{16}O$, $^{40}Ar^{16}O_2$, $^{40}Ca^{16}O_2$, $^{40}Ar^{32}S$

^{73}Ge $^{36}Ar_2{}^{74}Ge$ $^{40}Ar^{34}S$, $^{36}Ar^{38}Ar$, $^{37}Cl_2$, $^{38}Ar^{36}S$

^{76}Ge $^{35}Ar^{40}Ar$, $^{38}Ar_2$, $^{40}Ar^{36}S$

Hafnium

^{177}Hf $^{161}Dy^{16}O$

Holium

^{165}Ho $^{149}Sm^{16}O$

Iron

^{54}Fe $^{37}Cl^{16}O^1H$, $^{40}Ar^{14}N$, $^{38}Ar^{15}N^1H$, $^{36}Ar^{18}O$, $^{38}Ar^{16}O$, ^{36}Ar $^{17}O^1H$, $^{36}S^{18}O$, $^{35}Cl^{18}O^1H$, $^{37}Cl^{17}O$

^{56}Fe $^{40}Ar^{16}O$, $^{40}Ca^{16}O$, $^{40}Ar^{15}N^1H$, $^{38}Ar^{18}O$, ^{38}Ar $^{17}O^1H$, $^{37}Cl^{18}O^1H$

^{57}Fe $^{40}Ar^{16}O^1H$, $^{40}Ca^{16}O^1H$, $^{40}Ar^{17}O$, $^{38}Ar^{18}O^1H$, $^{40}Ar^{18}O$, $^{40}Ar^{17}O^1H$

^{58}Fe $^{40}Ar^{18}O$, $^{40}Ar^{17}O^1H$

Krypton

^{78}Kr $^{40}Ar^{38}Ar$

^{80}Kr $^{40}Ar_2$

Luticium

^{175}Lu $^{159}Tb^{16}O$

(continues)

APPENDIX 3 *(continued)*

Magnesium

    ^{24}Mg $^{12}C_2$

    ^{25}Mg $^{12}C_2{}^{1}H$

    ^{26}Mg $^{12}C^{14}N, {}^{12}C^{13}C^{1}H$

Manganese

    ^{55}Mn $^{40}Ar^{14}N^{1}H, {}^{39}K^{16}O, {}^{37}Cl^{18}O, {}^{40}Ar^{15}N, {}^{38}Ar^{17}O,$
                               $^{36}Ar^{18}O^{1}H, {}^{38}Ar^{16}O^{1}H, {}^{37}Cl^{17}O^{1}H, {}^{23}Na^{32}S$

Molybdenum

    ^{94}Mo $^{39}K_2{}^{16}O$

    ^{95}Mo $^{40}Ar^{39}K^{16}O, {}^{79}Br^{16}O$

    ^{96}Mo $^{39}K^{41}K^{16}O, {}^{79}Br^{17}O$

    ^{97}Mo $^{40}Ar^{41}K^{16}O, {}^{81}Br^{16}O$

    ^{98}Mo $^{81}Br^{17}O, {}^{41}K_2{}^{16}O$

Nickel

    ^{58}Ni $^{23}Na^{35}Cl, {}^{40}Ar^{18}O, {}^{40}Ca^{18}O, {}^{40}Ca^{17}O^{1}H,$
                               $^{42}Ca^{16}O, {}^{29}Si_2, {}^{40}Ar^{17}O^{1}H$

    ^{60}Ni $^{44}Ca^{16}O, {}^{23}Na^{37}Cl, {}^{43}Ca^{16}O^{1}H$

    ^{61}Ni $^{44}Ca^{16}O^{1}H, {}^{45}Sc^{16}O$

    ^{62}Ni $^{46}Ti^{16}O, {}^{23}Na^{39}K, {}^{46}Ca^{16}O$

    ^{64}Ni $^{32}S^{16}O_2, {}^{32}S_2$

Palladium

    ^{195}Pd $^{40}Ar^{65}Cu$

Phosphorus

    ^{31}p $^{14}N^{16}O_2, {}^{15}N_2{}^{1}H, {}^{15}N^{16}O, {}^{14}N^{17}O, {}^{13}C^{18}O,$
                               $^{12}C^{18}O^{1}H$

Potassium

    ^{39}K $^{38}Ar^{1}H$

    ^{41}K $^{40}Ar^{1}H$

Rhodium

    ^{103}Rh $^{40}Ar^{63}Cu$

Ruthenium

    ^{101}Ru $^{40}Ar^{61}Ni, {}^{64}Ni^{37}Cl$

    ^{33}S $^{15}N^{18}O, {}^{14}N^{18}O^{1}H, {}^{15}N^{17}O^{1}H, {}^{16}O^{17}O,$
                               $^{16}O_2{}^{1}H\ {}^{32}S^{1}H$

    ^{34}S $^{15}N^{18}O^{1}H, {}^{16}O^{18}O, {}^{17}O_2, {}^{16}O^{17}O^{1}H, {}^{33}S^{1}H$

Scandium

    ^{45}Sc $^{28}Si^{16}O^{1}H, {}^{29}Si^{16}O, {}^{13}C^{16}O_2$

Selenium

    ^{74}Se $^{40}Ar^{34}S, {}^{36}Ar^{38}Ar, {}^{37}Cl_2, {}^{38}Ar^{36}S$

    ^{76}Se $^{35}Ar^{40}Ar, {}^{38}Ar_2, {}^{40}Ar^{36}S$

    ^{77}Se $^{40}Ar^{37}Cl, {}^{36}Ar^{40}Ar^{1}H, {}^{38}Ar_2{}^{1}H$

    ^{78}Se $^{40}Ar^{38}Ar, {}^{38}Ar40Ca$

    ^{80}Se $^{40}Ar_2$

    ^{82}Se $^{12}C^{35}Cl_2$

(continues)

APPENDIX 3 *(continued)*

Silicon		
	^{28}Si	^{14}N$_2$, ^{12}C^{16}O
	^{29}Si	^{14}N^{15}N, ^{14}N$_2$^{1}H, ^{13}C^{16}O, ^{12}C^{17}O, ^{12}C^{16}O^{1}H
	^{30}Si	^{15}N$_2$, ^{14}N^{15}N^{1}H, ^{14}N^{16}O, ^{12}C^{18}O, ^{13}C^{17}O, ^{13}C ^{16}O^{1}H, ^{12}C^{17}O^{1}H
Silver		
	^{107}Ag	^{91}Zr^{16}O
	^{109}Ag	^{92}Zr^{16}O^{1}H
Strontium		
	^{86}Sr	^{85}Rb^{1}H
Sulfur		
	^{32}S	^{16}O$_2$, ^{14}N^{18}O, ^{15}N^{17}O, ^{14}N^{17}O^{1}H, ^{15}N^{16}O^{1}H
Tantalum		
	^{181}Ta	^{165}Ho^{16}O
Terbium		
	^{159}Tb	^{143}Nd^{16}O
Thulium		
	^{169}Tm	^{153}Eu^{16}O
Titanium		
	^{46}Ti	^{32}S^{14}N, ^{14}N^{16}O$_2$, ^{14}N$_2$^{16}O
	^{47}Ti	^{32}S^{14}N^{1}H, ^{30}Si^{16}O^{1}H, ^{32}S^{15}N, ^{33}S^{14}N, ^{15}N^{16}O$_2$, ^{12}C^{35}Cl, ^{31}P^{16}O
	^{48}Ti	^{32}S^{16}O, ^{32}S^{14}N, ^{14}N^{16}O^{18}O, ^{14}N^{17}N$_2$, ^{36}Ar^{12}C
	^{49}Ti	^{32}S^{17}O, ^{32}S^{16}O^{1}H, ^{35}Cl^{14}N, ^{34}S^{15}N, ^{33}S^{16}O, ^{14}N ^{35}Cl, ^{35}Ar^{13}C, ^{36}Ar^{12}C^{1}H, ^{12}C^{37}Cl, ^{31}P^{18}O
	^{50}Ti	^{32}S^{18}O, ^{32}S^{17}O^{1}H, ^{36}Ar^{14}N, ^{35}Cl^{15}N, ^{36}S ^{14}N, ^{33}S^{17}O, ^{34}S^{16}O, ^{14}N^{35}Cl^{1}H
Tungsten		
	^{182}W	^{166}Er^{16}O
Vanadium		
	^{50}V	^{34}S^{16}O, ^{36}Ar^{14}N, ^{35}Cl^{15}N, ^{36}S^{14}N, ^{32}S^{18}O, ^{36}Ar^{14}N^{1}H
	^{51}V	^{34}S^{16}O^{1}H, ^{35}Cl^{16}O, ^{38}Ar^{13}C, ^{36}Ar^{15}N, ^{36}Ar^{14}N^{1}H, ^{37}Cl ^{14}N, ^{36}S^{15}N, ^{33}S^{18}O, ^{34}S^{17}O
Ytterbium		
	^{172}Yb	^{156}Gd^{16}O
	^{173}Yb	^{157}Gd^{16}O
Zinc		
	^{64}Zn	^{32}S^{16}O$_2$, ^{48}Ti^{16}O, ^{48}Ca^{16}O, ^{32}S$_2$, ^{36}Ar^{14}N$_2$
	^{66}Zn	^{50}Ti^{16}O, ^{34}S^{16}O$_2$, ^{32}S^{16}O^{18}O, ^{32}S^{17}O$_2$, ^{33}S^{16}O^{17}O, ^{32}S^{34}S, ^{33}S$_2$
	^{67}Zn	^{35}Cl^{16}O$_2$, ^{33}S^{34}S, ^{34}S^{16}O^{17}O, ^{33}S^{16}O^{18}O, ^{32}S^{17}O ^{18}O, ^{33}S^{17}O$_2$, ^{35}Cl^{16}O$_2$
	^{68}Zn	^{36}S^{16}O$_2$, ^{34}S^{16}O^{18}O, ^{40}Ar^{14}N$_2$, ^{35}Cl^{16}O^{17}O, ^{34}S$_2$, ^{36}Ar^{32}S, ^{34}S^{17}O$_2$, ^{33}S^{17}O^{18}O, ^{32}S^{18}O$_2$, ^{32}S^{36}S
	^{70}Zn	^{35}Cl$_2$, ^{40}Ar^{14}N^{16}O, ^{35}Cl^{17}O^{18}O, ^{37}Cl^{16}O^{17}O, ^{34}S^{18}O$_2$, ^{36}S^{16}O^{18}O, ^{36}S^{17}O$_2$, ^{34}S^{36}S, ^{36}Ar^{34}S, ^{38}Ar^{32}S

Table of Student's t Distribution

Confidence interval[a] (%)	90	95	98	99
n				
1	6.31	12.71	31.82	63.66
2	2.92	4.30	6.96	9.92
3	2.35	3.18	4.54	5.84
4	2.13	2.78	3.75	4.60
5	2.02	2.57	3.36	4.03
6	1.94	2.45	3.14	3.71
7	1.89	2.36	3.00	3.50
8	1.86	2.31	2.90	3.36
9	1.83	2.26	2.82	3.25
10	1.81	2.23	2.76	3.17
12	1.78	2.18	2.68	3.05
14	1.76	2.14	2.62	2.98
16	1.75	2.12	2.58	2.92
18	1.73	2.10	2.55	2.88
20	1.72	2.09	2.53	2.85
30	1.70	2.04	2.46	2.75
50	1.68	2.01	2.40	2.68
∞	1.64	1.96	2.33	2.58

[a]Confidence intervals are for a two-tailed test.

Certified Standard Reference Materials Available from the U.S. National Institute of Standards and Technology and the National Research Council of Canada

---

U.S. National Institute Standards and Technology

Air Particulates

1648	Urban air particulate matter
1649	Urban dust/organics
1650	Diesel particulate matter

Biological and Clinical

1400	Bone ash
1486	Bone meal
1515	Apple leaves
1547	Peach leaves
1548	Total diet
1549	Non-fat milk
1566	Oyster tissue
1566a	Oyster tissue
1567	Wheat flour
1567a	Wheat flour
1568	Rice flour
1568a	Rice flour
1569	Brewer's yeast
1570	Spinach
1570a	Spinach
1571	Orchard leaves
1572	Citrus leaves
1573	Tomato leaves
1573a	Tomato leaves
1575	Pine needles

(continues)

APPENDIX 5 *(continued)*

	1577	Bovine liver
	1577a	Bovine liver
	1577b	Bovine liver
	1589	PCBs in human serum
	1590	Stabilized wine
	1598	Inorganic constituents in bovine serum
	1845	Cholesterol in whole egg powder
	1846	Infant formula
	1945	Organics in whale blubber
	2670	Trace metals in urine
	2670e	Trace metals in urine—elevated
	2670n	Trace metals in urine—normal
	2671	Fluoride in urine
	2671a	Freeze-dried urine certified for fluoride
	2671e	Fluoride in urine—elevated
	2671n	Fluoride in urine—normal
	2672	Mercury in urine
	2672a	Freeze-dried urine certified for mercury
	2672e	Mercury in urine—elevated
	2672n	Mercury in urine—normal
	2695-L	Fluoride in vegetation—low level
	2695-H	Fluoride in vegetation—high level
Glass		
	1411	Soft borosilicate glass
	1412	Multicomponent glass
	1413	Glass sand
	1414	Lead-silica glass
	1416	Glass Al silicate
	1830	Soda lime
	1834	Fused ore
	1871-1	Lead-silicate glasses for microanalysis
	1871-2	Lead-silicate glasses for microanalysis
	1871-3	Lead-silicate glasses for microanalysis
	1872-1	Lead-germanate glasses for microanalysis
	1872-2	Lead-germanate glasses for microanalysis
	1872-3	Lead-germanate glasses for microanalysis
	1873-1	Barium-zinc-silicate glasses for microanalysis
	1873-2	Barium-zinc-silicate glasses for microanalysis
	1873-3	Barium-zinc-silicate glasses for microanalysis
	1874-1	Lithium-aluminum-borate glasses for microanalysis
	1874-2	Lithium-aluminum-borate glasses for microanalysis
	1874-3	Lithium-aluminum-borate glasses for microanalysis
	1875-1	Aluminum-magnesium-phosphate glasses for microanalysis
	1875-2	Aluminum-magnesium-phosphate glasses for microanalysis
	1875-3	Aluminum-magnesium-phosphate glasses for microanalysis

(continues)

APPENDIX 5 *(continued)*

Minerals and Ores		
	19	Calcite
	25c	Manganese ore
	25d	Manganese ore
	104c	Magnesite
	113	Zinc concentrate
	113a	Zinc concentrate
	113b	Zinc concentrate
	120	Native sulfur
	137	Tin ore
	181	Lithium ore (spodumene)
	182	Lithium ore (petalite)
	183	Lithium ore (lepidolite)
Coal		
	1630	Mercury in coal
	1631a	Sulfur in coal
	1631b	Sulfur in coal
	1631c	Sulfur in coal
	1632	Trace elements in coal
	1632a	Trace elements in coal
	1632b	Trace elements in coal
	1635	Trace elements in coal
	2682	Sulfur in coal
	2682a	Sulfur in coal
	2683	Sulfur in coal
	2683a	Sulfur in coal
	2684	Sulfur in coal
	2684a	Sulfur in coal
	2685	Sulfur in coal
	2685a	Sulfur in coal
	2689	Coal fly ash
	2690	Coal fly ash
	2691	Coal fly ash
	2692	Sulfur in coal
	2692a	Sulfur in coal
Oil		
	1083	Wear metals in lubricating oil
	1084	Wear metals in lubricating oil
	1084a	Wear metals in lubricating oil
	1085	Wear metals in lubricating oil
	1085a	Wear metals in lubricating oil
	1818	Chlorine in lubricating base oil
	1818-1	Chlorine in lubricating base oil
	1818-2	Chlorine in lubricating base oil
	1818-3	Chlorine in lubricating base oil

(continues)

APPENDIX 5 *(continued)*

	1818-4	Chlorine in lubricating base oil
	1818-5	Chlorine in lubricating base oil
	1818a	Chlorine in lubricating base oil
	1819	Sulfur in lubricating base oil
	1819-1	Sulfur in lubricating base oil
	1819-2	Sulfur in lubricating base oil
	1819-3	Sulfur in lubricating base oil
	1819-4	Sulfur in lubricating base oil
	1819-5	Sulfur in lubricating base oil
	1819a	Sulfur in lubricating base oil
Fuel		
	1616	Sulfur in kerosene
	1617	Sulfur in kerosene
	1618	Vanadium and nickel in residual fuel oil
	1619	Sulfur in residual fuel oil
	1619a	Sulfur in residual fuel oil
	1620	Sulfur in residual fuel oil
	1620a	Sulfur in residual fuel oil
	1620b	Sulfur in residual fuel oil
	1621	Sulfur in residual fuel oil
	1621a	Sulfur in residual fuel oil
	1621b	Sulfur in residual fuel oil
	1621d	Sulfur in residual fuel oil
	1622	Sulfur in residual fuel oil
	1622a	Sulfur in residual fuel oil
	1622b	Sulfur in residual fuel oil
	1622c	Sulfur in residual fuel oil
	1622d	Sulfur in residual fuel oil
	1623	Sulfur in residual fuel oil
	1623a	Sulfur in residual fuel oil
	1623b	Sulfur in residual fuel oil
	1624	Sulfur in distillate oil
	1624a	Sulfur in distillate oil
	1624b	Sulfur in distillate oil
	1634	Trace metals in fuel oil
	1634a	Trace metals in fuel oil
	1634b	Trace metals in fuel oil
	1634c	Trace metals in fuel oil
	2717	Sulfur in residual fuel oil
	2724	Sulfur in diesel fuel oil
Silicate, Soil, and Rock		
	1a	Argillaceous limestone
	1b	Argillaceous limestone
	1c	Argillaceous limestone
	102	Silica brick

(continues)

APPENDIX 5 *(continued)*

120a	Phosphate rock
120b	Phosphate rock
143r	Trace elements in a sewage sludge
198	Silica brick
199	Silica brick
278	Obsidian rock
1645	River sediment
1646	Estuarine sediment
1646a	Estuarine sediment
1939	PCBs in river sediment A
2704	Buffalo river sediment
2709	San Joaquin soil
2709-L	EPA method 3050 leachable concentrations
2710	Montana soil
2710-L	EPA method 3050 leachable concentrations
2711	Montana soil
2711-L	EPA method 3050 leachable concentrations

Cement

1880	Portland cement
1881	Portland cement
1882	Calcium aluminate cement
1883	Calcium aluminate cement
1884	Portland cement
1885	Portland cement
1886	Portland cement
1887	Portland cement
1888	Portland cement
1889	Portland cement

Water and Precipitation

1641	Mercury in water
1641a	Mercury in water
1641b	Mercury in water
1641c	Mercury in water
1642	Mercury in water
1642a	Mercury in water
1642b	Mercury in water
1642c	Mercury in water
1643	Trace elements in water
1643a	Trace elements in water
1643b	Trace elements in water
1643c	Trace elements in water
1643d	Trace elements in water
2694-1	Simulated rainwater concentration I
2694-2	Simulated rainwater concentration II
2694a	Simulated rainwater

(continues)

APPENDIX 5 *(continued)*

2694a–1	Simulated rainwater concentration I
2694a–2	Simulated rainwater concentration II
Fly Ash	
1633	Trace elements in coal fly ash
1633a	Trace elements in coal fly ash
1633b	Trace elements in coal fly ash
Sludge	
145r	Trace elements in a sewage sludge
2781	Domestic sludge
Metallurgical	
8162	Fine silver[a]
8165	Fine silver[a]
8168	Fine silver[a]
8171	Fine silver[a]
8050	Fine gold[a]
8053	Fine gold[a]
8056	Fine gold[a]
8059	Fine gold[a]
8062	Fine gold[a]
8065	Fine gold[a]
8068	Gold bullion[a]
8071	Gold bullion[a]
8074	Gold bullion[a]
8077	Gold bullion[a]
8080	Gold bullion[a]
897	Nickel base superalloy (Tracealloy A)
898	Nickel base superalloy (Tracealloy B)
899	Nickel base superalloy (Tracealloy C)

National Research Council of Canada

CASS-4	Nearshore seawater
NASS-5	Open ocean seawater
SLEW-2	Estuarine water
SLEW-3	Estuarine water
SLRS-4	Riverine water
HISS-1	Marine sediment
MESS-3	Marine sediment
PACS-2	Marine sediment
DOLT-2	Dogfish liver
DORM-2	Dogfish muscle
LUTS-1	Non-defatted lobster hepatopancreas
TORT-2	Lobster hepatopancreas

[a]Prepared and certified by the Royal Canadian Mint (distributed by NIST).

Supplemental References

(1990). Semi-quantitative analysis by inductively coupled plasma mass spectrometry. *J. Anal. At. Spectrom.* **5**(6), 457–463.

(1990). Effect of ion-iens tuning and flow injection on non-spectroscopic matrix interferences in inductively coupled plasma mass spectrometry. *J. Anal. At. Spectrom.* **5**(6), 445–451.

(1990). Minimisation of sample matrix effects and signal enhancement for trace analytes using anodic stripping voltammetry with detection by inductively coupled plasma atomic emission spectrometry and inductively coupled plasma mass spectrometry. *J. Anal. At. Spectrom.* **5**(6), 437–445.

(1990). Effect of organic solvents and molecular gases on polyatomic ion interferences in inductively coupled plasma mass spectrometry. *J. Anal. At. Spectrom.* **5**(6), 425–431.

(1994). Advances in fatigue lifetime predictive techniques/applications of ICP-MS for determining radionuclides. *Stand. News* **22**(4), 14.

(1997). Dissolved states of trace metal ions in natural water as elucidated by ultrafiltration /size exclusion chromatography /ICP-MS. *Anal. Sci.: Int. J. Jpn. Soc. Anal. Chem.*, p. 393.

(1997). Determination of Cd, Cu, Zn and Pb in rice flour reference materials by isotope dilution inductively coupled plasma mass spectrometry. *Anal. Sci.: Int. J. Jpn. Soc. Anal. Chem.*, p. 429.

Abou-Shakra, F. R., Rayman, M. P., Ward, N. I., and Hotton, V. (1997). Enzymatic digestion for the determination of trace elements in blood serum by inductively coupled plasma mass spectrometry. *J. Anal. At. Spectrom.* **12**(4), 429.

Aggarwal, J. K., Shabani, M. B., Palmer, M. R., and Vala Ragnarsdottir, K. (1996). Determination of the rare earth elements in aqueous samples at subppt levels by inductively coupled plasma mass spectrometry and flow injection ICPMS. *Anal. Chem.* **68**(24), 4418.

Ahmed, K. O., Al-Swaidan, H. M., and Davies, B. E. (1993). Simultaneous elemental analysis in dust of the city of Riyadh, Saudi Arabia by inductively coupled plasma-mass spectrometry (ICP/MS). *Sci. Total Environ.* **138**(1–3), 207.

Akatsuka, K., McLaren, J. W., Lam, J. W., and Berman, S. S. (1992). Determination of iron and ten other trace elements in the ocean seawater reference material NASS-3 by inductively coupled plasma mass spectrometry. *J. Anal. At. Spectrom.* **7**(6), 889–894.

Akatsuka, K., Hoshi, S., McLaren, J. W., and Berman, S. S. (1994). Ion-exchange separation of nanogram platinum in environmental dust samples for isotope dilution ICP-MS. *Bunseki Kagaku* **43**(1), 61.

Akatsuka, K., Hoshi, S., Katoh, T., Willie, S. N., and McLaren, J. W. (1995). An improved anion-exchange separation for the determination of platinum in environmental samples by inductively coupled plasma mass spectrometry. *Chem. Lett.*, p. 817.

Akatsuka, K., Suzuki, T., Nobuyama, N., and Hoshi, S. (1998). Determination of trace elements in sea-water by inductively coupled plasma mass spectrometry after preconcentration by formation of watersoluble complexes and their adsorption on C18-bonded silica gel. *J. Anal. At. Spectrom.* **13**(4), 271.

Alaimo, R., and Censi, P. (1992). Quantitative determination of major, minor, and trace elements in USGS rock standards by inductively coupled plasma-mass spectrometry. *At. Spectrosc.* **13**(4), 113–117.

Al-Ammar, A., Gupta, R. K., and Barnes, R. M. (1999). Elimination of boron memory effect in inductively coupled plasma-mass spectrometry by addition of ammonia. *Spectrochim. Acta, Part B* **54**(13), 1077.

Al-Ammar, A. S., Reitznerova, E., and Barnes, R. M. (1999). Feasibility of using beryllium as internal reference to reduce non-spectroscopic carbon species matrix effect in the inductively coupled plasma-mass spectrometry (ICP-MS) determination of boron in biological samples. *Spectrochim. Acta, Part B* **54**(13), 1813–1820.

Al-Ammar, A. S., Gupta, R. K., and Barnes, R. M. (1999). Correction for non-spectroscopic matrix effects in inductively coupled plasma-mass spectrometry by common analyte internal standardization. *Spectrochim. Acta, Part B* **54**(13), 1849–1860.

Alary, J. F., and Salin, E. D. (1998). Quanitation of water plasma diagnosis for electrothermal vaporization—inductively coupled plasma—mass spectrometry: The use of argon and argide polyatomics as probing species. *Spectrochim. Acta, Part B* **53**(12), 1705.

Albrecht, A., Hall, G. S., and Herzog, G. F. (1992). Determination of trace element concentrations in meteorites by inductively coupled plasma-mass spectrometry. *J. Radioanal. Nucl. Chem.* **164**(1), 13–22.

Aldstadt, J. H., Kuo, J. M., Smith, L. L., and Erickson, M. D. (1996). Determination of uranium by flow injection inductively coupled plasma mass spectrometry. *Anal. Chim. Acta* **319**(1–2), 135.

Alexander, M. L., Smith, M. R., Hartman, J. S., Mendoza, A., and Koppenaal, D. W. (1998). Laser ablation inductively coupled plasma mass spectrometry. *Appl. Surf. Sci.*, p. 255.

Alibo, D. S., Amakawa, H., and Nozaki, Y. (1998). Determination of indium in natural waters by flow injection inductively coupled plasma mass spectrometry. *Acad. Proc. Earth Planet. Sci.* **107**(4), 359.

Alimonti, A., Petrucci, F., Santucci, B., Cristaudo, A., and Caroli, S. (1995). Determination of chromium and nickel in human blood by means of inductively coupled plasma mass spectrometry. *Anal. Chim. Acta* **306**(1), 35.

Alimonti, A., Petrucci, F., Fioravanti, S., Laurenti, F., and Caroli, S. (1997). Assessment of the content of selected trace elements in serum of term and preterm newborns by inductively coupled plasma mass spectrometry. *Anal. Chim. Acta* **342**(1), 75.

Alkanani, T., Friel, J. K., Jackson, S. E., and Longerich, H. P. (1994). Comparison between digestion procedures for the multielemental analysis of milk by inductively coupled plasma mass spectrometry. *J. Agric. Food Chemi.* **42**(9), 1965.

Allain, P., Berre, S., Premel-Cabic, A., Mauras, Y., and Delaporte, T. (1990). Concentrations of rare earth elements in plasma and urine of healthy subjects determined by inductively coupled plasma mass spectrometry. *Clin. Chem. (Winston-Salem, N.C.)* **36**(11), 2011.

Allain, P., Berre, S., Premel-Cabic, A., Mauras, Y., Delaporte, T., and Cournot, A. (1991). Investigation of the direct determination of uranium in plasma and urine by inductively coupled plasma mass spectrometry. *Anal. Chim. Acta* **251**(1–2), 183–186.

Allain, P., Berre, S., Premel-Cabic, A., Mauras, Y., Cledes, A., and Cournot, A. (1991). Urinary elimination of molybdenum by healthy subjects as determined by inductively coupled plasma mass spectrometry. *Magnesium Trace Elem.* **10**(1), 47–50.

Allain, P., Mauras, Y., Douge, C., Jaunault, L., Delaporte, T., and Beaugrand, C. (1991). Determination of iodine and bromine in plasma and urine by inductively coupled plasma mass spectrometry. *Analyst (London)* **115**(6), 813.

Allain, P., Berre, S., Mauras, Y., and le Bouil, A. (1992). Evaluation of inductively coupled mass spectrometry for the determination of platinum in plasma. *Biol. Mass Spectrom.* **21**(3), 141–143.

Allen, L. A. (1996). Inductively coupled plasma mass spectrometry with a twin quadrupole instrument using laser ablation sample introduction and monodisperse dried microparticulate injection. Unpublished Ph.D. Thesis, Iowa State University, Ames.

Allen, L. A., Leach, J. J., Pang, H. M., and Houk, R. S. (1997). Precise measurement of ion ratios in solid samples using laser ablation with a twin quadrupole inductively coupled plasma mass spectrometer. *J. Anal. At. Spectrom.* **12**(2), 171.

Allen, L., St, G., Myers, D. P., and Brushwyler, K. (1998). Trace elemental analysis of metals by laser ablation inductively coupled plasma time-of-flight mass spectrometry. *Phys. Status Solidi A* **167**(2), 357.

Allibone, J., Fatemian, E., and Walker, P. J. (1999). Determination of mercury in potable water by ICP-MS using gold as a stabilising agent. *J. Anal. At. Spectrom.* **14**(2), 235.

Al-Maawali, S. (1992). Studies on electrothermal atomization in graphite furnace atomic absorption spectrometry and electrothermal vaporization inductively coupled plasma mass spectrometry. Unpublished M.Sc. Thesis, Carleton University, Ottawa.

Almeida, C. M. R., and Vasconcelos, M. T. S. D. (1999). Determination of lead isotope ratios in port wine by inductively coupled plasma mass spectrometry after pre-treatment by UV-irradiation. *Anal. Chim. Acta* **396**(1), 45.

Almeida, C. M. R., Teresa, M., and Vasconcelos, S. D. (1999). UV-irradiation and MW-digestion pre-treatment of Port wine suitable for the determination of lead isotope ratios by inductively coupled plasma mass spectrometry. *J. Anal. At. Spectrom.* **14**(12), 1815–1822.

Alonso, J. I. G., Babelot, J. F., Glatz, J. P., Cromboom, O., and Koch, L. (1993). Applications of a glove-box ICP-MS for the analysis of nuclear materials. *Radiochim. Acta* **62**(1–2), 71–80.

Al-Swaidan, H. M. (1993). Trace determination of vanadium and nickel in Saudi Arabian petroleum and petroleum products by microemulsion ICP-MS. *At. Spectrosc.* **14**(6), 170–173.

Al-Swaidan, H. M. (1993). Determination of vanadium and nickel in oil products from Saudi Arabia by inductively coupled plasma mass spectrometry (ICP/MS). *Anal. Lett.* **26**(1), 141.

Al-Swaidan, H. M. (1994). Microemulsion determination of lead and cadmium in Saudi Arabian petroleum products by inductively coupled plasma mass spectrometry (ICP/MS). *Sci. Total Environ.* **145**(1–2), 157.

Alteyrac, J., Augagneur, S., Medina, B., Vivas, N., and Glories, Y. (1995). Mise au point d'une méthode de dosage des éléments minéraux du bois de chêne par ablation laser ICP-MS. *Analusis* **23**(10), 523.

Alvarado, J. (1994). Hydride interference on the determination of minor actinide isotopes by inductively coupled plasma mass spectrometry. *J. Anal. At. Spectrom.* **9**(11), 1223.

Alvarado, J. S., Neal, T. J., Smith, L. L., and Erickson, M. D. (1996). Microwave dissolution of plant tissue and the subsequent determination of trace lanthanide and actinide elements by inductively coupled plasma-mass spectrometry. *Anal. Chim. Acta* **322**(1–2), 11.

Alves, L. C. (1993). Reduction of polyatomic ion interferences in inductively coupled plasma mass spectrometry with cryogenic desolvation. Unpublished Ph.D. Thesis, Iowa State University, Ames.

Alves, L. C., Wiederin, D. R., and Houk, R. S. (1992). Reduction of polyatomic ion interferences in inductively coupled plasma mass spectrometry by cryogenic desolvation. *Anal. Chem.* **64**(10), 1164–1169.

Alves, L. C., Allen, L. A., and Houk, R. S. (1993). Measurement of vanadium, nickel, and arsenic in seawater and urine reference materials by inductively coupled plasma mass spectrometry with cryogenic desolvation. *Anal. Chem.* **65**(18), 2468.

Alves, L. C., Minnich, M. G., Wiederin, D. R., and Houk, R. S. (1994). Removal of organic solvents by cryogenic desolvation in inductively coupled plasma mass spectrometry. *J. Anal. At. Spectrom.* **9**(3), 399–404.

Amarasiriwardena, D., Durrant, S. F., Lasztity, A., Krushevska, A., Argentine, M. D., and Barnes, R. M. (1997). Semiquantitative analysis of biological materials by inductively coupled plasma-mass spectrometry. *Microchem. J.* **56**(3), 352.

Amarasiriwardena, D., Sharma, K., and Barnes, R. M. (1998). Determination of lead concentration and lead isotope ratios in calcium supplements by inductively coupled plasma mass spectrometry after high pressure, high temperature digestion. *Fresenius' J. Anal. Chem.* **362**(5), 493.

Amarasiriwardena, C. J., Lupoli, N., Potula, V., and Korrick, S. (1998). Determination of the total arsenic concentration in human urine by inductively coupled plasma mass spectrometry: A comparison of the accuracy of three analytical methods. *Analyst* **123**(3), 441.

Amosse, J. (1998). Determination of platinum-group elements (PGEs) and gold in geological matrices by inductively coupled plasma-mass spectrometry (ICP-MS) after separation with selenium and tellurium carriers. *Geostand. Newsl.* **22**(1), 93.

Amouroux, D., Tessier, E., Pecheyran, C., and Donard, O. F. X. (1998). Sampling and probing volatile metal(loid) species in natural waters by in-situ purge and cryogenic trapping followed by gas chromatography and inductively coupled plasma mass spectrometry (P-CT-GC-ICP/MS). *Anal. Chim. Acta* **377**(2), 241.

Anderson, S. T. G., Robert, R. V. D., and Farrer, H. N. (1994). Determination of total and leachable arsenic and selenium in soils by continuous hydride generation inductively coupled plasma mass spectrometry. *J. Anal. At. Spectrom.* **9**(10), 1107.

Anderson, W. A. C., Thorpe, S. A., Owen, L. M., and Anderson, S. E. (1998). The analysis of cyhexatin residues in apples, pears and kiwi fruit using inductively coupled plasma mass spectrometry as an initial screen for total tin, with confirmation by gas chromatography-mass spectrometry. *Food Addit. Contam.* **15**(3), 288.

Angeles Quijano, M., Gutierrez, A. M., Perez Conde, M. C., and Camara, C. (1995). Optimization of flow injection hydride generation inductively coupled plasma mass spectrometry for the determination of selenium in water and serum samples. *J. Anal. At. Spectrom.* **10**(10), 871.

Appelblad, P. K., Rodushkin, I., and Baxter, D. C. (2000). The use of PT guard electrode in inductively coupled plasma sector field mass spectrometry: Advantages and limitations. *J. Anal. At. Spectrom.*, Vol: 359–364.

Arar, E. J., Long, S. E., Martin, T. D., and Gold, S. (1992). Determination of hexavalent-chromium in sludge incinerator emissions using ion chromatography and inductively coupled plasma mass spectrometry. *Environ. Sci. Technol.* **26**(10), 1944.

Aro, A., Amarasiriwardena, C., Lee, M.-L., Kim, R., and Hu, H. (2000). Validation of K x-ray fluorescence bone lead measurements by Inductively coupled plasma mass spectrometry in cadaver legs. *Med. Phys.* **27**(1), 119.

Arslan, Z. (2000). Development of analytical methods for determination of trace elements in marine plankton by atomic absorption and inductively coupled plasma mass spectrometry. Unpublished Ph.D. Thesis, University of Massachusetts, Amherst.

Arslan, Z., Ertas, N., Tyson, J. F., Uden, P. C., and Denoyer, E. R. (2000). Determination of trace elements in marine plankton by inductively coupled plasma mass spectrometry (ICP-MS). *Fresenius' J. Anal. Chem.* **366**(3), 273.

Arunachalam, J., Mohl, C., Ostapczuk, P., and Emons, H. (1995). ICP-MS evaluation of continuous-flow sample preparation for the determination of lead in environmental samples. Multielement characterization of soil samples with ICP-MS for environmental studies. *J. Res. Natl. Inst. Stand. Technol.* **100**(3), 316.

Ashley, D. (1992). Polyatomic interferences due to the presence of inorganic carbon in environmental samples in the determination of chromium at mass 52 by ICP-MS. *At. Spectrosc.* **13**(5), 169–173.

Augagneur, S., Medina, B., Szpunar, J., and Lobinski, R. (1996). Determination of rare earth elements in wine by inductively coupled plasma mass spectrometry using a microconcentric nebulizer. *J. Anal. At. Spectrom.* **11**(9), 713.

Augagneur, S., Medina, B., and Grousset, F. (1997). Measurement of lead isotope ratios in wine by ICP-MS and its applications to the determination of lead concentration by isotope dilution. *Fresenius' J. Anal. Chem.* **357**(8), 1149.

Averitt, D. W., and Wallace, G. F. (1992). Investigation into the validity of using ICP-MS with microwave dissolution for the analysis of phosphatic fertilizers and animal feedstuffs. *At. Spectrosc.* **13**(1), 7–10.

Bachmann, H. J. (1996). Special aspects in automatic analysis of environmental samples (soil extracts, fertilizers, plant material) with ICP-MS: Blank values, quality control, retrospective analysis. *Analusis* **24**(9–10), 32.

Baglan, N., Bérard, P., Trompier, F., Cossonnet, C., and Guen, B. L. (1998). How to reduce the uncertainties of committed does derived from urinary uranium measurements: Investigation of new protocols using ICP-MS. *Radiat. Prot. Dosimetry* **79**(1–4), 477.

Baglan, N., Cossonnet, C., Trompier, F., Ritt, J., and Bérard, P. (1999). Implementation of ICP-MS protocols for uranium urinary measurements in worker monitoring. *Health Phys.* **77**(4), 455.

Baglan, N., Cossonnet, C., Pilet, P., Cavadore, D., Exmelin, L., and Bérard, P. (2000). On the use of ICP-MS for measuring plutonium in urine. *J. Radioanal. Nucl. Chem.* **243**(2), 397.

Bailey, E. H., Kemp, A. J., and Ragnarsdottir, K. V. (1993). Determination of uranium and thorium in basalts and uranium in aqueous solution by inductively coupled plasma mass spectrometry. *J. Anal. At. Spectrom.* **8**(4), 551–556.

Baiocchi, C., Giacosa, D., Saini, G., and Cavalli, P. (1994). Determination of thallium in Antarctica snow by means of laser induced atomic fluorescence and high resolution inductively coupled plasma mass spectrometry. *Int. J. Environ. Anal. Chem.* **55**(1//4), 211.

Baker, S. A. (1998). Analysis of glass, ceramic, and soil samples using laser ablation inductively coupled plasma mass spectrometry. Unpublished Ph.D. Thesis, University of Florida, Gainesville.

Baker, S. A., Smith, B. W., and Winefordner, J. D. (1997). Laser ablation inductively coupled plasma mass spectrometry with a compact laser source. *Appl. Spectrosc.* **51**(12), 1918.

Baker, S. A., Smith, B. W., and Winefordner, J. D. (1997). Determination of trace amount of lead in natural waterby isotope dilution -inductively coupled plasma mass spectrometry. Investigation of light scattering for normalization of signals in laser ablation inductively coupled plasma mass spectrometry. *Anal. Sci.: Int. J. Jpn. Soc. Anal. Chem.* **52**(1), 7.

Baker, S. A., Dellavecchia, M. J., Smith, B. W., and Winefordner, J. D. (1997). Analysis of silicon nitride bearings with laser ablation inductively coupled plasma mass spectrometry. *Anal. Chim. Acta* **355**(2–3), 113.

Baker, S. A., Bradshaw, D. K., and Miller-Ihli, N. J. (1999). Trace element determinations in food and biological samples using inductively coupled plasma mass spectrometry. *At. Spectrosc.* **20**(5), 167.

Baker, S. A., Bi, M., Aucelio, R. Q., and Smith, B. W. (1999). Analysis of soil and sediment samples by laser ablation inductively coupled plasma mass spectrometry. *J. Anal. At. Spectrom.* **14**(1), 19.

Bakowska, E., and Gluodenis, T. J. (1999). Analysis of uranium in urine using inductively coupled plasma mass spectrometry. *Am. Clin. Lab.*, pp. 14–17.

Balaram, V. (1993). Characterization of trace elements in environmental samples by ICP-MS. *At. Spectrosc.* **14**(6), 174–179.

Balaram, V. (1995). Developments and trends in inductively coupled plasma mass spectrometry and its influence on the recent advances in trace element analysis. *Curr. Sci.* **69**(8), 640.

Balaram, V. (1997). Microwave dissolution techniques for the analysis of geological materials by ICP-MS. *Curr. Sci.* **73**(11), 1017.

Balaram, V. (1999). Determination of precious metal concentrations in a polymetallic nodule reference sample from the Indian Ocean by ICP-MS. *Mar. Georesour. Geotechnol.* **17**(1), 17.

Balaram, V., and Anjaiah, K. V. (1997). Direct estimation of gold in geological samples by inductively coupled plasma mass spectrometry. *J. Indian Chem. Soc.* **74**(7), 581.

Balaram, V., and Ramesh, S. L. (1999). Rapid dissolution of larger amounts of rock and ore samples by microwave digestion for the estimation of gold by ICP-MS for geochemical prospecting studies. *J. Indian Chem. Soc.* **76**(10), 491.

Balaram, V., Manikyamba, C., Ramesh, S. L., and Saxena, V. K. (1990). Determination of rare earth elements in Japanese rock standard by inductively coupled plasma-mass spectrometry. *At. Spectrosc.* **11**(1), 19.

Balaram, V., Manikyamba, C., Ramesh, S. L., and Anjaiah, K. V. (1992). Rare earth and trace element determination in iron-formation reference samples by ICP-MS. *At. Spectrosc.* **13**(1), 19–25.

Balaram, V., Ramesh, S. L., and Anjaiah, K. V. (1995). Comparitive study of the sample decomposition procedures in the determination of trace and rare earth elements in anorthosites and related rocks by ICP-MS. *Fresenius' J. Anal. Chem.* **353**(2), 176.

Balaram, V., Anjaiah, K. V., and Reddy, M. R. P. (1995). Comparative study on the trace and rare earth element analysis of an Indian polymetallic nodule reference sample by inductively coupled plasma atomic emission spectrometry and inductively coupled plasma mass spectrometry. *Analyst (London)* **120**(5), 1401.

Balaram, V., Ramesh, S. L., and Anjaiah, K. V. (1996). New trace element and REE data in thirteen GSF reference samples by ICP-MS. *Geostand. Newsl.* **20**(1), 71.

Balaram, V., Hussain, S. M., Uday Raij, B., Charan, S. N., Subba Rao, D. V., Anjaiah, K. V., Ramesh, S. L., and Ilangovan, S. (1997). Determination of gold, platinum, palladium, and silver in rocks and ores by ICP-MS for geochemical exploration studies. *At. Spectrosc.* **18**(1), 17.

Balarama Krishna, M. V., Shekhar, R., Karunasagar, D., and Arunachalam, J. (2000). Multi-element characterization of high purity cadmium using inductively coupled plasma quadrupole mass spectrometry and glow-discharge quadrupole mass spectrometry. *Anal. Chim. Acta* **408**(1), 199.

Baranov, V. I., and Tanner, S. D. (1999). A dynamic reaction cell for inductively coupled plasma mass spectrometry (ICP-DRC-MS). Part 1. The RF-field energy contribution in thermodynamics of ion-molecule reactions. *J. Anal. At. Spectrom.* **14**(8), 1133.

Barany, E., Bergdahl, I. A., Schuetz, A., and Skerfving, S. (1997). Inductively coupled plasma mass spectrometry for direct multi-element analysis of diluted human blood and serum. *J. Anal. At. Spectrom.* **12**(9), 1005.

Barbante, C., Bellomi, T., Mezzadri, G., and Cescon, P. (1997). Direct determination of heavy metals at picogram per gram levels in Greenland and Antarctic snow by double focusing inductively coupled plasma mass spectrometry. *J. Anal. At. Spectrom.* **12**(9), 925.

Barbante, C., Cozzi, G., Capodaglio, G., Van de Velde, K., Ferrari, C., Veysseyre, A., Boutron, C. F., Scarponi, G., and Cescon, P. (1999). Determination of Rh, Pd, and Pt in polar and alpine snow and ice by double-focusing ICPMS with microconcentric nebulization. *Anal. Chem.* **71**(19), 4125.

Barbante, C., Cozzi, G., Capodaglio, G., and Van de Velde, K. (1999). Trace element determination in alpine snow and ice by double focusing inductively coupled plasma mass spectrometry with microconcentric nebulization. *J. Anal. At. Spectrom.* **14**(9), 1433.

Barbaro, M., Passariello, B., Quaresima, S., Casciello, A., and Marabini, A. (1995). Analysis of rare earth elements in rock samples by inductively coupled plasma-mass spectrometry (ICP-MS). *Microchem. J.* **51**(3), 312.

Barefoot, R. R. (1998). Determination of the precious metals in geological materials by inductively coupled plasma mass spectrometry. *J. Anal. At. Spectrom.* **13**(10), 1077.

Barinaga, C. J., and Koppenaal, D. W. (1994). Ion-trap mass spectrometry with an inductively coupled plasma source. *Rapid Commun. Mass Spectrom.* **8**(1), 71.

Barnes, R. M. (1993). Advances in inductively coupled plasma mass spectrometry: Human nutrition and toxicology. *Anal. Chim. Acta* **283**(1), 115.

Barnowski, C., Jakubowski, N., Stuewer, D., and Broekaert, J. A. C. (1997). Speciation of chromium by direct coupling of ion exchange chromatography with inductively coupled plasma mass spectrometry. *J. Anal. At. Spectrom.* **12**(10), 1155.

Barrat, J. A., Keller, F., Amosse, J., Taylor, R. N., Nesbitt, R. W., and Hirata, T. (1996). Determination of rare earth elements in sixteen silicate reference samples by ICP-MS after Tm addition and ion exchange separation. *Geostand. Newsl.* **20**(1), 133.

Barwick, V. J., Ellison, S. L. R. and Fairman, B. (1999). Estimation of uncertainties in ICP-MS analysis—a practical methodology. *Anal. Chim. Acta* **394**(2), 281.

Batterham, G. J., Munksgaard, N. C., and Parry, D. L. (1997). Determination of trace metals in seawater by inductively coupled plasma mass spectrometry after off-line dithiocarbamate solvent extraction. *J. Anal. At. Spectrom.* **12**(11), 1277.

Baumann, H. (1990). Rapid and sensitive determination of iodine in fresh milk and milk powder by inductively coupled plasma-mass spectrometry (ICP-MS). *Fresenius' J. Anal. Chem.* **338**(7), 809.

Baumann, H. (1992). Solid sampling with inductively coupled plasma-mass spectrometry—a survey. *Fresenius' J. Anal. Chem.* **342**(12), 907–916.

Bayon, M. M., Alonso, J. I. G., and Medel, A. S. (1998). Enhanced semiquantitative multi-analysis of trace elements in environmental samples using inductively coupled plasma mass spectrometry. *J. Anal. At. Spectrom.* **13**(4), 277.

Bayon, M. M., Garcia, A. R., Alonso, J. I. G., and Sanz-Medel, A. (1999). Indirect determination of trace amounts of fluoride in natural waters by ion chromatography: A comparison of on-line post- column fluorimetry and ICP-MS detectors. *Fluoride* **32**(2), 116.

Beals, D. M. (1996). Determination of technetium-99 in aqueous samples by isotope dilution inductively coupled plasma-mass spectrometry. *J. Radioanal. Nucl. Chem.* **204**(2), 253.

Beary, E. S., Paulsen, P. J., Jassie, L. B., and Fassett, J. D. (1997). Determination of environmental lead using continuous-flow microwave digestion isotope dilution inductively coupled plasma mass spectrometry. *Anal. Chem.* **69**(4), 758.

Beauchemin, D. (1991). Inductively coupled plasma mass spectrometry in hyphenation: A multielemental analysis technique with almost unlimited potential. *Trends Anal. Chem.* **10**(2), 71.

Beauchemin, D. (1993). Preliminary characterization of inductively coupled plasma mass spectrometry with flow injection Into a gaseous (air) carrier. *Analyst (London)* **118**(7), 815.

Beauchemin, D. (1998). Hydride generation interface for the determination of inorganic arsenic and organoarsenic by inductively coupled plasma mass spectrometry using open-focused microwave digestion to enhance the pre-reduction process. *J. Anal. At. Spectrom.* **13**(1), 1.

Beauchemin, D., and Craig, J. M. (1991). Investigations on mixed-gas plasmas produced by adding nitrogen to the plasma gas in ICP-MS. *Spectrochim. Acta, Part B* **46**(5), 603.

Beauchemin, D., and Kisilevsky, R. (1998). A method based on ICP-MS for the analysis of Alzheimer's amyloid plaques. *Anal. Chem.* **70**(5), 1026.

Beauchemin, D., and Specht, A. A. (1997). On-line isotope dilution analysis with ICPMS using reverse flow injection. *Anal. Chem.* **69**(16), 3183.

Beauchemin, D., Micklethwaite, R. K., vanLoon, G. W., and Hay, G. W. (1992). Determination of metal-organic associations in soil leachates by inductively coupled plasma–mass spectrometry. *Chem. Geol.* **95**(1–2), 191.

Becker, J. S., and Dietze, H. J. (1998). Ultratrace and precise isotope analysis by double-focusing sector field inductively coupled plasma mass spectrometry. *J. Anal. At. Spectrom.* **13**(9), 1057.

Becker, J. S., and Dietze, H.-J. (1999). Precise isotope ratio measurements for uranium, thorium and plutonium by quadrupole-based inductively coupled plasma mass spectrometry. *Fresenius' J. Anal. Chem.* **364**(5), 482.

Becker, J. S., and Dietze, H. J. (1999). Determination of trace elements in geological samples by ablation inductively coupled plasma mass spectrometry. *Fresenius' J. Anal. Chem.* **365**(5), 429.

Becker, J. S., Dietze, H.-J., Lean, J. A. M., and Montaser, A. (1999). Ultratrace and isotope analysis of long-lived radionuclides by inductively coupled plasma quadrupole mass spectrometry using a direct injection high efficiency nebulizer. *Anal. Chem.* **71**(15), 3077.

Becker, J. S., Kerl, W., and Dietze, H.-J. (1999). Nuclide analysis of an irradiated tantalum target of a spallation neutron source using high performance ion chromatography and inductively coupled plasma mass spectrometry. *Anal. Chim. Acta* **387**(2), 145.

Becker, J. S., Soman, R. S., Sutton, K. L., and Caruso, J. A. (1999). Determination of long-lived radionuclides by inductively coupled plasma quadrupole mass spectrometry using different nebulizers. *J. Anal. At. Spectrom.* **14**(6), 933.

Becker, S., and Hirner, A. V. (1994). Coupling of inductively coupled plasma mass spectrometry (ICP-MS) with electrothermal vaporisation (ETV). *Fresenius' J. Anal. Chem.* **350**(4/5), 260.

Becotte-Haigh, P., Tyson, J. F., Denoyer, E., and Hinds, M. W. (1996). Determination of arsenic in gold by flow injection inductively coupled plasma mass spectrometry with matrix removal by reductive precipitation. *Spectrochim. Acta, Part B* **51**(14), 1823.

Becotte-Haigh, P., Tyson, J. F., and Denoyer, E. (1998). Flow injection manifold for matrix removal in inductively coupled plasma mass spectrometry by solid phase extraction: determination of Al, Be, Li and Mg in a uranium matrix. *J. Anal. At. Spectrom.* **13**(12), 1327.

Becotte-Haigh, P. E. (1997). Methods for the reduction of matrix interferences in trace element determinations. Unpublished Ph.D. Thesis, University of Massachusetts, Amherst.

Bedini, R. M., and Bodinier, J. L. (1999). Distribution of incompatible trace elements between the constituents of spinel peridotite xenoliths: ICP-MS data from the East African rift—Evidence from the poikiloblastic peridotite xenoliths from Boree (Massif Central, France). *Geochim. Cosmoch. Acta* **63**(22), 3883–3900.

Begerow, J., and Dunemann, L. (1996). Mass spectral interferences in the determination of trace levels of precious metals in human blood using quadrupole magnetic sector field and inductively coupled plasma mass spectrometry. *J. Anal. At. Spectrom.* **11**(4), 303.

Begerow, J., Turfeld, M., and Dunemann, L. (1996). Determination of physiological platinum levels in human urine using magnetic sector field inductively coupled plasma mass spectrometry in combination with ultraviolet photolysis. *J. Anal. At. Spectrom.* **11**(10), 913.

Begerow, J., Turfeld, M., and Dunemann, L. (1997). Determination of physiological palladium and platinum levels in urine using double focusing magnetic sector field ICP-MS. *Fresenius' J. Anal. Chem.* **359**(4–5), 427.

Begerow, J., Turfeld, M., and Dunemann, L. (1997). Determination of physiological palladium, platinum, iridium and gold levels in human blood using double focusing magnetic sector field inductively coupled plasma mass spectrometry. *J. Anal. At. Spectrom.* **12**(9), 1095.

Begerow, J., Turfeld, M., and Dunemann, L. (2000). New horizons in human biomonitoring of environmentally and occupationally relevant metals-sector-field ICP-MS versus electrothermal AAS. *J. Anal. At. Spectrom.* **15**, 347–352.

Begley, I. S., and Sharp, B. L. (1994). Occurrence and reduction of noise in inductively coupled plasma mass spectrometry for enhanced precision in isotope ratio measurement. *J. Anal. At. Spectrom.* **9**(3), 171–176.

Begley, I. S., and Sharp, B. L. (1997). Characterisation and correction of instrumental bias in inductively coupled plasma quadrupole mass spectrometry for accurate measurement of lead isotope ratios. *J. Anal. At. Spectrom.* **12**(4), 395.

Behrens, A. (1995). Determination of gadolinium in geological samples by inductively coupled plasma mass spectrometry and multivariate calibration. *Spectrochim. Acta, Part B* **50**(12), 1521.

Behrens, A. (1997). Multivariate calibration standardization used for the reduction of standards needed for multivariate calibration in inductively coupled plasma mass spectrometry. *Spectrochim. Acta, Part B* **52**(4), 445.

Bei, W., Xiao-Quan, S., and Shu-Guang, X. (1999). Preconcentration of ultratrace rare earth elements in seawater with 8-hydroxyquinoline immobilized polyacrylonitrile hollow fiber membrane for determination by inductively coupled plasma mass spectrometry. *Analyst (London)* **124**(4), 621.

Bendahl, L., Sidenius, U., and Gammelgaard, B. (2000). Determination of selenoprotein P in human plasma by solid phase extraction and inductively coupled plasma mass spectrometry. *Anal. Chim. Acta* **411**(1), 103.

Bensimon, M., Leuenberger, M., and Parriaux, A. (1996). Application of high resolution plasma source mass spectrometry (HR-ICP-MS) to the determination of trace elements having spectral interferences in natural and polluted waters. *Analusis* **24**(9–10), 8.

Bensted, J. (1993). Use of inductively coupled plasma mass spectrometry (ICPMS) for heavy metal trace analysis of API class G oilwell cement. *Cem. Concr. Res.* **23**(4), 993–994.

Beres, S. A., Bruckner, P. H., and Denoyer, E. (1994). Performance evaluation of a cyclonic spray chamber for ICP-MS. *At. Spectrosc.* **15**(2), 96–99.

Berg, T., Royset, O., and Steinnes, E. (1993). Blank values of trace eleemnts in aerosol filters determined by ICP-MS. *Atmo. Environ.* **27A**(15), 2435.

Berg, T., Roeyset, O., and Steinnes, E. (1994). Trace elements in atmospheric precipitation at Norwegian background stations (1989–1990) measured by ICP-MS. *Atmos. Environ.* **28**(21), 3519.

Berg, T., Roeyset, O., Steinnes, E., and Vadset, M. (1995). Atmospheric trace element deposition: Principal component analysis of ICP-MS data from moss samples. *Environ. Pollu.* **88**(1), 67.

Berger, S., Webb, K., and Hams, G. (1997). Routine ICP-MS determinations in biomedical fluids. Part 2: Performance evaluation of the Varian ultramass ICP-MS. *Clin. Biochem. Rev.* **18**(3), 93.

Berryman, N. G., and Probst, T. U. (1996). Parameter optimization of electrothermal vaporization inductively coupled plasma mass spectrometry for oligoelement determination in standard reference materials. *Fresenius' J. Anal. Chem.* **355**(7–8), 783.

Bersier, P. M., and Howell, J. (1994). Tutorial review. Advanced electroanalytical techniques versus atomic absorption spectrometry, inductively coupled plasma atomic emission spectrometry and inductively coupled plasma mass spectrometry in environmental analysis. *Analyst (London)* **119**(2), 219.

Bertucci, M., and Zydowicz, P. (1996). Analyse de traces dans le PVDF par fluorescence X et ICP-MS avec ablation laser. *J. Phys. IV* **6**(4), 853.

Besson, T., and Stroh, A. (1996). Performances et caractéristiques d'un ICP-MS moderne challenges et nouveaux champs d'application. *Analusis* **24**(9–10), 12.

Bettinelli, M., and Baroni, U. (1995). ICP-MS multielemental characterization of coal fly ash. *At. Spectrosc.* **16**(5), 203.

Bettinelli, M., Spezia, S., Baroni, U., and Bizzarri, G. (1995). Determination of trace elements in fuel oils by inductively coupled plasma mass spectrometry after acid mineralization of the sample in a microwave. *J. Anal. At. Spectrom.* **10**(8), 555.

Bettinelli, M., Spezia, S., and Bizzarri, G. (1996). Trace element determination in lichens by ICP-MS. *At. Spectrosc.* **17**(3), 133.

Bettinelli, M., Baroni, U., Bilei, F., and Bizzarri, G. (1997). Characterization of SCR-DENOx materials by ICP-MS and ICP-AES: Comparison with XRF and NAA. *At. Spectrosc.* **18**(3), 71.

Bettinelli, M., Spezia, S., Baroni, U., and Bizzarri, G. (1998). Microwave oven digestion of power plant emissions and ICP-MS determination of trace elements. *At. Spectrosc.* **19**(3), 73.

Bettinelli, M., Spezia, S., Baroni, U., and Bizzarri, G. (1998). Determination of trace elements in power plant emissions by inductively coupled plasma mass spectrometry: Comparison with other spectrometric techniques. *Microchem. J.* **59**(2), 203.

Bhandari, S. A., and Amarasiriwardena, D. (2000). Closed-vessel microwave acid digestion of commercial maple syrup for the determination of lead and seven other trace elements by inductively coupled plasma-mass spectrometry. *Microchem. J.* **64**(1), 73.

Bibak, A., Behrens, A., Sturup, S., Knudsen, L., and Gundersen, V. (1998). Concentrations of 63 major and trace elements in Danish agricultural crops measured by inductively coupled plasma mass spectrometry. 1. Onion (*Allium cepa. J. Agric. Food Chem.* **46**(8), 3139.

Bibak, A., Behrens, A., Sturup, S., Knudsen, L., and Gundersen, V. (1998). Concentrations of 55 major and trace elements in Danish agricultural crops measured by inductively coupled plasma mass spectrometry. 2. Pea (*Pisum sativum* Ping Pong). *J. Agric. Food Chem.* **46**(8), 3146.

Bibak, A., Sturup, S., Haahr, V., Gundersen, P., and Gundersen, V. (1999). Concentrations of 50 major and trace elements in Danish agricultural crops measured by inductively coupled plasma mass spectrometry. 3. Potato (*Solanum tuberosum* Folva). *J. Agric. Food Chem.* **47**(7), 2678.

Bird, S. M. (1998). Elemental speciation for bioanalytical applications HPLC-ICP-MS and CE-ICP-MS. Unpublished Ph.D Thesis, University of Massachusetts, Amherst.

Bird, S. M., Ge, H., Uden, P. C., Tyson, J. F., Block, E., and Denoyer, E. (1998). High-performance liquid chromatography of selenoamino acids and organo selenium compounds—Speciation by inductively coupled plasma mass spectrometry. *J. Chromatogr.* **798**(1–2), 349.

Bitterli, B. A., Cousin, H., and Magyar, B. (1997). Determination of metals in airborne particles by electrothermal vaporization inductively coupled plasma mass spectrometry after accumulation by electrostatic precipitation. *J. Anal. At. Spectrom.* **12**(9), 957.

BjornBjorn, E., BjornFrech, W., BjornHoffmann, E., and LudkeLudke, C. (1998). Investigation and quantification of spectroscopic interferences from polyatomic species in inductively coupled plasma mass spectrometry using electrothermal vaporization or pneumatic nebulization for sample introduction. *Spectrochim. Acta, Part B* **53**(13), 1765.

Bloxham, M. J., Hill, S. J., and Worsfold, P. J. (1993). Analysis of trace metals in sea-water by inductively coupled plasma mass spectrometry and related techniques. *Anal. Proc.* **30**(3), 159.

Bloxham, M. J., Worsfold, P. J., and Hill, S. J. (1994). Matrix suppression in sea-water analysis using inductively coupled plasma mass spectrometry with mixed gas plasmas. *Anal. Proc.* **31**(3), 95.

Bollinger, D. S., and Schleisman, A. J. (1999). Analysis of high purity acids using a dynamic reaction cell ICP-MS. *At. Spectrosc.* **20**(2), 60.

Bonchin-Cleland, S. L., Cleland, T. J., Olson, L. K., and Caruso, J. A. (1996). Evaluation of a modified ion lens system for inductively coupled plasma-mass spectrometry. *Am. Lab.* **28**(3), 34P.

Boonen, S., Vanhaecke, F., Moens, L., and Dams, R. (1996). Direct determination of Se and As in solid certified reference materials using electrothermal vaporization inductively coupled plasma mass spectrometry. *Spectrochim. Acta, Part B* **51**(2), 271.

Borisov, O.V., Coleman, D. M., and Carter, R. O. (1997). Determination of vanadium, rhodium and platinum in automotive catalytic converters using inductively coupled plasma mass spectrometry with spark ablation. *J. Anal. At. Spectrom.* **12**(2), 231.

Borisov, O.V., Coleman, D. M., Oudsema, K. A., and Carter, R. O. (1997). Determination of platinum, palladium, rhodium and titanium in automotive catalytic converters using inductively coupled plasma mass spectrometry with liquid nebulization. *J. Anal. At. Spectrom.* **12**(2), 239.

Bortoli, A., Gerotto, M., Marchiori, M., Palonta, R., and Troncon, A. (1992). Applications of inductively coupled plasma mass spectrometry to the early detection of potentially toxic elements. *Microchem. J.* **46**(2), 167–173.

Bowins, R. J., and McNutt, R. H. (1994). Electrothermal isotope dilution inductively coupled plasma mass spectrometry method for the determination of sub-ng levels of lead in human plasma. *J. Anal. At. Spectrom.* **9**(11), 1233.

Branch, S., Ebdon, L., Ford, M., Foulkes, M., and O'Neill, P. (1991). Determination of arsenic in samples with high chloride content by inductively coupled plasma mass spectrometry. *J. Anal. At. Spectrom.* **6**(2), 151–155.

Branch, S., Corns, W. T., Ebdon, L., Hill, S., and O'Neill, P. (1991). Determination of arsenic by hydride generation inductively coupled plasma mass spectrometry using a tubular membrane gas-liquid separator. *J. Anal. At. Spectrom.* **6**(2), 155–159.

Brandl, W., and Hunter, R. (1999). Benchtop ICP-MS breaks the bottleneck in inorganic elemental analysis: Case studies from a contract analytical laboratory. *Spectroscopy* **14**(1), 30.

Brenner, I. B., and Taylor, H. E. (1992). A critical review of inductively coupled plasma-mass spectrometry for geoanalysis, geochemistry, and hydrology. Part I. Analytical performance. *CRC Crit. Rev. Anal. Chem.*, pp. 355–368.

Brenner, I. B., Zander, A., Plantz, M., and Zhu, J. (1997). Characterization of an ultrasonic nebulizer-membrane separation interface with inductively coupled plasma mass spectrometry for the determination of trace elements by solvent extraction. *J. Anal. At. Spectrom.* **12**(3), 273.

Bricker, T. M. (1994). Studies of selenium and xenon in inductively coupled plasma mass spectrometry. Unpublished M.S. Thesis, Iowa State University, Ames.

Bricker, T. M., and Houk, R. S. (1995). Speciation of selenium in human serum by size exclusion chromatography and inductively coupled plasma mass spectrometry. *Spec. Pub.—R. Soc. Chem.*, p. 109.

Briggs, P. H. (1999). The Determination of forty two elements in geological materials by inductively coupled plasma-mass spectrometry. *Geol. Surv. Open-File Rep. (U.S.),* **99–166.**

Broadhead, M. (1991). Laser Sampling ICP-MS: Determination of rhenium in molybdenum concentrates, cooper concentrates and other geological materials. *At. Spectrosc.* **12**(2), 45.

Broadhead, M., Broadhead, R., and Hager, J. W. (1990). Laser sampling ICP- MS: Semiquantitative determination of sixty-six elements in geological samples. *At. Spectrosc.* **11**(6), 205.

Broekaert, J. A. C., Brandt, R., Leis, F., and Pilger, C. (1994). Analysis of aluminium oxide and silicon carbide ceramic materials by inductively coupled plasma mass spectrometry. *J. Anal. At. Spectrom.* **9**(9), 1063.

Brown, A. A., Ebdon, L., and Hill, S. J. (1994). Development of a coupled liquid chromatography-isotope dilution inductively coupled plasma mass spectrometry method for lead speciation. *Anal. Chim. Acta* **286**(3), 391.

Brown, J. A., Kunz, F. W., and Belitz, R. K. (1991). Characterization of automotive catalysts using inductively coupled plasma mass spectrometry: Sample preparation. *J. Anal. At. Spectrom.* **6**(5), 393.

Brown, P. H., Picchioni, G., Jenkin, M., and Mu, H. (1992). Use of ICP-MS and 10B to trace the movement of boron in plants and soil. *Commun. Soil Sci. Plant Analy.* **23**(17–20), 2781.

Brown, R., Gray, D. J., and Tye, D. (1995). Hydride generation ICP-MS (HG- ICP-MS) for the ultra lowlevel determination of mercury in biota. *Water, Air, Soil Pollut.* **80**(1–4), 1237.

Brunk, S. (1994). A simplified method for serum aluminum by ICP-MS. *At. Spectrosc.* **15**(4), 145.

Buckley, W. T., and Ihnat, M. (1993). Determination of copper, molybdenum and selenium in biological reference materials by inductively coupled plasma mass spectrometry. *Fresenius' J. Anal. Chem.* **345**(2–4), 217–220.

Buckley, W. T., Budac, J. J., Godfrey, D. V., and Koenig, K. M. (1992). Determination of selenium by inductively coupled plasma mass spectrometry utilizing a new hydride generation sample introduction system. *Anal. Chem.* **64**(7), 724–728.

Buseth, E., Wibetoe, G., and Martinsen, I. (1998). Determination of endogenous concentrations of the lanthanides in body fluids and tissues using electrothermal vaporization inductively coupled plasma mass spectrometry. *J. Anal. At. Spectrom.* **13**(9), 1039.

Butler, I. B., and Nesbitt, R. W. (1999). Trace element distributions in the chalcopyrite wall of a black smoker chimney: Insights from laser ablation coupled plasma mass spectrometry (LA-ICP-MS). *Earth Planet. Sci. Lett.* **167**(3), 335.

Byrne, J. P., and Chapple, G. (1998). Direct determination of trace metals in seawater by electrothermal vaporization ICP-MS with Pd-HNO3 modifier. *At. Spectrosc.* **19**(4), 116.

Byrne, J. P., Chakrabarti, C. L., Gregoire, D. C., Lamoureux, M., and Ly, T. (1992). Mechanisms of chloride interferences in atomic absorption spectrometry using a graphite furnace atomizer investigated by electrothermal vaporization inductively coupled plasma mass spectrometry. Part 1. Effect of magnesium chloride matrix and ascorbic acid chemical modifier on manganese. *J. Anal. At. Spectrom.* **7**(2), 371–382.

Byrne, J. P., Lamoureux, M. M., Chakrabarti, C. L., Tam, L., and Grégoire, D. C. (1993). Mechanisms of chloride interferences in atomic absorption spectrometry using a graphite furnace atomizer investigated by electrothermal vaporization inductively coupled plasma mass spectrometry. Part 2. Effect of sodium chloride matrix and ascorbic acid chemical modifier on manganese. *J. Anal. At. Spectrom.* **8**(4), 599–610.

Byrne, J. P., Hughes, D. M., Chakrabarti, C. L., and Grégoire, D. C. (1994). Mechanism of volatilization of tungsten in the graphite furnace investigated by electrothermal vaporization inductively coupled plasma mass spectrometry. *J. Anal. At. Spectrom.* **9**(9), 913.

Byrne, J. P., Grégoire, D. C., Goltz, D. M., and Chakrabarti, C. L. (1994). Vaporization and atomization of boron in the graphite furnace investigated by electrothermal vaporization inductively coupled plasma mass spectrometry. *Spectrochim. Acta, Part B* **49B**(5), 433.

Caddia, M., and Iversen, B. S. (1998). Determination of uranium in urine by inductively coupled plasma mass spectrometry with pneumatic nebulization. *J. Anal. At. Spectrom.* **13**(4), 309.

Cairns, W. R. L., Hill, S. J., and Ebdon, L. (1996). Directly coupled high performance liquid chromatography—Inductively coupled plasma-mass spectrometry for the determination of organometallic species in tea. *Microchem. J.* **54**(2), 88.

Cairns, W. R. L., McLeod, C. W., and Hancock, B. (1997). Atomic spectroscopy perspectives: Determination of platinum in human serum by flow injection inductively coupled plasma-mass spectrometry. *Spectroscopy* **12**(4), 16.

Campana, S. E., Fowler, A. J., and Jones, C. M. (1994). Otolith elemental fingerprinting for stock identification of Atlantic cod *(Gadus morhua)* using laser ablation ICPMS. *Can. J. Fish. Aqua. Sci.* **51**(9), 1942.

Campbell, A. J., and Humayun, M. (1999). Trace element microanalysis in iron meteorites by laser ablation ICPMS. *Anal. Chem.* **71**(5), 939.

Campbell, M. (1995). Development of ICPMS and ID-ICPMS with the determination of Pb and Hg in environmental matrices as an example. *Tech. Instrum. Anal. Chem.*, p. 28.

Campbell, M. J., and Toervenyi, A. (1999). Non-spectroscopic suppression of zinc in ICP-MS in a candidate biological reference material (IAEA 392 Algae). *J. Anal. At. Spectrom.* **14**(9), 1313.

Campbell, M. J., Vermeir, G., Dams, R., and Quevauviller, P. (1992). Influence of chemical species on the determination of mercury in a biological matrix (cod muscle) using inductively coupled plasma mass spectrometry. *J. Anal. At. Spectrom.* **7**(4), 617–622.

Campbell, M. J., Demesmay, C., and Olle, M. (1994). Determination of total arsenic concentrations in biological matrices by inductively coupled plasma mass spectrometry. *J. Anal. At. Spectrom.* **9**(12), 1379.

Cao, X., Zhao, G., Yin, M., and Li, J. (1998). Determination of ultratrace rare earth elements in tea by inductively coupled plasma mass spectrometry with microwave digestion and AG50W-x8 cation exchange chromatography. *Analyst (London)* **123**(5), 1115.

Cao, X., Yin, M., and Li, B. (1999). Determination of rare earth impurities in high purity gadolinium oxide by inductively coupled plasma mass spectrometry after 2-ethylhexyl-hydrogen-ethylhexy phosphonate extraction chromatographic separation. *Talanta* **48**(3), 517.

Carey, J. M., Evans, E. H., Caruso, J. A., and Wei-Lung, S. (1991). Evaluation of a modified commercial graphite furnace for reduction of isobaric interferences in argon inductively coupled plasma mass spectrometry. *Spectrochim. Acta, Part B* **46B**(13), 1711.

Carlo, E. H. D., and Pruszkowski, E. (1995). Laser ablation ICP-MS determination of alkaline and rare earth elements in marine ferro-manganese deposits. *At. Spectrosc.* **16**(2), 65.

Cary, E. E., Wood, R. J., and Schwartz, R. (1990). Stable Mg isotopes as tracers using ICP-MS. *J. Micronutrient Anal.* **8**(1), 13.

Casetta, B. (1990). An interactive reference program for spectral interferences forecast in ICP-MS. *At. Spectrosc.* **11**(3), 102.

Casiot, C., Szpunar, J., Lobinski, R., and Potin-Gautier, M. (1999). Sample preparation and HPLC separation approaches to speciation analysis of selenium in yeast by ICP-MS. *J. Anal. At. Spectrom.* **14**(4), 645.

Castillano, M. T. M. (1994). Solution nebulization device (ultrasonic nebulizer) for inductively coupled plasma spectrometry and alternate plasma source (low pressure inductively coupled plasma) for mass spectrometry. Unpublished Ph.D. Thesis, University of Cincinnati, Cincinnati, OH.

Castillano, T. M., Giglio, J. J., Hywel Evans, E., and Caruso, J. A. (1994). Evaluation of low pressure inductively coupled plasma mass spectrometry for the analysis of gaseous samples. *J. Anal. At. Spectrom.* **9**(12), 1335.

Castle, J. E., and Qiu, J. H. (1990). The application of ICP-MS and XPS to studies of ion selectivity during passivation of stainless steels. *J. Electrochem. Soci.* **137**(7), 2031.

Catterick, T., Handley, H., and Merson, S. (1995). Analytical accuracy in ICP-MS using isotope dilution and its application to reference materials. *At. Spectrosc.* **16**(6), 229.

Catterick, T., Fairman, B., and Harrington, C. (1998). Structured approach to achieving high accuracy measurements with isotope dilution inductively coupled plasma mass spectrometry. *J. Anal. At. Spectrom.* **13**(9), 1009.

Chang, C. C., and Jiang, S. J. (1997). Determination of Hg and Bi by electrothermal vaporization inductively coupled plasma mass spectrometry using vapor generation with in situ preconcentration in a platinum-coated graphite furnace. *Anal. Chim. Acta* **353**(2–3), 173.

Chang, C. C., and Jiang, S. J. (1997). Determination of copper, cadmium and lead in biological samples by electrothermal vaporization isotope dilution inductively coupled plasma mass spectrometry. *J. Anal. At. Spectrom.* **12**(1), 75.

Chapple, G., and Byrne, J. P. (1996). Direct determination of trace metals in sea-water using electrothermal vaporization inductively coupled plasma mass spectrometry. *J. Anal. At. Spectrom.* **11**(8), 549.

Chartier, F., Aubert, M., and Pilier, M. (1999). Determination of Am and Cm in spent nuclear fuels by isotope dilution inductively coupled plasma mass spectrometry and isotope dilution thermal ionization mass spectrometry after separation by high-performance liquid chromatography. *Fresenius' J. Anal. Chem.* **364**(4), 320.

Chassaigne, H., and Szpunar, J. (1998). The coupling of reversed-phase HPLC with ICP-MS in bioinorganic analysis. *Analusis* **26**(6), M48.

Chassery, S., Grousset, F. E., Lavaux, G., and Quetel, C. R. (1998). 87Sr/86Sr measurements on marine sediments by inductively coupled plasma-mass spectrometry. *Fresenius' J. Anal. Chem.* **360**(2), 230.

Chaudhary-Webb, M., Paschal, D. C., Crawford Elliott, W., Hopkins, H. P., Ghazi, A. M., Ting, B. C., and Romieu, I. (1998). ICP-MS determination of lead isotope ratios in whole blood, pottery, and leaded gasoline: Lead sources in Mexico City. *At. Spectrosc.* **19**(5), 156.

Cheatham, M. M., Sangrey, W. M., and White, W. M. (1993). Sources of error in external calibration ICP-MS analysis of geological samples and an improved non-linear drift correction procedure. *Spectrochim. Acta, Part B* **48B**(3), E487.

Chen, C. S., and Jiang, S. J. (1996). Determination of As, Sb, Bi and Hg in water samples by flow-injection inductively coupled plasma mass spectrometry with an in-situ nebulizer/ hydride generator. *Spectrochim. Acta, Part B* **51**(14), 1813.

Chen, K.-L. B., Amarasiriwardena, C. J., and Christiani, D. C. (1999). Determination of total arsenic concentrations in nails by inductively coupled plasma mass spectrometry. *Biol. Trace Elem. Res.* **67**(2), 109.

Chen, S. F., and Jiang, S. J. (1998). Determination of arsenic, selenium and mercury in fish samples by slurry sampling electrothermal vaporization inductively coupled plasma mass spectrometry. *J. Anal. At. Spectrom.* **13**(7), 673.

Chen, S. F., and Jiang, S. J. (1998). Determination of cadmium, mercury and lead in soil samples by slurry sampling electrothermal vaporization inductively coupled plasma mass spectrometry. *J. Anal. At. Spectrom.* **13**(10), 1113.

Chen, W. H., Lin, S. Y., and Liu, C. Y. (2000). Capillary electrochromatographic separation of metal ion species with on-line detection by inductively coupled plasma mass spectrometry. *Anal. Chim. Acta* **410**(1), 25.

Chen, X. (1995). Matrix effects in inductively coupled plasma mass spectrometry. Unpublished Ph.D. thesis, Iowa State University, Ames.

Chen, X., and Houk, R. S. (1995). Polyatomic ions as internal standards for matrix corrections in inductively coupled plasma mass spectrometry. *J. Anal. At. Spectrom.* **10**(10), 837.

Chen, Z. (1999). Inter-element fractionation and correction in laser ablation inductively coupled plasma mass spectrometry. *J. Anal. At. Spectrom.* **14**(12), 1823–1828.

Chen, Z., Fryer, B. J., Longerich, H. P., and Jackson, S. E. (1996). Determination of the precious metals in milligram samples of sulfides and oxides using inductively coupled plasma mass spectrometry after ion exchange preconcentration. *J. Anal. At. Spectrom.* **11**(9), 803.

Chen, Z., Doherty, W., and Grégoire, D. C. (1997). Application of laser sampling microprobe inductively coupled plasma mass spectrometry to the in situ trace element analysis of selected geological materials. *J. Anal. At. Spectrom.* **12**(6), 653.

Cheol-Su, K., Chang-Kyu, K., Jong-In, L., and Kun-Jai, L. (2000). Rapid determination of Pu isotopes and atom ratios in small amounts of environmental samples by an on-line sample pre- treatment system and isotope dilution high resolution inductively coupled plasma mass spectrometry. *J. Anal. At. Spectrom.* **15**, 247–256.

Chicarelli, M. I., Eckardt, C. B., Owen, C. R., Maxwell, J. R., Eglinton, G., Hutton, R. C., and Eaton, A. N. (1990). Application of inductively coupled plasma-mass spectrometry in the detection of organometallic compounds in chromatographic fractions from organic rich shales. *Org. Geochem.* **15**(3), 267.

Chin, C. J., Wang, C. F., and Jeng, S. L. (1999). Multi-element analysis of airborne particulate matter collected on PTFE-membrane filters by laser ablation inductively coupled plasma mass spectrometry. *J. Anal. At. Spectrom.* **14**(4), 663.

Chisum, M. E. (1991). Applications of negative ion analysis on the ELAN 250 ICP-MS. *Atomic Spectrosc.* **12**(5), 155.

Christodoulou, J., Kashani, M., Keohane, B. M., and Sadler, P. J. (1996). Determination of gold and platinum in the presence of blood plasma proteins using inductively coupled plasma mass spectrometry with direct injection nebulization. *J. Anal. At. Spectrom.* **11**(11), 1031.

Cidu, R. (1996). Comparison of ICP-MS and ICP-OES in the determination of trace elements in water. *At. Spectrosc.* **17**(4), 155.

Cleland, S. L., Olson, L. K., Caruso, J. A., and Carey, J. M. (1994). Optimization of arsenic supercritical fluid extraction with detection by inductively coupled plasma mass spectrometry. *J. Anal. At. Spectrom.* **9**(9), 975.

Cleland, T. J. (1995). I. Ion studies in inductively coupled plasma mass spectrometry: II. A generalized expression for generating quantum collision cross sections. Unpublished Ph.D. Thesis, University of Cincinnati, Cincinnati, OH.

Cleland, T. J., Bonchin-Cleland, S. L., Olson, L. K., and Meeks, F. R. (1995). More efficient ion transport for inductively coupled plasma mass spectrometry. *Spectrochim. Acta, Part B* **50**(8), 873.

Clemons, P. S. (1996). Development and evaluation of a graphite injector for inductively coupled plasma mass spectrometry. Unpublished Ph.D. Thesis, Iowa State University, Ames.

Cocherie, A., Negrel, P., Roy, S., and Guerrot, C. (1998). Direct determination of lead isotope ratios in rainwater using inductively coupled plasma mass spectrometry. *J. Anal. At. Spectrom.* **13**(9), 1069.

Coedo, A. G., and Dorado, T. (1995). Evaluation of the analytical performance of inductively coupled plasma mass spectrometry for the simultaneous determination of major and minor elements in basic slags. *Mikrochim. Acta* **118**(1–2), 75.

Coedo, A. G. and Dorado, T. (1995). Determination of trace elements in unalloyed steels by flow injection inductively coupled plasma mass spectrometry. *J. Anal. At. Spectrom.* **10**(6), 449.

Coedo, A. G. and Dorado, M. T. (1995). Use of flow injection inductively coupled plasma-mass spectrometry for the determination of niobium, vanadium, and titanium in microalloyed steels. *Appl. Spectrosc.* **49**(1), 115.

Coedo, A. G., Dorado, T., Fernandez, B. J., and Alguacil, F. J. (1996). Isotope dilution analysis for flow injection ICPMS determination of microgram per gram levels of boron in iron and steel after matrix removal. *Anal. Chem.* **68**(6), 991.

Coedo, A. G., Dorado, T., and Alguacil, F. J. (1996). Inductively coupled plasma mass spectrometry analysis of electric arc furnace flue dusts: Optimization or sample dissolution procedure. *J. Trace Microprobe tech.* **14**(4), 739.

Coedo, A. G., Lopez, T. D., and Alguacil, F. (1996). On-line ion-exchange separation and determination of niobium, tantalum, tungsten, zirconium and hafnium in high-purity iron by flow injection inductively coupled plasma mass spectrometry. *Anal. Chim. Acta* **315**(3), 331.

Coedo, A. G., Dorado, M. T., Padilla, I., and Alguacil, F. J. (1996). Study of the application of air-water flow injection inductively coupled plasma mass spectrometry for the determination of calcium in steels. *J. Anal. At. Spectrom.* **11**(11), 1037.

Coedo, A. G., Dorado, M. T., Padilla, I., and Alguacil, F. (1997). Preconcentration and matrix separation of precious metals in geological and related materials using metalfix-chelamine resin prior to inductively coupled plasma mass spectrometry. *Anal. Chim. Acta* **340**(1–3), 31.

Coedo, A. G., Dorado, T., Pagilla, I., and Alguacil, F. J. (1997). Determination of phosphorus content in steels using flow injection into an argon-water carrier for inductively coupled plasma-mass spectrometry. *ISIJ Int.* **37**(9), 878.

Coedo, A. G., Dorado, M. T., Padilla, I., and Alguacil, F. J. (1998). Use of boric acid to improve the microwave-assisted dissolution process to determine fluoride forming elements in steels by flow injection inductively coupled plasma mass spectrometry. *J. Anal. At. Spectrom.* **13**(10), 1193.

Coedo, A. G., Dorado, T., and Padilla, I. (1999). Submitted papers—Evaluation of a desolvating microconcentric nebulizer in inductively coupled plasma mass spectrometry to improve the determination of arsenic in steels. *Appl. Spectrosc.* **53**(8), 974.

Coedo, A. G., Padilla, I., Dorado, T., and Alguacil, F. J. (1999). A micro-scale mercury cathode electrolysis procedure for on- line flow injection inductively coupled plasma mass spectrometry trace elements analysis in steel samples. *Anal. Chim. Acta* **389**(1), 247.

Coedo, A. G., Dorado, T., Padilla, I., Maibusch, R., and Kuss, H. M. (2000). Slurry sampling electrothermal vaporization inductively coupled plasma mass spectrometry for steelmaking flue dust analysis. *Spectrochim. Acta, Part B* **55**(2), 185–196.

Colodner, D. C., Boyle, E. A., and Edmond, J. M. (1993). Determination of rhenium and platinum in natural waters and sediments, and iridium in sediments by flow injection isotope dilution inductively coupled plasma mass spectrometry. *Anal. Chem.* **65**(10), 1419.

Contado, C., Blo, G., Fagioli, F., Dondi, F., and Beckett, R. (1997). Characterisation of River Po particles by sedimentation field- flow fractionation coupled to GFAAS and ICP-MS. *Colloids Surf. A* **120**(1–3), 47.

Cook, J. M., Robinson, J. J., Chenery, S. R. N., and Miles, D. L. (1997). Determining cadmium in marine sediments by inductively coupled plasma mass spectrometry: Attacking the problems or the problems with the attack? *Analyst (London)* **122**(11), 1207.

Courtney, A. J., Die, D. J., and Holmes, M. J., (1994). Descriminating populations of the eastern king prawn, *Panaeus plebejus*, from different estuaries using ICP-MS trace element analysis. *At. Spectrosc.* **15**(1), 1–6.

Cousin, H., Weber, A., Magyar, B., and Abell, I. (1995). An auto-focus system for reproducible focusing in laser ablation inductively coupled plasma mass spectrometry. *Spectrochim. Acta, Part B* **50**(1), 63.

Cox, R. J., Pickford, C. J., and Thompson, M. (1992). Determination of iodine- 129 in vegetable samples by inductively coupled plasma mass spectrometry. *J. Anal. At. Spectrom.* **7**(4), 635–640.

Craig, J. M., and Beauchemin, D. (1992). Reduction of the effects of concomitant elements in inductively coupled plasma mass spectrometry by adding nitrogen to the plasma gas. *J. Anal. At. Spectrom.* **7**(6), 937–942.

Crain, J. S. (1996). Atomic spectroscopy perspectives: Applications of inductively coupled plasma–mass spectrometry in environmental radiochemistry. *Spectroscopy* **11**(2), 30.

Crain, J. S. (1998). A study of isotopic homogeneity in "blended uranium" using ICP-MS. *ASTM Spe. Tech. Pub.*, p. 16.

Crain, J. S., and Gallimore, D. L. (1992). Determination of trace impurities in uranium oxides by laser ablation inductively coupled plasma mass spectrometry. *J. Anal. At. Spectrom.* **7**(4), 605–610.

Crain, J. S., and Kiely, J. T. (1996). Waste reduction in inductively coupled plasma mass spectrometry using flow injection and a direct injection nebulizer. *J. Anal. At. Spectrom.* **11**(7), 525.

Crain, J. S., and Mikesell, B. L. (1992). Detection of sub-ng/L actinides in industrial wastewater matrices by inductively coupled plasma-mass spectrometry. *Appl. Spectrosc.* **46**(10), 1498.

Crain, J. S., Smith, L. L., Yaeget, J. S., and Alvarado, J. A. (1995). Determination of long-lived actinides in soil leachates by inductively coupled plasma-mass spectrometry. *J. Radioanal. Nuc. Chem.* **194**(1), 133.

Creed, J. T., Martin, T. D., and Brockhoff, C. A. (1995). Ultrasonic nebulization and arsenic valence state considerations prior to determination via inductively coupled plasma mass spectrometry. *J. Anal. At. Spectrom.* **10**(6), 443.

Creed, J. T., Magnuson, M. L., Brockhoff, C. A., and Chamberlain, I. (1996). Arsenic determination in saline waters utilizing a tubular membrane as a gas-liquid separator for hydride generation inductively coupled plasma mass spectrometry. *J. Anal. At. Spectrom.* **11**(7), 505.

Crews, H. M. (1993). A decade of ICP-MS analysis. *Am. Lab.* **25**(4), 34Z.

Crews, H. M., Baxter, M. J., Bigwood, T., Burrell, J. A., and Owen, L. M. (1992). Lead in feed incident—multi-element analysis of cattle feed and tissues by inductively coupled plasma-mass spectrometry and co-operative quality assurance scheme for lead analysis of milk. *Food Addit. Contam.* **9**(4), 365.

Cromwell, E. F., and Arrowsmith, P. (1995). Fractionation effects in laser ablation inductively coupled plasma mass spectrometry. *Appl. Spectrosc.* **49**(11), 1652.

Cromwell, E. F., and Arrowsmith, P. (1995). Semiquantitative analysis with laser ablation inductively coupled plasma mass spectrometry. *Anal. Chem.* **67**(1), 131.

Dadfarnia, S., and McLeod, C. W. (1994). On-line trace enrichment and determination of uranium in waters by flow injection inductively coupled plasma mass spectrometry. *Appl. Spectrosc.* **48**(11), 1331.

Da-Hai, S., Ren-Li, M., McLeod, C. W., Xiao-Ru, W., and Cox, A. G. (2000). Determination of boron in serum, plasma and urine by inductively coupled plasma mass spectrometry (ICPMS). Use of mannitol-ammonia as diluent and for eliminating memory effect. *J. Anal. At. Spectrom.* **15**(3), 257–262.

Dalpe, C., Baker, D. R., and Sutton, S. R. (1995). Synchrotron x-ray-fluorescence and laser-ablation ICP-MS microprobes: Useful instruments for analysis of experimental run-products. *Can. Mineral.* **33**(2), 481.

Das, A. K., Chakarborty, R., Cervera, M. L., and Guardia, M. D. I. (1998). A rapid method for the determination of trace element impurities in silicone oils by ICP-MS after microwave-assisted digestion. *Spectrosc. Lette.* **31**(6), 1245.

Date, A. R. (1991). Inductively coupled plasma-mass spectrometry. *Spectrochim. Acta Rev.* **14**(1–2), 3–32.

Date, A. R., and Gray, A. L. (1990). Applications of inductively coupled plasma mass spectrometry. *Spectrochim. Acta, Part B* **45**(1–2), 219.

Dauchy, X., Cottier, R., Batel, A., Jeannot, R., and Borsier, M. (1993). Speciation of butyltin compounds by high-performance liquid chromatography with inductively coupled plasma mass spectrometry detection. *J. Chromatogr. Sci.* **31**(10), 416.

Dauchy, X., Cottier, R., Batel, A., and Borsier, M. (1994). Application of butyltin speciation by HPLC/ICP-MS to marine sediments. *Environ. Techno.* **15**(6), 569.

De Boer, J. L. M. (1997). Possibilities and limitations of spectral fitting to reduce polyatomic ion interferences in inductively coupled plasma quadrupole mass spectrometry in the mass range 51–88. *Spectrochim. Acta, Part B* **52**(3), 389.

De Boer, J. L. M., Verweij, W., Van der Velde-Koerts, T., and Mennes, W. (1996). Levels of rare earth elements in Dutch drinking water and its sources. Determination by inductively coupled plasma mass spectrometry and toxicological implications. A Pilot Study. *Water Res.* **30**(1), 190.

Debrah, E., and Alary, J.-F. (1999). Improved sample throughput using flow injection rapid microsampling for ICP-MS. *At. Spectrosc.* **20**(1), 1.

Debrah, E., and Legere, G. (1999). Improving sensitivity and detection limits in ICP-MS with a novel high-efficiency sample introduction system. *At. Spectrosc.* **20**(2), 73.

Debrah, E., Beres, S. A., Gluodenis T. J., Jr. Thomas, R. J., and Denoyer, E. R. (1995). Benefits of a microconcentric nebulizer for multielement analysis of small sample volumes by ICP-MS. *At. Spectrosc.* **16**(5), 197.

Debrah, E., Denoyer, E. R., and Tyson, J. F. (1996). Flow injection determination of mercury with preconcentration by amalgamation on a gold-platinum gauze by inductively coupled plasma mass spectrometry. *J. Anal. At. Spectrom.* **11**(2), 127.

Delves, H. T., Sieniawska, C. E., Fell, G. S., and Lyon, T. D. B. (1997). Determination of antimony in urine, blood and serum and in liver and lung tissues of infants by inductively coupled plasma mass spectrometry. *Analyst (London)* **122**(11), 1323.

Deng, B., and Zhu, B. (1998). Determination of micro-amounts of impurities in tungsten oxide by inductively coupled plasma-mass spectrometry. *Phys. Test. Chem. anal.* **34**(1), 7.

Dennis, M. J., Burrell, A., Mathieson, K., and Willetts, P. (1994). The determination of the flour improver potassium bromate in bread by gas chromatographic and ICP-MS methods. *Food Addit. Contam.* **11**(6), 633.

Denoyer, E. (1992). An evaluation of spectral integration in ICP-MS. *At. Spectrosc.* **13**(3), 93–98.

Denoyer, E. R. (1994). Optimization of transient signal measurements in ICP-MS. *At. Spectrosc.* **15**(1), 7–16.

Denoyer, E. R. (1995). An advanced ICP-MS instrument. *Int. Lab. Eur. Ed.* **25**(3), 8.

Denoyer, E. R., and Qinghong, L. (1993). Characterization of operating parameters in flow injection ICP-MS. *At. Spectrosc.* **14**(6), 162–169.

Denoyer, E. R., and Stroh, A. (1992). Expanding ICP-MS capabilities using flow injection. *Am. Lab.* **24**(3), 74–82.

Denoyer, E. R., Stroh, A., and Qinghong, L. (1993). High sample throughput with rapid microsampling flow injection ICP-MS. *At. Spectrosc.* **14**(2), 55–59.

Denoyer, E. R., Jacques, D., Debrah, E., and Tanner, S. D. (1995). Determination of trace elements in uranium: Practical benefits of a new ICP-MS lens system. *At. Spectrosc.* **16**(1), 1.

Denoyer, E. R., Brueckner, P., and Debrah, E. (1995). Determination of trace impurities in semiconductor-grade hydrofluoric acid and hydrogen peroxide by ICP-MS. *At. Spectrosc.* **16**(1), 12.

Denoyer, E. R., Tanner, S. D., and Voellkopf, U. (1999). A new dynamic reaction cell for reducing ICP-MS interferences using chemical resolution—Through the process of chemical resolution in a dynamic reaction cell, interference in ICP-MS can be reduced or even eliminated. *Spectroscopy* **14**(2), 43.

De Smaele, T., Verrept, P., Moens, L., and Dams, R. (1995). A flexible interface for the coupling of capillary gas chromatography with inductively coupled plasma mass spectrometry. *Spectrochim. Acta, Part B* **50**(11), 1409.

De Smaele, T., Moens, L., Dams, R., and Sandra, P. (1996). ICP-MS—A sensitive detector for metal speciation with capillary GC. *Semicond. Int.* **19**(6), 90.

Devos, W., Moor, C., and Lienemann, P. (1999). Determination of impurities in antique silver objects for authentication by laser ablation inductively coupled plasma mass spectrometry (LA-ICP-MS). *J. Anal. At. Spectrom.* **14**(4), 621.

Devos, W., Senn-Luder, M., Moor, C., and Salter, C. (2000). Laser ablation inductively coupled plasma mass spectrometry (LA-ICP-MS) for spatially resolved trace analysis of early-medieval archaeological iron finds. *Fresenius' J. Anal. Chem.* **366**(8), 873.

Diaz-Somoano, M., and Martinez-Tarazona, M. R. (1999). Application of ICP-MS to arsenic determination in solid samples containing silica. *J. Anal. At. Spectrom.* **14**(9), 1439.

Diemer, J., and Heumann, K. G. (1997). Bromide/bromate speciation by NTI-IDMS and ICP-MS coupled with ion exchange chromatography. *Fresenius' J. Anal. Chem.* **357**(1), 74.

Ding, H. (1996). Chromatography with inductively coupled plasma mass spectrometry for elemental speciation, quantification, and fragmentation studies. Unpublished Ph.D. Thesis, University of Cincinnati, Cincinnati, OH.

Ding, H., Goldberg, M. M., Raymer, J. H., Holmes, J., Stanko, J., and Charney, S. G. (1999). Determination of platinum in rat dorsal root ganglion using ICP-MS. *Bio. Trace Elem. Res.* **67**(1), 1.

Divjak, B., and Goessler, W. (1999). Ion chromatographic separation of sulfur-containing inorganic anions with an ICP-MS as element-specific detector. *J. Chromatogr.* **844**(1), 161.

Doering, T., Schwikowski, M., and Gaeggeler, H. W. (1997). Determination of lead concentrations and isotope ratios in recent snow samples from high alpine sites with a double focusing ICP-MS. *Fresenius' J. Anal. Chem.* **359**(4–5), 382.

Donachie, A., Walmsley, A. D., and Haswell, S. J. (1999). Application and comparisons of chemometric techniques for calibration modeling using electrochemical/ICP-MS data for trace elements in UHQ water and. *Anal. Chim. Acta* **378**(1), 237.

D'Orazio, M. (1999). Boron determination in twenty one silicate rock reference materials by isotope dilution ICP-MS. *Geostand. Newsl.* **23**(1), 21.

D'Orazio, M., and Tonarini, S. (1997). Simultaneous determination of neodymium and samarium in silicate rocks and inerals by isotope dilution inductively coupled plasma mass spectrometry. *Anal. Chim. Acta* **351**(1–3), 325.

Dressier, V. L., Pozebon, D., and Curtius, A. J. (1998). Determination of heavy metals by inductively coupled plasma mass spectrometry after on-line separation and pre- concentration. *Spectrochim. Acta, Part B* **53**(11), 1527.

Dressler, V. L., Pozebon, D., and Curtius, A. J. (1999). Introduction of alcohols in inductively coupled plasma mass spectrometry by a flow injection system. *Anal. Chim. Acta* **379**(1), 175.

Du, Z. (1999). Attenuation of metal oxide ions in inductively coupled plasma mass spectrometry with a hexapole collision cell. Unpublished M. S. Thesis, Iowa State University, Ames.

Du, Z., and Houk, R. S. (2000). Attenuation of metal oxide ions in inductively coupled plasma mass spectrometry with hydrogen in a hexapole collision cell. *J. Anal. At. Spectrom.* 383–388.

Du, Z., Olney, T. N., and Douglas, D. J. (1997). Inductively coupled plasma mass spectrometry with a quadrupole mass filter operated in the third stability region. *J. Am. Soc. Mass Spectrom.* **8**(12), 1230.

Duane, M. J., and Facchetti, S. (1995). On-site environmental water analyses by ICP-MS. *Sci. Total Environ.* **172**(2–3), 133.

Duane, M. J., Facchetti, S., and Pigozzi, G. (1996). Site characterization of polluted soils and comparison of screening techniques for heavy metals by mobile ICP-MS, GFAAS/ICP-AES (fixed laboratory) and EDXRF (fixed laboratory). *Sci. Total Environ.* **177**(1–3), 195.

Dubinin, A. V. (1994). Mass spectrometry with an inductively coupled plasma: Determining the REE in standard ocean-floor specimens. *Geochem. Int.* **31**(6), 81.

Ducros, V., Peoc'h, M., Moulin, C., Ruffieux, D., Amosse, J., Favier, A., and Pasquier, B. (1996). Titanium identification and measurement by ICP-MS and ICP-OES in a human spleen fragment. *Clin. Chem. (Winsten-Salem, N.C.)* **42**(11), 1875.

Dufosse, T., and Touron, P. (1998). ICP-MS Changes increase use comparison of bullet alloys by chemical analysis: Use of ICP-MS method. *Res./Dev.* **40**(2), 53.

Dulski, P. (1992). Determination of minor and trace elements in four Canadian iron formation standard samples FeR-1, FeR-2, FeR-3 and FeR-4 by INAA and ICP-MS. *Geostand. Newsl.* **16**(2), 325–332.

Dulski, P. (1994). Interferences of oxide, hydroxide and chloride analyte species in the determination of rare earth elements in geological samples by inductively coupled plasma-mass spectrometry. *Fresenius' J. Anal. Chem.* **350**(4/5), 194.

Durrant, S. F. (1992). Multi-elemental analysis of environmental matrices by laser ablation inductively coupled plasma mass spectrometry. *Analyst (London)* **117**(10), 1585.

Durrant, S. F. (1993). Alternatives to all-argon plasmas in inductively coupled plasma mass spectrometry (ICP-MS): An overview. *Fresenius' J. Anal. Chem.* **347**(10–11), 389–392.

Durrant, S. F. (1994). Feasibility of improvement in analytical performance in laser ablation inductively coupled plasma-mass spectrometry (LA-ICP-MS) by addition of nitrogen to the argon plasma. *Fresenius' J. Anal. Chem.* **349**(10/11), 768.

Durrant, S. F. (1996). Laser ablation inductively coupled plasma mass spectrometry. *AIP Conf. Proc.*, p. 94.

Durrant, S. F. (1999). Laser ablation inductively coupled plasma mass spectrometry: Achievements, problems, prospects. *J. Anal. At. Spectrom.* **14**(9), 1385.

Durrant, S. F. (2000). Laser ablation inductively coupled plasma mass spectrometry: Achievements, problems, prospects. *J. Anal. At. Spectrom.* 237.

Durrant, S. F., Krushevska, A., Amarasiriwardena, D., Argentine, M. D., Romon-Guesnier, S., and Barnes, R. M. (1994). Matrix separation by chelation to prepare biological materials for isotopic zinc analysis by inductively coupled plasma mass spectrometry. *J. Anal. At. Spectrom.* **9**(3), 199–204.

Dziewatkoski, M. P. (2000). Speciation of metal-containing compounds found in the environment by micellar liquid chromatography interfaced to inductively coupled plasma mass spectrometry. *ACS Symp. Ser.*, p. 276.

Dziewatkoski, M. P., Daniels, L. B., and Olesik, J. W. (1996). Time-resolved inductively coupled plasma mass spectrometry measurements with individual, monodisperse drop sample introduction. *Anal. Chem.* **68**(7), 1101.

Eastwood, W. J., Pearce, N. J. G., Westgate, J. A., and Perkins, W. T. (1998). Recognition of Santorini (Minoan) Tephra in lake sediments from Goelhisar Goelue, southwest turkey by laser ablation ICP-MS. *J. Archaeol. Sci.* **25**(7), 677.

Eaton, A. N., Hutton, R. C., and Holland, J. G. (1992). Application of flow injection sample introduction to inductively coupled plasma-mass spectrometry for geochemical analysis. *Chem. Geol.* **95**(1–2), 63–72.

Ebdon, L., Fisher, A. S., Worsfold, P. J., Crews, H., and Baxter, M. (1993). On-line removal of interferences in the analysis of biological materials by flow injection inductively coupled plasma mass spectrometry. *J. Anal. At. Spectrom.* **8**(5), 691–696.

Ebdon, L., Fisher, A., Handley, H., and Jones, P. (1993). Determination of trace metals in concentrated brines using inductively coupled plasma mass spectrometry on-line preconcentration and matrix elimination with flow injection. *J. Anal. Atomic Spectrom.* **8**(7), 979–982.

Ebdon, L., Fisher, A. S., and Worsfold, P. J. (1994). Determination of arsenic, chromium, selenium and vanadium in biological samples by inductively coupled plasma mass spectrometry using on-line elimination of interference and preconcentration by flow injection. *J. Anal. At. Spectrom.* **9**(5), 611.

Ebdon, L., Evans, E. H., Pretorius, W. G., and Rowland, S. J. (1994). Analysis of geoporphyrins by high-temperature gas chromatography inductively coupled plasma mass spectrometry and high-performance liquid chromatography inductively coupled plasma mass spectrometry. *J. Anal. At. Spectrom.* **9**(9), 939.

Eckhoff, K. M., and Maage, A. (1997). Iodine content in fish and other food products from East Africa analyzed by ICP-MS. *J. Food Compos. Analy.* **10**(3), 270.

Edmonds, J. S., Shibata, Y., Francesconi, K. A., Yoshinaga, J., and Morita, M. (1992). Arsenic lipids in the digestive gland of the western rock lobster *Panulirus cygnus*: An investigation by HPLC ICP-MS. *Sci. Total Environ.* **122**(3), 321.

Eggins, S. M., Woodhead, J. D., Kinsley, L. P. J., Mortimer, G. E., Sylvester, P., McCulloch, M. T., Hergt, J. M., and Handler, M. R. (1996). A simple method for the precise determination of 2 40 trace elements in geological samples by ICPMS using enriched isotope internal standardisation. *Chem. Geol.* **134**(4), 311.

Eggins, S. M., Kinsley, L. P. J., and Shelley, J. M. G. (1998). Deposition and element fractionation processes during atmospheric pressure laser sampling for analysis by ICP-MS. *Appl. Surf. Sci.* 278.

Eggins, S. M., Rudnick, R. L., and McDonough, W. F. (1998). The composition of peridotites and their minerals: A laser- ablation ICP-MS study. *Earth Planet. Sci. Lett.* **154**(1–4), 53.

Eiser, W., and Beck, H. P. (1999). Trace analysis of impurities in sol-gel prepared BaTiO3-powders with ICP-MS. *Fresenius' J. Anal. Chem.* **364**(5), 417.

Ejima, A., Watanabe, C., Koyama, H., and Satoh, H. (1996). Analysis of trace elements in the central nerve tissues with inductively coupled plasma-mass spectrometry. *Tôhoku J. Exp. Med.* **178**(1), 1.

Ejima, A., Watanabe, C., Koyama, H., and Satoh, H. (1999). Matrix interferences in the analysis of digested biological tissues with inductively coupled plasma-mass spectrometry. *Bio. Trace Elem. Res.* **69**(2), 99.

Ejnik, J. W., Carmichael, A. J., Hamilton, M. M., McDiarmid, M., Squibb, K., Boyd, P., and Tardiff, W. (2000). Determination of the isotopic composition of uranium in urine by inductively coupled plasma mass spectrometry. *Health Phys.* **78**(2), 143.

Elliot, S. (1998). Analysis of fused basalt and granite samples by laser ablation ICP-MS. *Chem. N. Z.* **62**(4), 15.

Elokhin, V. A., Protopopov, S. V., Retivykh, S. N., and Chernetskii, S. M. (1997). PQ simplex software for data processing in the determination of rare-earth elements by inductively coupled plasma mass spectrometry. *J. Anal. Chem.* **52**(11), 1030.

Ely, J. C., Neal, C. R., O'Neill, J. A., Jr. and Jain, J. C. (1999). Quantifying the platinum group elements (PGEs) and gold in geological samples using cation exchange pretreatment and ultrasonic nebulization inductively coupled plasma-mass spectrometry (USN-ICP-MS). *Chem. Geol.* **157**(3), 219.

Emteborg, H., Tian, X., and Adams, F. C. (1999). Quality assurance of arsenic, lead, tin and zinc in copper alloys using axial inductively coupled plasma time-of-flight mass spectrometry (ICP-TOF-MS). *J. Anal. At. Spectrom.* **14**(10), 1567.

Emteborg, H., Tian, X., Ostermann, M., Berglund, M., and Adams, F. C. (2000). Isotope ratio and isotope dilution measurements using axial inductively coupled plasma time of flight mass spectrometry. *J. Anal. At. Spectrom.* 239–246.

Entwistle, J. A., and Abrahams, P. W. (1997). Multi-element analysis of soils and sediments from Scottish historical sites. The potential of inductively coupled plasma-mass spectrometry for rapid site investigation. *J. Archaeol. Sci.* **24**(5), 407.

Enzweiler, J., Potts, P. J., and Jarvis, K. E. (1995). Determination of platinum, palladium, ruthenium and iridium in geological samples by isotope dilution inductively coupled plasma mass spectrometry using a sodium peroxide fusion and tellurium coprecipitation. *Analyst (London)* **120**(5), 1391.

Epov, V. N., Vasil'eva, I. E., Lozhkin, V. I., Epova, E. N., Paradina, L. F., and Suturin, A. N. (1999). Determination of macroelements in Baikal water using inductively coupled plasma mass spectrometry. *J. Anal. Chem.* **54**(9), 837.

Erickson, B. (1999). ICPMS—Beyond quadrupole. New designs combat interferences and reduce costs. *Anal. Chem.* **71**(23), 811A.

Eroglu, A. E., McLeod, C. W., Leonard, K. S., and McCubbin, D. (1998). Determination of plutonium in seawater using co-precipitation and inductively coupled plasma mass spectrometry with ultrasonic nebulisation. *Spectrochim. Acta, Part B* **53**(6–8), 1221.

Eroglu, A. E., McLeod, C. W., Leonard, K. S., and McCubbin, D. (1998). Determination of technetium in sea-water using ion exchange and inductively coupled plasma mass spectrometry with ultrasonic nebulisation. *J. Anal. At. Spectrom.* **13**(9), 875.

Escobar, M. P. (1995). Determination of wear metals in lubricating oil by electrothermal vaporization inductively coupled plasma mass spectrometry. Unpublished Ph.D. Thesis, University of Florida, Tallahassee.

Escobar, M. P., Smith, B. W., and Winefordner, J. D. (1996). Determination of metallo-organic species in lubricating oil by electrothermal vaporization inductively coupled plasma mass spectrometry. *Anal. Chim. Acta* **320**(1), 11.

Evans, E. H., and Caruso, J. A. (1992). Optimization strategies for the reduction of non-spectroscopic interferences in inductively coupled plasma mass spectrometry. *Spectrochim. Acta, Part B* **47B**(8), 1001.

Evans, E. H., and Caruso, J. A. (1993). Low pressure inductively coupled plasma source for mass spectrometry. *J. Anal. At. Spectrom.* **8**(3), 427–432.

Evans, E. H., and Ebdon, L. (1991). Comparison of normal and low-flow torches for inductively coupled plasma mass spectrometry using optimized operating conditions. *J. Anal. At. Spectrom.* **6**(6), 421.

Evans, E. H., Giglio, J. J., Castillano, T. M., and Caruso, J. A. (1995). "Inductively Coupled and Microwave Induced Plasma Sources for Mass Spectrometry." Royal Society of Chemistry, Cambridge; UK.

Evans, E. H., Gigtio, J. J., Castellano, T. M., Caruso, J. A., and Uden, P. C. (1996). Inductively coupled and microwave induced plasma sources for mass spectrometry. *J. Chromatogr.* **746**(2), 304.

Evans, R. D., and Villeneuve, J. Y. (2000). A method for characterization of humic and fulvic acids by gel electrophoresis laser ablation inductively coupled plasma mass spectrometry. *J. Anal. At. Spectrom.*, 157–162.

Evans, R. D., Outridge, P. M., and Richner, P. (1994). Applications of laser ablation inductively coupled plasma mass spectrometry to the determination of environmental contaminants in calcified biological structures. *J. Anal. At. Spectrom.* **9**(9), 985.

Evans, R. D., Richner, P., and Outridge, P. M. (1995). Micro-spatial variations of heavy metals in the teeth of walrus as determined by laser ablation ICP-MS: The potential for reconstructing a history of metal exposure. *Arch. Environ. Contam. Toxicol.* **28**(1), 55.

Evans, S., and Kraehenbuehl, U. (1994). Improved boron determination in biological material by inductively coupled plasma mass spectrometry. *J. Anal. At. Spectrom.* **9**(11), 1249.

Faahan-Smith, T., and Woodford, D. (1995). The relative merits of the analysis of high purity gold by glow discharge mass spectrometry and inductively coupled plasma optical emission spectrometry. *Precious Met.*, p. 205.

Fadda, S., Rivoldini, A., and Cau, I. (1995). ICP-MS determination of 45 trace elements in whole coal using microwave oven acid digestion for sample preparation. *Geostand. Newsl.* **19**(1), 41.

Faires, L. M. (1993). Methods of Analysis by the U.S. Geological Survey National Water Quality Laboratory determination of metals in water by inductively coupled Plasma-mass spectrometry. *Geol. Surv. Open-File Rep. (U.S.)* **92–634**.

Fairman, B., and Catterick, T. (1997). Simultaneous determination of arsenic, antimony and selenium in aqueous matrices by electrothermal vaporization inductively coupled plasma mass spectrometry. *J. Anal. At. Spectrom.* **12**(8), 863.

Fairman, B., Catterick, T., Wheals, B., and Polinina, E. (1997). Reversed-phase ion-pair chromatography with inductively coupled plasma-mass spectrometry detection for the determination of organo-tin compounds in waters. *J. Chromatogr.* **758**(1), 85.

Fairman, B., Sanz-Medel, A., Jones, P., and Evans, E. H. (1998). Comparison of fluorimetric and inductively coupled plasma mass spectrometry detection systems for the determination of aluminium species in waters by high-performance liquid chromatography. *Analyst (London)* **123**(4), 699.

Falk, H. F., Geerling, R., Hattendorf, B., Krengel-Rothensee, K., and Schmidt, K. P. (1997). Capabilities and limits of ICP-MS for direct determination of element traces in saline solutions. *Fresenius' J. Anal. Chem.* **359**(4–5), 352.

Falk, H. F., Hattendorf, B., Krengel-Rothensee, K., Wieberneit, N., and Dannen, S. L. (1998). Calibration of laser-ablation ICP-MS. Can we use synthetic standards with pneumatic nebulization? *Fresenius' J. Anal. Chem.* **362**(5), 468.

Falter, R., and Wilken, R.-D. (1998). Determination of carboplatinum and cisplatinum by interfacing HPLC with ICP-MS using ultrasonic nebulisation. *Sci. Total Environ.* **225**(1), 167.

Falter, R., and Wilken, R.-D. (1998). Determination of rare earth elements by ICP-MS and ultrasonic nebulization in sludges of water treatment facilities. *Vom Wasser* 57.

Fangshi, L., Goessler, W., and Irgolic, K. J. (1999). Determination of selenium compounds by HPLC with ICP-MS or FAAS as selenium-specific detector. *Chin. J. Chromatogr.* **17**(3), 240.

Fannin, H. B., Evans, E. H., Giglio, J. J., Castillano, T. M., and Caruso, J. A. (1996). Inductively coupled and microwave induced plasma sources for mass spectrometry. *J. Am. Chem. Soc.* **118**(50), 12871.

Fardy, J. J., and Warner, I. M. (1992). A comparison of neutron activation analysis and inductively coupled plasma mass spectrometry for trace element analysis of biological materials. *J. Radioanal. Nucl. Chem.* **157**(2), 239.

Fatemian, E., Allibone, J., and Walker, P. J. (1999). Use of gold as a routine and long term preservative for mercury in potable water, as determined by ICP-MS. *Analyst (London)* **124**(8), 1233.

Fecher, P. A., Goldmann, I., and Nagengast, A. (1998). Determination of iodine in food samples by inductively coupled plasma mass spectrometry after alkaline extraction. *J. Anal. At. Spectrom.* **13**(9), 977.

Fedorowich, J. S., Richards, J. P., Jain, J. C., Kerrich, R., and Fan, J. (1993). A rapid method for REE and trace-element analysis using laser sampling ICP-MS on direct fusion whole-rock glasses. *Chem. Geol.* 229.

Fedorowich, J. S., Jain, J. C., Kerrich, R., and Sopuck, V. (1995). Trace-element analysis of garnet by laser-ablation microprobe ICP-MS. *Can. Mineral.* **33**(2), 469.

Feldmann, I., Tittes, W., Jakubowski, N., and Stuewer, D. (1994). Performance characteristics of inductively coupled plasma mass spectrometry with high mass resolution. *J. Anal. At. Spectrom.* **9**(9), 1007.

Feldmann, I., Jakubowski, N., Thomas, C., and Stuewer, D. (1999). Application of a hexapole collision and reaction cell in ICP-MS. Part II: Analytical figures of merit and first applications. *Fresenius' J. Anal. Chem.* **365**(5), 422.

Feldmann, I., Jakubowski, N., and Stuewer, D. (1999). Application of a hexapole collision and reaction cell in ICP-MS. Part I: Instrumental aspects and operational optimization. *Fresenius' J. Anal. Chem.* **365**(5), 415.

Feldmann, I., Jakubowski, N., Stuewer, D., and Thomas, C. (2000). Speciation of organic selenium compounds by reversed-phase liquid chromatography and inductively coupled plasma mass spectrometry. Part II. Sector field instrument with high mass resolution. *J. Anal. At. Spectrom.* **15**, 371–376.

Feldmann, J., Gruemping, R., and Hirner, A. V. (1994). Determination of volatile metal and metalloid compounds in gases from domestic waste deposits with GC/ICP-MS. *Fresenius' J. Anal. Chem.* **350**(4/5), 228.

Feldmann, J., Riechmann, T., and Hirner, A. V. (1996). Determination of organometallics in intra-oral air by LT-GC/ICP-MS. *Fresenius' J. Anal. Chem.* **354**(5–6), 620.

Fellows, K., and Pickston, L. (1994). ICP-MS a new era in elemental analysis. *Chem. N. Z.* **58**(1), 24.

Feng, R. (1994). In situ trace element determination of carbonates by Laserprobe inductively coupled plasma mass spectrometry using nonmatrix matched standardization. *Geochim. Cosmochim. Acta* **58**(6), 1615.

Feng, X., and Horlick, G. (1994). Analysis of aluminium alloys using inductively coupled plasma and glow discharge mass spectrometry. *J. Anal. At. Spectrom.* **9**(8), 823.

Feng, Y.-L., Chen, H.-Y., Tian, L.-C., and Narasaki, H. (1998). Off-line separation and determination of inorganic arsenic species in natural water by high resolution inductively coupled plasma mass spectrometry with hydride generation combined with reaction of arsenic(V) and L-cysteine. *Anal. Chim. Acta* **375**(1), 167.

Feng, Y.-L., Narasaki, H., Tian, L.-C., and Chen, H.-Y. (2000). Speciation of Sb(III) and Sb(V) by hydride generation high-resolution ICP-MS combined with prereduction of Sb(V) with L-cysteine. *At. Spectrosc.* **21**(1), 30.

Fernandez, R. G., Bayon, M. M., Alonso, J. I., and Sanz-Medel, A. (2000). Comparison of different derivatisation approaches for mercury speciation in biological tissues by gas chromatography/inductively coupled plasma mass spectrometry. *J. Mass Spectrom.*, pp. 639–646.

Field, A., and Taylor, P. D. P. (1999). A calculation method based on isotope ratios for the determination of dead time and its uncertainty in ICP-MS and application of the method to investigating some features of a continuous dynode multiplier. *J. Anal. At. Spectrom.* **14**(7), 1075.

Field, M. P., and Sherrell, R. M. (1998). Magnetic sector ICPMS with desolvating micronebulization: Interference-free subpicogram determination of rare earth elements in natural samples. *Anal. Chem.* **70**(21), 4480.

Field, M. P., Cullen, J. T., and Sherrell, R. M. (1999). Direct determination of 10 trace metals in 50 L samples of coastal seawater using desolvating micronebulization sector field ICP-MS. *J. Anal. At. Spectrom.* **14**(9), 1425.

Figg, D., and Kahr, M. S. (1997). Elemental fractionation of glass using laser ablation inductively coupled plasma mass spectrometry. *Appl. Spectrosc.* **51**(8), 1185.

Finkeldei, S., and Staats, G. (1997). ICP-MS—A powerful analytical technique for the analysis of traces of Sb, Bi, Pb, Sn and P in steel. *Fresenius' J. Anal. Chem.* **359**(4–5), 357.

Fischer, T. P., Shuttleworth, S., and O'Day, P. A. (1998). Determination of trace and platinum-group elements in high ionic-strength volcanic fluids by sector-field inductively coupled plasma mass spectrometry (ICP-MS). *Fresenius' J. Anal. Chem.* **362**(5), 457.

Fisher, C. G. (1991). Where to next? The future of ICP-MS. *At. Spectrosc.* **12**, 239–246.

Fitzgerald, N. (1999). Improvements in sample preparation and introduction for inductively coupled plasma-mass spectrometry incorporating microwave energy. Unpublished Ph.D. Thesis, University of Massachusetts, Amherst.

Fonseca, R. W., and Miller-Ihli, N. J. (1995). Analyte transport studies of aqueous solutions and slurry samples using electrothermal vaporization ICP-MS. *Appl. Spectrosc.* **49**(10), 1403.

Fonseca, R. W., and Miller-Ihli, N. J. (1996). Influence of sample matrix components on the selection of calibration strategies in electrothermal vaporization inductively coupled plasma mass spectrometry. *Spectrochim. Acta, Part B* **51**(13), 1591.

Ford, M., Ebdon, L., and Hill, S. J. (1992). Further investigations into the addition of nitrogen in inductively coupled plasma mass spectrometry to reduce the argon chloride interference on arsenic. *Anal. Proc.* **29**(3), 104–105.

Fordham, P. J., Gramshaw, J. W., Castle, L., and Crews, H. M. (1995). Determination of trace elements in food contact polymers by semi-quantitative inductively coupled plasma mass spectrometry. Performance evaluation using alternative multi-element techniques and in-house polymer reference materials. *J. Anal. At. Spectrom.* **10**(4), 303.

Fowler, A. J., Campana, S. E., Jones, C. M., and Thorrold, S. R. (1995). Experimental assessment of the effect of temperature and salinity on elemental composition of otoliths using solution-based ICPMS. *Can. J. Fish. Aqua. Sci.* **52**(7), 1421.

Fowler, A. J., Campana, S. E., Jones, C. M., and Thorrold, S. R. (1996). Experimental assessment of the effect of temperature and salinity on elemental composition of otoliths using laser ablation ICPMS. *Anal. Chem.* **68**(1), 46A.

Frengstad, B., Midtgard Skrede, A. K., Banks, D., Reidar Krog, J., and Siewers, U. (2000). The chemistry of Norwegian groundwaters: III. The distribution of trace elements in 476 crystalline bedrock groundwaters, as analysed by ICP-MS techniques. *Sci. Total Environ.* **246**(1), 21.

Friel, J. K., Skinner, C. S., Jackson, S. E., and Longerich, H. P. (1990). Analysis of biological reference materials, prepared by microwave dissolution, using inductively coupled plasma mass spectrometry. *Analyst (London)* **115**(3), 269.

Friel, J. K., Longerich, H. P., and Jackson, S. E. (1993). Determination of isotope ratios in human tissues enriched with zinc stable isotope tracers using inductively coupled plasma-mass spectrometry (ICP-MS). *Biol. Trace Elem. Res.* 123–136.

Fuchtjohann, L., Jakubowski, N., Gladtke, D., Barnowski, C., Klockow, D., and Broekaert, J. A. C. (2000). Determination of soluble and insoluble nickel compounds in airborne particulate matter by graphite furnace atomic absorption spectrometry and inductively coupled plasma mass spectrometry. *Fresenius' J. Anal. Chem.* **366**(2), 142.

Fucsko, J., Tan, S. S., and Balazs, M. K. (1993). Measurement of trace metallic contaminents on silicon wafer surfaces in native and dielectric silicon oxides by vapor-phase decomposition flow injection inductively coupled plasma-mass spectrometry. *J. Electrochem. Soc.* **140**(4), 1105.

Fuge, R., Palmer, T. J., Pearce, N. J. G., and Perkins, W. T. (1993). Minor and trace element chemistry of modern shells: A laser ablation inductively coupled plasma mass spectrometry study. *Appl. Geochem.* **8**(3), 111–116.

Fujimori, E., Sawatari, H., Hirose, A., and Haraguchi, H. (1994). Simultaneous multielement analysis of rock samples by inductively coupled plasma mass spectrometry using discrete microsampling technique. *Chem. Lett.*, p. 1467.

Fujimori, E., Wei, R., Sawatari, H., Chiba, K., and Haraguchi, H. (1996). Multielement determination of trace elements in sediment sample by inductively coupled plasma mass spectrometry with microsampling technique. *Bull. Chem. Soc. Jpn.* **69**(12), 3505.

Fujimori, E., Tomosue, Y., and Haraguchi, H. (1996). Determination of rare earth elements in blood serum reference sample by chelating resin preconcentration and inductively coupled plasma mass spectrometry. *Tohoku J. Exp. Med.* **178**(1), 63.

Fujimori, E., Sawatari, H., Chiba, K., and Haraguchi, H. (1996). Determination of minor and trace elements in urine reference sample by a combined system of inductively coupled plasma mass spectrometry and inductively coupled plasma atomic emission spectrometry. *Anal. Sci.: Int. J. Jpn. Soc. Anal. Chem.* **12**(3), 465.

Fujimori, E., Hayashi, T., Inagaki, K., and Haraguchi, H. (1999). Determination of lanthanum and rare earth elements in bovine whole blood reference material by ICP-MS after coprecipitation preconcentration with heme-iron as coprecipitant. *Fresenius' J. Anal. Chem.* **363**(3), 277.

Fujino, O., and Terada, E. (1996). Determination and distribution of lanthanoids in hard tissue (shell and pearl) of shellfish by ICP-MS. *Anal. Sci.: Int. J. Jpn. Soc. Anal. Chem.* **12**(6), 963.

Fujino, O., Umetani, S., Matsui, M., Nishida, S., Sanada, K., and Orimi, K. (1996). Determination of rare earth elements, uranium and thorium in apatite minerals by inductively coupled plasma-mass spectrometry. *Nippon Kagaku Kaishi.*, p. 650.

Fukuchi, T., and Imai, N. (1998). ESR isochron dating of the Nojima fault gouge, southwest Japan, using ICP-MS: An approach to fluid flow events in the fault zone. *Spec. Publ.— Geol. Soc. London* 261.

Fukuda, M., and Sayama, Y. (1995). Determination of traces uranium and thorium in antimony(III) oxide by ICP-MS. *Bunseki Kagaku* **44**(9), 719.

Fukuda, M., and Sayama, Y. (1997). Determination of traces of uranium and thorium in (Ba, Sr)TiO3 ferroelectrics by inductively coupled plasma mass spectrometry. *Fresenius' J. Anal. Chem.* **357**(6), 647.

Galbacs, G., Vanhaecke, F., Moens, L., and Dams, R. (1996). Determination of cadmium in certified reference materials using solid sampling electrothermal vaporization inductively coupled plasma mass spectrometry supplemented with thermogravimetric studies. *Microchem. J.* **54**(3), 272.

Gallagher, P. A., Wei, X., Shoemaker, J. A., Brockhoff, C. A., and Creed, J. T. (1997). Applications in ICP-AES and ICP-MS, detection of arsenosugars from kelp extracts via IC-electrospray ionization-MS-MS and IC membrane hydride generation ICP-MS. *Int. Labmate* **22**(1), 28.

Gallus, S. M., and Heumann, K. G. (1996). Development of a gas chromatography inductively coupled plasma isotope dilution mass spectrometry system for accurate determination of volatile element species. Part 1. Selenium speciation. *J. Anal. At. Spectrom.* **11**(9), 887.

Gammelgaard, B., and Joens, O. (1999). Determination of selenium in urine by inductively coupled plasma mass spectrometry: Interferences and optimization. *J. Anal. At. Spectrom.* **14**(5), 867.

Gammelgaard, B., and Larsen, E. H. (1998). Sensitivities of selenite, selenate, selenomethionine and trimethylselenonium ion in aqueous solution and in blood plasma—ETAAS compared with ICP-MS. *Talanta* **47**(2), 503.

Gammelgaard, B., Jessen, K. D., Kristensen, F. H., and Jons, O. (2000). Determination of trimethylselenonium ion in urine by ion chromatography and inductively coupled plasma mass spectrometry detection. *Anal. Chim. Acta* **404**(1), 47.

Garbarino, J. R. (1999). Methods of analysis by the U.S. Geological Survey National Water Quality Laboratory: Determination of Dissolved Arsenic, Boron, Lithium, Selenium, Strontium, Thallium, and Vanadium Using Inductively Coupled Plasma-Mass Spectrometry. *Geol. Surv. Open-File Rep. (U.S.)* **94–358**.

Garbarino, J. R. (2000). Methods of analysis by the U.S. Geological Survey National Water Quality Laboratory: Determination of Whole-Water Recoverable Arsenic, Boron and Vanadium Using Inductively Coupled Plasma-Mass Spectrometry. *Geol. Surv. Open-File Rep. (U.S.)* **99–464.**

Garbarino, J. R., and Taylor, H. E. (1996). Inductively coupled plasma-mass spectrometric method for the determination of dissolved trace elements in natural water. *Geol. Surv. Open-File Rep. (U.S.)* **94–358.**

Garbe-Schonberg, C. D. (1993). Simultaneous determination of thirty-seven trace elements in twenty-eight international rock standards by ICP-MS. *Geostand. Newsl.* **17**(1), 81–98.

Garcia Alonso, J. I., Thoby-Schultzendorff, D., Giovanonne, B., and Glatz, J. P. (1994). Characterization of spent nuclear fuel dissolver solutions and dissolution residues by inductively coupled plasma mass spectrometry. *J. Anal. At. Spectrom.* **9**(11), 1209.

Garcia Alonso, J. I., Sena, F., and Koch, L. (1994). Determination of 99Tc in nuclear samples by inductively coupled plasma mass spectrometry. *J. Anal. At. Spectrom.* **9**(11), 1217.

Garcia-Alonso, J. I., Thoby-Schultzendorff, D., Giovannone, B., and Koch, L. (1996). Analysis of long-lived radionuclides by ICP-MS. *J. Radioanal. Nucl. Chem.* **203**(1), 19.

Garcia Alonso, J. I., Gutierrez Camblor, M., Montes Bayon, M., and Marchante-Gayon, Gayon, J. M. (1997). Different quantification approaches for the analysis of biological and environmental samples using inductively coupled plasma mass spectrometry. *J. Mass Spectrom.* **32**(5), 556.

Garraud, H., Robert, M., Quetel, C. R., Szpunar, J., and Donard, O. F. X. (1996). Focused microwave-assisted digestion of biological reference materials for the determination of trace metals by ICP-MS. *At. Spectrosc.* **17**(5), 183.

Gastel, M., Becker, J. S., Kueppers, G., and Dietze, H. J. (1997). Determination of long-lived radionuclides in concrete matrix by laser ablation inductively coupled plasma mass spectrometry. *Spectrochim. Acta, Part B* **52**(14), 2051.

Ge, H. (1997). A study of on-line sample introduction to inductively coupled plasma spectrometry. Unpublished Ph.D. Thesis, University of Massachusetts, Amherst.

Gelinas, Y., Youla, M., Beliveau, R., Schmit, J. P., and Ferraris, J. (1992). Multi-element analysis of biological tissues by inductively coupled plasma mass spectrometry: Healthy Sprague Dawley rats. *Anal. Chim. Acta* **269**(1), 115.

Gelinas, Y., Lenjer, R., and Schmit, J.-P. (1996). Optimally interfacing an ICP-MS with a conventional autosampler for the analysis of solutions with a high dissolved solids content using continuous nebulization. *At. Spectrosc.* **17**(4), 143.

Gelinas, Y., Lafond, J., and Schmit, J.-P. (1997). Multielement analysis of human fetal tissues using inductively coupled plasma-mass spectrometry. *Biol. Trace Elem. Res.* **59**(1–3), 63.

Gelinas, Y., Krushevska, A., and Barnes, R. M. (1998). Determination of total iodine in nutritional and biological samples by ICP-MS following their combustion within an oxygen stream. *Anal. Chem.* **70**(5), 1021.

Gelinas, Y., Iyengar, G.V., and Barnes, R. M. (1998). Total iodine in nutritional and biological reference materials using neutron activation analysis and inductively coupled plasma mass spectrometry. *Fresenius' J. Anal. Chem.* **362**(5), 483.

Gerotto, M., Dell'Andrea, E., Bortoli, A., and Marchiori, M. (1995). Interference effects and their control in ICP-MS analysis of serum and saline solutions. *Microchem. J.* **51**(1/2), 73.

Ghazi, A. M. (1994). Lead in archaeological samples: An isotopic study by ICP-MS. *Appl. Geochem.: J. Int. Assoc. Geochem. Cosmochem.* **9**(6), 627.

Ghazi, A. M., and Shuttleworth, S. (2000). Trace element determination of single fluid inclusions by laser ablation ICP-MS: Applications for halites from sedimentary basins. *Analyst (London)* 205–210.

Ghazi, A. M., Vanko, D. A., Roedder, E., and Seeley, R. C. (1993). Determination of rare earth elements in fluid inclusions by inductively coupled plasma-mass spectrometry (ICP-MS). *Geochim. Cosmoch. Acta* **57**(18), 4513.

Giessmann, U., and Greb, U. (1994). High resolution ICP-MS—a new concept for elemental mass spectrometry. *Fresenius' J. Anal. Chem.* **350**(4/5), 186.

Gillyon, E. C. P., Hunter, J., Tye, C. T., and Alavosus, T. (1994). Improving ICP and ICP-MS instrument uptime and reducing service costs by use of a remote diagnostic facility. *Int. Labmate* **19**(6), 33.

Gilmutdinov, A. K., Staroverov, A. E., Grégoire, D. C., and Sturgeon, R. E. (1994). Kinetics of release of carbon and carbon monoxide from a graphite furnace investigated by electrothermal vaporization inductively coupled plasma mass spectrometry. *Spectrochim. Acta, Part B* **49**(10), 1007.

Giussani, A., Roth, P., Werner, E., Schramel, P., Wendler, I., and Nusslin, F. (1997). Thermal ionization and inductively coupled plasma mass spectrometry: Potential for application to studies on the biokinetics of molybdenum in humans. *Isotopenpraxis* **33**(1–2), 207.

Gluodenis, T. J., Jr. (1998). ICP-MS: The new standard for inorganic analysis. *Am. Lab.* **30**(23), 24S.

Gluodenis, T. J., Jr. (1999). ICP-MS: The new standard for inorganic analysis. *Int. Lab., Eur. Ed.* **29**(4//A), 26.

Gluodenis, T. J., Jr. Sakata, K. i., and McCurdy, E. (1999). Minimizing polyatomic interferences in ICP-MS—The authors review the three commonly used methods for minimizing interferences in ICP-MS and discuss their use of a cool-plasma-based system involving a collision/reaction cell. *Spectroscopy* **14**(3), 16.

Goenaga Infante, H., Fernandez Sanchez, M. L., and Sanz-Medel, A. (1998). Vesicle-assisted determination of ultratrace amounts of cadmium in urine by electrothermal atomic absorption spectrometry and inductively coupled plasma mass spectrometry. *J. Anal. At. Spectrom.* **13**(9), 899.

Goergen, M. G., Murshak, V. F., Roettger, P., Murshak, I., and Edelman, D. (1992). ICP-MS analysis of toxic characteristic leaching procedure (TCLP) extract: Advantages and disadvantages. *At. Spectrosc.* **13**(1), 11–18.

Goguel, R. (1992). Group separation by solvent extraction from silicate rock matrix of Nb, Mo, Ta, and W at low levels for ICP-MS. *Fresenius' J. Anal. Chem.* **344**(7–8), 326–333.

Goltz, D. M., and Salin, E. D. (1997). Inductively heated vaporizer for inductively coupled plasma mass spectrometry. *J. Anal. At. Spectrom.* **12**(10), 1175.

Goltz, D. M., Chakrabarti, C. L., Grégoire, D. C., and Byrne, J. P. (1995). Vaporization and atomization of uranium in a graphite tube electrothermal vaporizer: A mechanistic study using electrothermal vaporization inductively coupled plasma mass spectrometry and graphite furnace atomic absorption spectrometry. *Spectrochim. Acta, Part B* **50**(8), 803.

Goltz, D. M., Grégoire, D. C., and Chakrabarti, C. L. (1995). Mechanism of vaporization of yttrium and rare earth elements in electrothermal vaporization inductively coupled plasma mass spectrometry. *Spectrochim. Acta, Part B* **50**(11), 1365.

Gomes Neto, J. A., Borba Silva, J. B., Goncalves Souza, I., and Curtius, A. J. (1999). Use of FIA systems for on-line dilution in multielement determination by ICP-MS. *Lab. Robot. Autom.* **11**(4), 240.

Gomez, M. B., Gomez, M. M., and Palacios, M. A. (2000). Control of interferences in the determination of Pt, Pd and Rh in airborne particulate matter by inductively coupled plasma mass spectrometry. *Anal. Chim. Acta* **404**(2), 285.

Gomez-Ariza, J. L., Sanchez-Rodas, D., Giraldez, I., and Morales, E. (2000). A comparison between ICP-MS and AFS detection for arsenic speciation in environmental samples. *Talanta* **51**(2), 257–268.

Gomez Gomez, M. M., and McLeod, C. W. (1995). Trace enrichment and determination of gold by flow injection inductively coupled plasma spectrometry. Part II. Inductively coupled plasma mass spectrometry. *J. Anal. At. Spectrom.* **10**(2), 89.

Goossens, J., Vanhaecke, F., Moens, L., and Dams, R. (1993). Elimination of interferences in the determination of arsenic and selenium in biological samples by inductively coupled plasma mass spectrometry. *Anal. Chim. Acta* **280**(1), 137.

Goossens, J., de Smaele, T., Moens, L., and Dams, R. (1993). Accurate determination of lead in wines by inductively coupled plasma mass spectrometry. *Fresenius' J. Anal. Chem.* **347**(3–4), 119–125.

Goossens, J., Moens, L., and Dams, R. (1994). Determination of lead by flow-injection inductively coupled plasma mass spectrometry comparing several calibration techniques. *Anal. Chim. Acta* **293**(1/2), 171.

Goossens, J., Moens, L., and Dams, R. (1994). A mathematical correction method for spectral interferences on selenium in inductively coupled plasma mass spectrometry. *Talanta* **41**(2), 187.

Graham, S. M., and Robert, R. V. D. (1994). The analysis of high-purity noble metals and their salts by ICP-MS. *Talanta* **41**(8), 1369.

Gratuze, B. (1999). Obsidian characterization by laser ablation ICP-MS and its application to prehistoric trade in the Mediterranean and the Near East: Sources and distribution of obsidian within the Aegean and Anatolia. *J. Archaeol. Sci.* **26**(8), 869.

Gray, A. L., Williams, J. G., Ince, A. T., and Liezers, M. (1994). Noise sources in inductively coupled plasma mass spectrometry: An investigation of their importance to the precision of isotope ratio measurements. *J. Anal. At. Spectrom.* **9**(10), 1179.

Greenough, J. D., Longerich, H. P., and Jackson, S. E. (1996). Trace element concentrations in wines by ICP-MS: Evidence for the role of solubility in determining uptake by plants. *Can. J. Appl. Spectrosc.* **41**(3), 76.

Grégoire, D. C. (1990). Determination of boron in fresh and saline waters by inductively coupled plasma mass spectrometry. *J. Anal. At. Spectrom.* **5**(7), 623.

Grégoire, D. C. (1992). Electrothermal vaporization for inductively coupled plasma mass spectrometry: New applications in mass spectrometry and graphite furnace atomic absorption spectrometry. *Anal. Proc.* **29**(7), 276.

Grégoire, D. C. (1999). Atomic spectroscopy perspectives: analysis of geological materials by inductively coupled plasma mass spectrometry. *Spectroscopy* **14**(4), 14.

Grégoire, D. C., and De Lourdes Ballinas, M. (1997). Direct determination of arsenic in fresh and saline waters by electrothermal vaporization inductively coupled plasma mass spectrometry. *Spectrochim. Acta, Part B* **52**(1), 75.

Grégoire, D. C., and Lee, J. (1994). Determination of cadmium and zinc isotope ratios in sheep's blood and organ tissue by electrothermal vaporization inductively coupled plasma mass spectrometry. *J. Anal. At. Spectrom.* **9**(3), 393–398.

Grégoire, D. C., and Naka, H. (1995). Mechanism of vaporization of sulfur in electrothermal vaporization inductively coupled plasma mass spectrometry. *J. Anal. At. Spectrom.* **10**(10), 823.

Grégoire, D. C., and Sturgeon, R. E. (1993). Background spectral features in electrothermal vaporizaiton inductively coupled plasma mass spectrometry: Molecular ions resulting from the use of chemical modifiers. *Spectrochim. Acta, Part B* **48B**(11), 1347.

Grégoire, D. C., and Sturgeon, R. E. (1999). Analyte transport efficiency with electrothermal vaporization inductively coupled plasma mass spectrometry. *Spectrochim. Acta, Part B* **54**(5), 773.

Grégoire, D. C., Al-Maawali, S., and Chakrabarti, C. L. (1992). Use of Mg/Pd chemical modifiers for the determination of volatile elements by electrothermal vaporization ICP-MS: Effect on mass transport efficiency. *Spectrochim. Acta, Part B* **47B**(9), 1123.

Grégoire, D. C., Lamoureux, M., Chakrabarti, C. L., Al-Maawali, S., and Byrne, J. P. (1992). Electrothermal vaporization for inductively coupled plasma mass spectrometry and atomic absorption spectrometry: Symbiotic analytical techniques. *J. Anal. At. Spectrom.* **7**(4), 579–586.

Grégoire, D. C., Miller-Ihli, N. J., and Sturgeon, R. E. (1994). Direct analysis of solids by ultrasonic slurry electrothermal vaporization inductively coupled plasma mass spectrometry. *J. Anal. At. Spectrom.* **9**(5), 605.

Grégoire, D. C., Acheson, B. M., and Taylor, R. P. (1996). Measurement of lithium isotope ratios by inductively coupled plasma mass spectrometry: Application to geological materials. *J. Anal. At. Spectrom.* **11**(9), 765.

Grotti, M., Gnecco, C., and Bonfiglioli, F. (1999). Multivariate quantification of spectroscopic interferences caused by sodium, calcium, chlorine and sulfur in inductively coupled plasma mass spectrometry. *J. Anal. At. Spectrom.* **14**(8), 1171.

Gruenke, K., Staerk, H. J., Wennrich, R., Ortner, H. M., and Broekaert, J. A. C. (1997). An investigation of different modifiers in electrothermal vaporization inductively coupled plasma mass spectrometry (ETV-ICP-MS). *Fresenius' J. Anal. Chem.* **359**(4–5), 465.

Gruenke, K., Staerk, H. J., and Ortner, H. M. (1997). Removal of water and hydrogen from graphite tubes investigated by electrothermal vaporization inductively coupled plasma mass spectrometry. *Spectrochim. Acta, Part B* **52**(9/10), 1545.

Grunwald, E. J. (1998). Capillary electrophoresis inductively coupled plasma mass spectrometry: Detection limits, sample matrix, and metal-ligand considerations. Unpublished Ms. Thesis, Ohio State University, Columbus.

Guenther, D., Longerich, H. P., Jackson, S. E., and Forsythe, L. (1996). Effect of sampler orifice diameter on dry plasma inductively coupled plasma mass spectrometry (ICP-MS) backgrounds, sensitivities, and limits of detection using laser ablation sample introduction. *Fresenius' J. Anal. Chem.* **355**(7–8), 771.

Guérin, T., Astruc, A., and Astruc, M. (1997). Chromatographic ion-exchange simultaneous separation of arsenic and selenium species with inductively coupled plasma-mass spectrometry on-line detection. *J. Chromatogr. Sci.* **35**(5), 213.

Gunther, D., Longerich, H. P., and Jackson III, S. E., (1995). A new enhanced sensitivity quadrupole inductively coupled plasma-mass spectrometer (ICP-MS). *Can. J. Appl. Spectrosc.* **40**(4)

Gunther, K., von Bohlen, A., Paprott, G., and Klockenkamper, R. (1992). Multielement analysis of biological reference materials by total- reflection X-ray fluorescence and inductively coupled plasma mass spectrometry in the semiquant mode. *Fresenius' J. Anal. Chem.* **342**(4–5), 444–448.

Gwiazda, R., Woolard, D., and Smith, D. (1998). Improved lead isotope ratio measurements in environmental and biological samples with a double focussing magnetic sector inductively coupled plasma mass spectrometer (ICP-MS). *J. Anal. At. Spectrom.* **13**(11), 1233.

Haichen, L., Ying, L., and Zhanxia, Z. (1998). Determination of ultra-trace rare earth elements in chon- dritic meteorites by inductively coupled plasma mass spectrometry. *Spectrochim. Acta, Part B* **53**(10), 1399.

Haldimann, M., Zimmerli, B., Als, C., and Gerber, H. (1998). Direct determination of urinary iodine by inductively coupled plasma mass spectrometry using isotope dilution with iodine-129. *Clin. Chem. (Winston-Salem, N.C.)* **44**(4), 817.

Halicz, L., Lam, J. W. H., and McLaren, J. W. (1994). An on-line method for the determination of lead and lead isotope ratios in fresh and saline waters by inductively coupled plasma mass spectrometry. *Spectrochim. Acta, Part B* **49**(7), 637.

Halicz, L., Erel, Y., and Veron, A. (1996). Lead isotope ratio measurements by ICP-MS: Accuracy, precision, and long-term drift. *At. Spectrosc.* **17**(5), 186.

Halicz, L., Segal, I., and Yoffe, O. (1999). Direct REE determination in fresh waters using ultrasonic nebulization ICP-MS. *J. Anal. At. Spectrom.* **14**(10), 1579.

Halicz, L., Galy, A., Belshaw, N. S., and O'Nions, R. K. (1999). High-precision measurement of calcium isotopes in carbonates and related materials by multiple collector inductively coupled plasma mass spectrometry (MC-ICP-MS). *J. Anal. At. Spectrom.* **14**(12), 1835–1838.

Hall, G. E. M. (1992). Inductively coupled plasma mass spectrometry in geoanalysis. *J. Geochem. Explor.* **55**, 201–250.

Hall, G. E. M., and Pelchat, J. C. (1990). Analysis of standard reference materials for Zr, Nb, Hf and Ta by ICP-MS after lithium metaborate fusion and cupferron separation. *Geostand. Newsl.* **14**(1), 197.

Hall, G. E. M., and Pelchat, J. C. (1993). Determination of palladium and platinum in fresh waters by inductively coupled plasma mass spectrometry and activated charcoal preconcentration. *J. Anal. At. Spectrom.* **8**(8), 1059–1066.

Hall, G. E. M., and Pelchat, J.-C. (1997). Determination of As, Bi, Sb, Se and Te in fifty five geochemical reference materials by hydride generation ICP-MS. *Geostand. Newsl.* **21**(1), 85.

Hall, G. E. M., and Pelchat, J. C. (1997). Analysis of geological materials for bismuth, antimony, selenium and tellurium by continuous flow hydride generation inductively coupled plasma mass spectrometry. Part 1. Mutual hydride interferences. *J. Anal. At. Spectrom.* **12**(1), 97.

Hall, G. E. M., and Pelchat, J. C. (1997). Analysis of geological materials for bismuth, antimony, selenium and tellurium by continuous flow hydride generation inductively coupled plasma mass spectrometry. Part 2. Methodology and results. *J. Anal. At. Spectrom.* **12**(1), 103.

Hall, G. E. M., Pelchat, J. C., and Loop, J. (1990). Determination of zirconium, niobium, hafnium and tantalum at low levels in geological materials by inductively coupled plasma mass spectrometry. *J. Anal. At. Spectrom.* **5**(5), 339.

Hall, G. S. (1993). "Determination of Sources of Pb and Cr in Soil and Dust Samples by ICP-MS." New Jersey Dept. of Environmental Protection Division of Science and Research, Trenton.

Hall, G. S., Yamaguchi, D. K., and Rettberg, T. M. (1990). Multielemental analyses of tree rings by inductively coupled plasma mass spectrometry. *J. Radioanal. Nucl. Chem.* **146**(4), 255.

Hall, M. E., Brimmer, S. P., Li, F. H., and Yablonsky, L. (1998). ICP-MS and ICP-OES studies of gold from a late Sarmatian burial. *J. Archaeol. Sci.* **25**(6), 545.

Halliday, A. N., Lee, D. C., Christensen, J. N., Walder, A. J., Freedman, P. A., Jones, C. E., Hall, C. M., Yi, W., and Teagle, D. (1995). Recent developments in inductively coupled plasma magnetic sector multiple collector mass spectrometry. *Int. J. Mass Spectrom. Ion Process.* p. 21.

Halliday, A. N., Christensen, J. N., Der-Chuen, L., Hall, C. M., Luo, X., and Rehkamper, M. (2000). Multiple-collector inductively coupled plasma mass spectrometry. *Pract. Spectrosc.*, pp. 291–328.

Hamanaka, T., Itoh, A., Itoh, S., Sawatari, H., and Haraguchi, H. (1995). Multielement determination of rare earth elements in rock sample by liquid chromatography Inductively Coupled Plasma Mass Spectrometry. *Chem. Lett.*, p. 363.

Hamester, M., Stechmann, H., Krause, P., and Dannecker, W. (1992). Characterization of selected sources for particulate emissions by inductively coupled plasma mass spectrometry. *Anal. Proce.* **29**(7), 293.

Hamester, M., Wiederin, D., Wills, J., Kerl, W., and Douthitt, C. B. (1999). Strategies for isotope ratio measurements with a double focusing sector field ICP-MS. *Fresenius' J. Anal. Chem.* **364**(5), 495.

Hams, G., Webb, K., and Berger, S. (1997). Routine ICP-MS determinations in biomedical fluids. Part 1: Operational parameters. *Clin. Biochem. Rev.* **18**(3), 93.

Hanada, K., Fujimoto, K., Shimura, M., and Yoshioka, K. (1998). Determination of trace amounts of Si and P in iron and steel using gel chromatographic separation followed by ICP-MS. *Phys. Status Solidi A* **167**(2), 383.

Hanada, T., Isobe, H., Saito, T., Ogura, S., Takekawa, H., Yamazaki, K., Tokuchi, Y., and Kawakami, Y. (1998). Intracellular accumulation of thallium as a marker of cisplatin cytotoxicity in nonsmall cell lung carcinoma: An application of inductively coupled plasma mass spectrometry. *Cancer (Philadelphia)* **83**(5), 930.

Hansen, S. H., Larsen, E. H., Pritzl, G., and Cornett, C. (1992). Separation of seven arsenic compounds by high-performance liquid chromatography with on-line detection by hydrogen-argon flame atomic absorption spectrometry and inductively coupled plasma mass spectrometry. *J. Anal. At. Spectrom.* **7**(4), 629–634.

Haraguchi, H., Itoh, A., Kimata, C., and Miwa, H. (1998). Speciation of yttrium and lanthanides in natural water by inductively coupled plasma mass spectrometry after preconcentration by ultrafiltration and with a chelating. *Analyst (London)* **123**(5), 773.

Haraldsson, C., Lyven, B., Oehman, P., and Munthe, J. (1994). Determination of mercury isotope ratios in samples containing sub-nanogram-amounts of mercury using inductively coupled plasma mass spectrometry. *J. Anal. At. Spectrom.* **9**(11), 1229.

Harrington, C. F., and Catterick, T. (1997). Problems encountered during the development of a method for the speciation of mercury and methylmercury by high-performance liquid chromatography coupled to inductively coupled plasma mass spectrometry. *J. Anal. At. Spectrom.* **12**(9), 1053.

Harrington, C. F., Elahi, S., Ponnampalavanar, P., and D'Silva, T. M. (1999). A protocol for the multielemental analysis of trace metals in food samples by flow injection coupled to ICP-MS. *At. Spectrosc.* **20**(5), 174.

Hartley, J. H. D., Ebdon, L., and Hill, S. J. (1992). Analysis of trimethylgallium etherate by flow injection inductively coupled plasma mass spectrometry. *Anal. Proc.* **29**(3), 94–95.

Hartley, J. H. D., Hill, S. J., and Ebdon, L. (1993). Analysis of slurries by inductively coupled plasma mass spectrometry using desolvation to improve transport efficiency and atomization efficiency. *Spectrochim. Acta, Part B* **48B**(11), 1421.

Hasegawa, S.-I, Kobayashi, T., Sato, K., Igarashi, S., and Naito, K. (1999). Determination of trace elements in high purity iron by chromazurol B separation/ICP-MS. *Mater. Trans. JIM* **40**(8), 1069.

Hassellov, M., Lyven, B., Haraldsson, C., and Sirinawin, W. (1999). Determination of continuous size and trace element distribution of colloidal material in natural water by on-line coupling of flow field-flow fractionation with ICPMS. *Anal. Chem.* **71**(16), 3497.

Hassler, D. R., Peucker-Ehrenbrink, B., and Ravizza, G. E. (2000). Rapid determination of Os isotopic composition by sparging OsO4 into a magnetic-sector ICP-MS. *Chem. Geolo.* **166**(1), 1.

Hastings, D. W., Emerson, S. R., and Nelson, B. K. (1996). Determination of picogram quantities of vanadium in calcite and seawater by isotope dilution inductively coupled plasma mass spectrometry with electrothermal vaporization. *Anal. Chem.* **68**(2), 371.

Hauptkorn, S., Krivan, V., Gercken, B., and Pavel, J. (1997). Determination of trace impurities in high-purity quartz by electrothermal vaporization inductively coupled plasma mass spectrometry using the slurry sampling technique. *J. Anal. At. Spectrom.* **12**(4), 421.

Hearn, R., and Wildner, H. (1998). Isotope ratios of uranium using high resolution inductively coupled plasma-mass spectrometry (ICP-MS). *ASTM Spec. Tech. Publ.*, p. 137.

Heisterkamp, M., De Smaele, T., Candelone, J. P., and Moens, L. (1997). Inductively coupled plasma mass spectrometry hyphenated to capillary gas chromatography as a detection system for the speciation analysis of organolead compounds in environmental waters. *J. Anal. At. Spectrom.* **12**(9), 1077.

Held, A., Taylor, P., Ingelbrecht, C., and De Bievre, P. (1995). Determination of scandium in high-purity titanium using inductively coupled plasma mass spectrometry and glow discharge mass spectrometry as part of its certification as a reference material. *J. Anal. At. Spectrom.* **10**(10), 849.

Held, A., Taylor, P. D. P., and Ingelbrecht, C. (1999). Measuring metal homogeneity in a matrix via the measurement of the ratio metal to matrix oxide using ICP-MS. *Fresenius' J. Anal. Chem.* **364**(5), 437.

Hepiegne, P., Dall'ava, D., Clement, R., and Degros, J. P. (1995). The separation of 0909Tc from low and medium-level radioactive wastes and its determination by inductively coupled plasma mass spectrometry. *Talanta* **42**(6), 803.

Heumann, K. G., Gallus, S. M., Raedlinger, G., and Vogl, J. (1998). Industry applications: Determination of lead and copper in drinking water by ICP-MS. Accurate determination of element species by on-line coupling of chromatographic systems with ICP-MS using isotope dilution technique. *Chem. N. Z.* **62**(4), 14.

Heydorn, K., Alfassi, Z., Damsgaard, E., Rietz, B., and Solgaard, P. (1995). Validation of methods for the determination of aluminium in fish gills by INAA and ICP-MS. *J. RadioAnal. Nucl. Chem.* **192**(2), 321.

Hidalgo, M. M., Gomez, M. M., and Palacios, M. A. (1996). Trace enrichment and measurement of platinum by flow injection inductively coupled plasma mass spectrometry. *Fresenius' J. Anal. Chem.* **354**(4), 420.

Higashiura, M., Uchida, H., Uchida, T., and Wada, H. (1995). Inductively coupled plasma mass spectrometric determination of gold in serum: Comparison with flame and furnace atomic absorption spectrometry. *Anal. Chim. Acta* **304**(3), 317.

High, K. A., Blais, J. S., Methven, B. A. J., and McLaren, J. W. (1995). Probing the characteristics of metal-binding proteins using high-performance liquid chromatography-atomic absorption spectroscopy and inductively coupled plasma mass spectrometry. *Analyst (London)* **120**(3), 629.

Hill, S. J. (1999). "Inductively Coupled Plasma Spectrometry and its Applications." Academic Press, Sheffield.

Hill, S. J., Hartley, J., and Ebdon, L. (1992). Determination of trace metals in volatile organic solvents using inductively coupled plasma atomic emission spectrometry and inductively coupled plasma mass spectrometry. *J. Anal. At. Spectrom.* **7**(1), 23–28.

Hill, S. J., Ford, M. J., and Ebdon, L. (1992). Simplex optimization of nitrogen-argon plasmas in inductively coupled plasma mass spectrometry for the removal of chloride-based interferences. *J. Anal. At. Spectrom.* **7**(5), 719–726.

Hill, S. J., Hartley, J., and Ebdon, L. (1992). Determination of impurities in organometallic compounds dissolved in diethyl ether by flow injection inductively coupled plasma mass spectrometry. *J. Anal. At. Spectrom.* **7**(6), 895–898.

Hill, S. J., Bloxham, M. J., and Worsfold, P. J. (1993). Chromatography coupled with inductively coupled plasma atomic emission spectrometry and inductively coupled plasma mass spectrometry. *J. Anal. At. Spectrom.* **8**(4), 499–516.

Hill, S. J., Brown, A., Rivas, C., and Sparkes, S. (1995). High performance liquid chromatography—isotope dilution—inductively coupled plasma-mass spectrometry for lead and tin speciation in environmental samples. *Tech. Instrum. Anal. Chem.*, p. 412.

Hill, S. J., Pitts, L. J., and Fisher, A. S. (2000). High-performance liquid chromatography-isotope dilution inductively coupled plasma mass spectrometry for speciation studies: An overview. *Trends Anal. Chem.* **19**(2), 120.

Himri, M. E., Pastor, A., and de la Guardia, M. (2000). Determination of uranium in tap water by ICP-MS. *Fresenius' J. Anal. Chem.* **367**(2), 151.

Hinds, M. W., Grégoire, D. C., and Ozaki, E. A. (1997). Direct determination of volatile elements in nickel alloys by electrothermal vaporization inductively coupled plasma mass spectrometry. *J. Anal. At. Spectrom.* **12**(2), 131.

Hinrichs, J., and Schnetger, B. (1999). A fast method for the simultaneous determination of 020300Th, 020304U and 020305U with isotope dilution sector field ICP-MS. *Analyst (London)* **124**(6), 927.

Hintelmann, H., and Evans, R. D. (1997). Application of stable isotopes in environmental tracer studies—Measurement of monomethylmercury (CH3Hg+) by isotope dilution ICP-MS and detection of species transformation. *Fresenius' J. Anal. Chem.* **358**(3), 378.

Hintelmann, H., Falter, R., Ilgen, G., and Evans, R. D. (1997). Determination of artifactual formation of monomethylmercury (CH3Hg+) in environmental samples using stable Hg2+ isotopes with ICP-MS detection: Calculation of contents applying species specific isotope addition. *Fresenius' J. Anal. Chem.* **358**(3), 363.

Hioki, A., Lam, J. W. H., and McLaren, J. W. (1997). On-line determination of dissolved silica in seawater by ion exclusion chromatography in combination with inductively coupled plasma mass spectrometry. *Anal. Chem.* **69**(1), 21.

Hirata, T. (1996). Evaluation of correction method for mass discrimination effect in multiple collector inductively coupled plasma mass spectrometry. *Bunseki Kagaku* **45**(6), 465.

Hirata, T. (1996). Lead isotopic analyses of NIST standard reference materials using multiple collector inductively coupled plasma mass spectrometry coupled with a modified external correction method for mass discrimination effect. *Analyst (London)* **121**(10), 1407.

Hirata, T., and Masuda, A. (1990). Determination of rhenium with enhanced sensitivity using inductively coupled plasma mass spectrometry. *J. Anal. At. Spectrom.* **5**(7), 627.

Hobbs, S. E., and Olesik, J. W. (1992). Inductively coupled plasma mass spectrometry signal fluctuations due to individual aerosol droplets and vaporizing particles. *Anal. Chem.* **64**(3), 274–282.

Hoelzl, R., Fabry, L., Kotz, L., and Pahlke, S. (2000). Routine analysis of ultra pure water by ICP-MS in the low- and sub-ng/L level. *Fresenius' J. Anal. Chem.* **366**(1), 64.

Hoffmann, E., Luedke, C., Scholze, H., and Stephanowitz, H. (1994). Analytical investigations of tree rings by laser ablation ICP-MS. *Fresenius' J. Anal. Chem.* **350**(4/5), 253.

Hoffmann, P., Karandashev, V. K., Sinner, T., and Ortner, H. M. (1997). Chemical analysis of rain and snow samples from Chernogolovka/Russia by IC, TXRF and ICP-MS. *Fresenius' J. Anal. Chem.* **357**(8), 1142.

Hollenbach, M., Grohs, J., Mamich, S., and Kroft, M. (1994). Determination of technetium-99, thorium-230 and uranium-234 in soils by inductively coupled plasma mass spectrometry using flow injection preconcentration. *J. Anal. At. Spectrom.* **9**(9), 927.

Hollenbach, M., Grohs, J., Kroft, M., and Mamich, S. (1995). Determination of 230Th, 234U, 239Pu, and 240Pu in soils by ICP-MS using flow-injection preconcentration. *ASTM Spec. Tech. Publ.*, p. 99.

Hollocher, K., and Ruiz, J. (1995). Major and trace element determinations on NIST glass standard reference materials 611, 612, 614 and 1834 by inductively coupled plasma-mass spectrometry. *Geostand. Newsl.* **19**(1), 27.

Hollocher, K., Fakhry, A., and Ruiz, J. (1995). Trace element determinations for USGS basalt BHVO-1 and NIST standard reference materials 278, 688 and 694 by inductively coupled plasma-mass spectrometry. *Geostand. Newsl.* **19**(1), 35.

Holmes, L. J., Robinson, V. J., Makinson, P. R., and Livens, F. R. (1996). Multi-element determination in complex matrices by inductively coupled plasma-mass spectrometry (ICP-MS). *Sci. Total Environ.* **180**, 345.

Hoppstock, K., Becker, J. S., and Dietze, H. J. (1997). Assessment of the determination of 79 selenium using double-focusing sector field ICP-MS after hydride generation. *At. Spectrosc.* **18**(6), 180.

Horlick, G. (1994). Inductively coupled plasma mass spectrometry: Has it matured already? *J. Anal. At. Spectrom.* **9**(5), 593.

Horn, I., Hinton, R. W., Jackson, S. E., and Longerich, H. P. (1997). Ultra-trace element analysis of NIST SRM 616 and 614 using laser ablation microprobe inductively coupled plasma-mass spectrometry (LAM-ICP-MS): A comparison with secondary ion mass spectrometry (SIMS). *Geostand. Newsl.* **21**(2), 191.

Horn, M. (1999). Applications of ICP-MS in semiconductor industry. *Fresenius' J. Anal. Chem.* **364**(5), 385.

Horn, M., and Heumann, K. G. (1994). Comparison of heavy metal analyses in hydrofluoric acid used in microelectronic industry by ICP-MS and thermal ionization isotope dilution mass spectrometry. *Fresenius' J. Anal. Chem.* **350**(4/5), 286.

Hoskin, P. W. O. (1998). Minor and trace element analysis of natural zircon (ZrSiO4) by SIMS and laser ablation ICPMS: A consideration and comparison of two broadly competitive techniques. *J. Trace Microprobe Tech.* **16**(3), 301.

Hou, H.-B., and Narasaki, H. (1999). Determination of antimony in river water by hydride generation high-resolution ICP-MS. *At. Spectrosc.* **20**(1), 20.

Hou, H. B., and Narasaki, H. (1999). Notes—Differential determination of antimony(III) and antimony(V) in river water by hydride-generation inductively coupled plasma mass spectrometry. *Anal. Sci.: Int. J. Jpn. Soci. Anal. Chem.* **15**(9), 911.

Hou, X., Li, C., Ding, W., Qian, Q., and Chai, C. (1998). Determination of 24 elements in four algae reference materials by neutron activation analysis and inductively coupled plasma mass spectrometry. *Fresenius' J. Anal. Chem.* **360**(3), 423.

Houk, R. S. (1994). Elemental and isotopic analysis by inductively coupled plasma mass spectrometry. *Acc. Chem. Res.* **27**(11), 333.

Houk, R. S. (2000). Atomic emission spectroscopy—Inductively coupled plasma-mass spectrometry and the European discovery of America. *J. Chem. Educ.* **77**(5), 598.

Howe, S. E., Davidson, C. M., and McCartney, M. (1999). Operational speciation of uranium in inter-tidal sediments from the vicinity of a phosphoric acid plant by means of the BCR sequential extraction procedure and ICP-MS. *J. Anal. At. Spectrom.* **14**(2), 163.

Hsiung, C.-S. (1997). Multielement analysis of biological materials: Optimization of inductively coupled plasma mass spectrometry. Unpublished Ph.D. Thesis, University of Utah, Salt Lake City.

Hsiung, C.-S., Andrade, J. D., Costa, K., and Ash, K. O. (1997). Minimizing interferences in the quantitative multielement analysis of trace elements in biological fluids by inductively coupled plasma mass spectrometry. *Clin. Chem. (Winston-Salem, N.C.)* **43**(12), 2303.

Hu, K. (1992). Enhancement of ion transmission and reduction of background and interferences in inductively coupled plasma mass spectrometry. Unpublished Ph.D. Thesis, Iowa State University, Ames.

Hu, K., and Houk, R. S. (1993). Inductively coupled plasma mass spectrometry with an enlarged sampling orifice and offset ion lens. II. Polyatomic Ion interferences and matrix effects. *J. Am. Soc. Mass Spectrom.* **4**(1), 28–37.

Hu, K., and Houk, R. S. (1993). Inductively coupled plasma mass spectrometry with an electrically floating sampling interface. *J. Am. Soc. Mass Spectrom.* **4**(9), 733–741.

Hu, K., and Houk, R. S. (1996). Ion deposition by inductively coupled plasma mass spectrometry. *J. Vac. Sci. Technol. A* **14**(2), 370.

Hu, K., Clemons, P. S., and Houk, R. S. (1993). Inductively coupled plasma mass spectrometry with an enlarged sampling orifice and offset ion lens. I. Ion trajectories and detector performance. *J. Am. Soc. Mass Spectrom.* **4**(1), 16–27.

Hu, S., Lin, S., Liu, Y., and Gao, S. (2000). Studies on the calibration of matrix effects and polyatomic ion for rare earth elements in geochemical samples by ICP-MS. *Kao Teng Hsueh Hsiao Hua Hsueh Hsueh Pao (Chem. J. Chin. Univ.)*, pp. 372–375.

Hu, Y., Vanhaecke, F., Moens, L., and Dams, R. (1997). Semi-quantitative panoramic analysis of industrial samples by inductively coupled plasma mass spectrometry. *Anal. Chim. Acta* **355**(2–3), 105.

Hu, Y., Vanhaecke, F., Moens, L., Dams, R., del Castilho, P., and Japenga, J. (1998). Determination of the aqua regia soluble content of rare earth elements in fertilizer, animal fodder phosphate and manure samples using inductively coupled plasma mass spectrometry. *Anal. Chim. Acta* **373**(1), 95.

Hu, Y., Vanhaecke, F., Moens, L., and Dams, R. (1999). Determination of ruthenium in photographic materials using solid sampling electrothermal vaporization inductively coupled plasma mass spectrometry. *J. Anal. At. Spectrom.* **14**(4), 589.

Huang, C.-C., Yang, M.-H., and Shih, T.-S. (1997). Automated on-line sample pretreatment system for the determination of trace metals in biological samples by inductively coupled plasma mass spectrometry. *Anal. Chem.* **69**(19), 3930.

Huang, K. S., and Jiang, S. J. (1993). Determination of trace levels of metal ions in water samples by inductively coupled plasma mass spectrometry after on-line preconcentration on SO3-oxine CM-cellulose. *Fresenius' J. Anal. Chem.* **347**(6–7), 238–242.

Huang, M. F., Jiang, S. J., and Hwang, C. J. (1995). Determination of arsenic in environmental and biological samples by flow injection inductively coupled plasma mass spectrometry. *J. Anal. At. Spectrom.* **10**(1), 31.

Hughes, D. M. Grégoire, D. C., Naka, H., and Chakrabarti, C. L. (1997). The vaporization of phosphorus compounds and the use of chemical modifiers for the determination of phosphorus by electrothermal vaporization inductively coupled plasma mass spectrometry. *Spectrochim. Acta, Part B* **52**(4), 517.

Hughes, D. M., Grégoire, D. C., Naka, H., and Chakrabarti, C. L. (1998). Determination of trace amounts of phosphorus in high purity iron by electrothermal vaporization inductively coupled plasma mass spectrometry. *Spectrochim. Acta, Part B* **53**(6–8), 1079.

Hutton, R. C., and Grote, B. (1996). Sample introduction systems for ICP-OES and ICP-MS. *Analusis* **24**(9–10), 29.

Hutton, R. C., Bridenne, M., Coffre, E., Marot, Y., and Simondet, F. (1990). Investigations into the direct analysis of semiconductor grade gases by inductively coupled plasma mass spectrometry. *J. Anal. At. Spectrom.* **5**(6), 463.

Huynh, M. P. T., Carrot, F., Ngoc, S. C. P., Dang Vu, M., and Revel, G. (1997). Determination of rare earth elements in rice by INAA and ICP-MS. *J. RadioAnal. Nucl. Chem.* **217**(1), 95.

Hwang, C. J., and Jiang, S. J. (1994). Determination of arsenic compounds in water samples by liquid chromatography—inductively coupled plasma mass spectrometry with an in situ nebulizer-hydride generator. *Anal. Chim. Acta* **289**(2), 205.

Hwang, T. J., and Jiang, S. J. (1996). Determination of copper, cadmium and lead in biological samples by isotope dilution inductively coupled plasma mass spectrometry after on-line pre-treatment by anodic stripping voltammetry. *J. Anal. At. Spectrom.* **11**(5), 353.

Hwang, T. J., and Jiang, A. J. (1997). Determination of cadmium by flow injection isotope dilution inductively coupled plasma mass spectrometry with vapour generation sample introduction. *J. Anal. At. Spectrom.* **12**(5), 579.

Hwang, T. J., and Jiang, S. J. (1997). Determination of trace amounts of zinc in water samples by flow injection isotope dilution inductively coupled plasma mass spectrometry. *Analyst (London)* **122**(3), 233.

Hywel Evans, E., and Giglio, J. J. (1993). Interferences in inductively coupled plasma mass spectrometry. A review. *J. Anal. At. Spectrom.* **8**(1), 1–18.

Hywel Evans, E., Pretorius, W., Ebdon, L., and Rowland, S. (1994). Low-pressure inductively coupled plasma ion source for molecular and atomic mass spectrometry. *Anal. Chem.* **66**(20), 3400.

Hywel Evans, E., Truscott, J. B., Bromley, L., and Jones, P. (1998). Evaluation of chelation preconcentration for the determination of actinide elements by flow injection ICP-MS. *ASTM Spec. Tech. Publ.*, p. 79.

Ignacio Garcia Alonso, J. (1995). Determination of fission products and actinides by inductively coupled plasma-mass spectrometry using isotope dilution analysis: A study of random and systematic errors. *Anal. Chim. Acta* **312**(1), 57.

Ignacio Garcia Alonso, J., Sena, F., Arbore, P., and Betti, M. (1995). Determination of fission products and actinides in spent nuclear fuels by isotope dilution ion chromatography inductively coupled plasma mass spectrometry. *J. Anal. At. Spectrom.* **10**(5), 381.

Ihnat, M., Gamble, D. S., and Gilchrist, G. F. R. (1993). Determination of trace element levels in natural fresh water by inductively coupled plasma mass spectrometry. *Int. J. Environ. Anal. Chem.* **53**(1), 63–78.

Ihsanullah. (1995). Significance and initial investigations for the separation of molybdenum from the anion exchange resin prior to technetium analysis by ICP-MS. *J. Radioanal. Nucl. Chem.* **191**(1), 67.

Imai, N. (1992). Microprobe analysis of geological materials by laser ablation inductively coupled plasma mass spectrometry. *Anal. Chim. Acta* **269**(2), 263.

Imai, S., Nishiyama, Y., Tanaka, T., and Hayashi, Y. (1995). Investigations of pyrolysed ascorbic acid in an electrothermal graphite furnace by inductively coupled argon plasma mass spectrometry and raman spectrometry. *J. Anal. At. Spectrom.* **10**(6), 439.

Imbert, J. L., and Telouk, P. (1993). Application of laser ablation ICP-MS to elemental analysis of glasses. *Mikrochim. Acta* **110**(4–6), 151.

Inagaki, K., and Haraguchi, H. (1997). Rare earth elements in human blood serum as determined by inductively coupled plasma mass spectrometry. *Chem. Lett.*, p. 775.

Inagaki, K., and Haraguchi, H. (2000). Determination of rare earth elements in human blood serum by inductively coupled plasma mass spectrometry after chelating resin preconcentration. *Analyst (London)* **125**, 191–196.

Inagaki, K., Mikuriya, N., Morita, S., Haraguchi, H., Nakahara, Y., Hattori, M., Kinosita, T., and Saito, H. (2000). Speciation of protein-binding zinc and copper in human blood serum by chelating resin pre-treatment and inductively coupled plasma mass spectrometry. *Analyst (London)* **125**, 197–204.

Inoue, Y., Kawabata, K., Takahashi, H., and Endo, G. (1995). Determination of inorganic and organic arsenic compounds in urine using ion chromatography with ICP-MS. *Bunseki Kagaku* **44**(3), 203.

Inoue, Y., Sakai, T., and Kumagai, H. (1995). Simultaneous determination of chromium(III) and chromium(VI) by ion chromatography with inductively coupled plasma mass spectrometry. *J. Chromatogr.* **706**(1–2), 127.

Inoue, Y., Date, Y., Yoshida, K., Chen, H., and Endo, G. (1996). Speciation of arsenic compounds in the urine of rats orally exposed to dime hylarsinic acid by ion chromatography with ICP-MS as an element-selective. *Appl. Organomet. Chem.* **10**(9), 707.

Inoue, Y. I., Kawabata, K., and Suzuki, Y. (1995). Speciation of organotin compounds using inductively coupled plasma mass spectrometry with micellar liquid chromatography. *J. Anal. At. Spectrom.* **10**(5), 363.

Ionov, D. A., Savoyant, L., and Dupuy, C. (1992). Application of the ICP-MS technique to trace element analysis of peridotites and their minerals. *Geostand. Newsl.* **16**(2), 311–316.

Itoh, A., Hamanaka, T., Rong, W., Ikeda, K., Sawatari, H., Chiba, K., and Haraguchi, H. (1999). Multielement determination of rare earth elements in geochemical samples by liquid chromatography/inductively coupled plasma mass spectrometry. *Anal. Sci.: Int. J. Jpn. Soci. Anal. Chem.* **15**(1), 17.

Itoh, A., Iwata, K., Chiba, K., and Haraguchi, H. (2000). Analytical and inorganic—articles—Chemical stability of large organic molecule-metal complexes dissolved in natural water as studied by size exclusion chromatography/inductively coupled plasma mass spectrometry. *Bull. Chem. Soc. Jpn.* **73**(1), 121.

Itoh, M., Ohmichi, M., and Suzuki, K. T. (1995). Detection of metabolites of selenite labeled with stable isotope by HPLC/ICP-MS. *Eisei Kagaku* **41**(1), 47.

Jackson, S. E., Fryer, B. J., Gosse, W., Healey, D. C., Longerich, H. P., and Strong, D. F. (1990). Determination of the precious metals in geological materials by inductively coupled plasma-mass spectrometry (ICP-MS) with nickel sulphide fire-assay collection and tellurium coprecipiation. *Chem. Geo.* **83**(1–2), 119–132.

Jackson, S. E., Longerich, H. P., Dunning, G. R., and Fryer, B. J. (1992). The application of laser-ablation microprobe—inductively coupled plasma—mass spectrometry (LAM-ICP-MS) to in situ trace-element determinations in minerals. *Can. Mineral.*, pp. 1049–1064.

Jakubowski, N., Feldmann, I., Sack, B., and Stuewer, D. (1992). Analysis of conducting solids by inductively coupled plasma mass spectrometry with spark ablation. *J. Anal. At. Spectrom.* **7**(2), 121–126.

Jakubowski, N., Feldmann, I., Stuewer, D., and Berndt, H. (1992). Hydraulic high pressure nebulization—application of a new nebulization system for inductively coupled plasma mass spectrometry. *Spectrochim. Acta, Part B* **47B**(1), 119.

Jakubowski, N., Feldmann, I., and Stuewer, D. (1993). Diagnostic investigations of aerosols with varying water content in inductively coupled plasma mass spectrometry. *J. Anal. At. Spectrom.* **8**(7), 969–978.

Jakubowski, N., Jepkens, B., Stuewer, D., and Berndt, H. (1994). Speciation analysis of chromium by inductively coupled plasma mass spectrometry with hydraulic high pressure nebulization. *J. Anal. At. Spectrom.* **9**(3), 193–198.

Jakubowski, N., Feldmann, I., and Stuewer, D. (1995). Comparison of ICP-MS with spark ablation and GDMS for direct element analysis of conductive solids. *Spectrochim. Acta, Part B* **50**(4/7), 639.

Jakubowski, N., Tittes, W., Pollmann, D., and Stuewer, D. (1996). Comparative analysis of aluminium oxide powders by inductively coupled plasma mass spectrometry with low and high mass resolution. *J. Anal. At. Spectrom.* **11**(9), 797.

Jakubowski, N., Thomas, C., Stuewer, D., and Dettlaff, I. (1996). Speciation of inorganic selenium by inductively coupled plasma mass spectrometry with hydraulic high pressure nebulization. *J. Anal. At. Spectrom.* **11**(11), 1023.

Jakubowski, N., Feldmann, I., and Stuewer, D. (1997). Grimm-type glow discharge ion source for operation with a high resolution inductively coupled plasma mass spectrometry instrument. *J. Anal. At. Spectrom.* **12**(2), 151.

Jakubowski, N., Moens, L., and Vanhaecke, F. (1998). Sector field mass spectrometers in ICP-MS. *Spectrochim. Acta, Part B* **53**(13), 1739.

Jakubowski, N., Thomas, C., Klueppel, D., and Stuewer, D. (1998). Speciation of metals in by use of inductively coupled plasma mass spectrometry with low and high mass resolution. *Analusis* **26**(6), M37.

Jakubowski, N., Brandt, R., Stuewer, D., Eschnauer, H. R., and Gortges, S. (1999). Analysis of wines by ICP-MS: Is the pattern of the rare earth elements a reliable fingerprint for the provenance? *Fresenius' J. Anal. Chem.* **364**(5), 424.

Jalkanen, L. M., and Haesaenen, E. K. (1996). Simple method for the dissolution of atmospheric aerosol samples for analysis by inductively coupled plasma mass spectrometry. *J. Anal. At. Spectrom.* **11**(5), 365.

Jarvis, K. E. (1990). A critical evaluation of two sample preparation techniques for low-level determination of some geologically incompatible elements by inductively coupled plasma-mass spectrometry. *Chem. Geol.* **83**(1–2), 89–104.

Jarvis, K. E., and Williams, J. G. (1993). Laser ablation inductively coupled plasma mass spectrometry (LA-ICP-MS): A rapid technique for the direct, quantitative determination of major, trace and rare-earth elements in geological samples. *Chem. Geol.* **104**, 251.

Jarvis, K. E., Williams, J. G., Alcantara, E., and Wills, J. D. (1996). Determination of trace and ultra-trace elements in saline waters by inductively coupled plasma mass spectrometry after off-line chromatographic separation and preconcentration. *J. Anal. At. Spectrom.* **11**(10), 917.

Jarvis, K. E., Gray, A. L., and Houk, R. S. (1996). "Handbook of Inductively Coupled Plasma Mass Spectrometry." Blackie Academic and Professional Chapman & Hall, London and New York.

Jarvis, I., Totland, M. M., and Jarvis, K. E. (1997). Determination of the platinum-group elements in geological materials by ICP-MS using microwave digestion, alkali fusion and cation-exchange chromatography. *Chem. Geo.* **143**(1–2), 27.

Jarvis, I., Totland, M. M., and Jarvis, K. E. (1997). Assessment of dowex 1-X8- based anion-exchange procedures for the separation and determination of ruthenium, rhodium, palladium, iridium, platinum and gold in geological samples by inductively coupled plasma mass spectrometry. *Analyst (London)* **122**(1), 19.

Jarvis, K. E., Mason, P., Platzner, T., and Williams, J. G. (1998). Critical assessment of the effects of skimmer cone geometry on spectroscopic and non-spectroscopic interference in inductively coupled plasma mass spectrometry. *J. Anal. At. Spectrom.* **13**(8), 689.

Jeffries, T. E., Perkins, W. T., and Pearce, N. J. G. (1995). Measurements of trace elements in basalts and their phenocrysts by laser probe microanalysis inductively coupled plasma mass spectrometry (LPMA-ICP-MS). *Chem. Geol.* **121**(1–4), 131.

Jeffries, T. E., Perkins, W. T., and Pearce, N. J. G. (1995). Comparisons of infrared and ultraviolet laser probe microanalysis inductively coupled plasma mass spectrometry in mineral analysis. *Analyst (London)* **120**(5), 1365.

Jeong, S. H., Borisov, O. V., Yoo, J. H., Mao, X. L., and Russo, R. E. (1999). Effects of particle size distribution on inductively coupled plasma mass spectrometry signal intensity during laser ablation of glass samples. *Anal. Chem.* **71**(22), 5123.

Jian, L., Goessler, W., and Irgolic, K. J. (2000). Mercury determination with ICP-MS: Signal suppression by acids. *Fresenius' J. Anal. Chem.* **366**(1), 48.

Jiansheng, W., Evans, E. H., and Caruso, J. A. (1991). Minimization of non-spectroscopic matrix interferences for the determination of trace elements in fusion samples by flow injection inductively coupled plasma mass spectrometry. *J. Anal. At. Spectrom.* **6**(8), 605–608.

Jiansheng, W., Evans, E. H., and Caruso, J. A. (1992). Addition of molecular gases to argon gas flows for the reduction of polyatomic-ion interferences in inductively coupled plasma mass spectrometry. *J. Anal. At. Spectrom.* **7**(6), 929–936.

Joannon, S., Telouk, P., and Pin, C. (1997). Determination of U and Th at ultra-trace levels by isotope dilution inductively coupled plasma mass spectrometry using a geyser-type ultrasonic nebulizer: Application to geological samples. *Spectrochim. Acta, Part B* **52**(12), 1783.

Jochum, K. P., and Jenner, G. (1994). Trace element analysis of Geological Survey of Japan silicate reference materials: Comparison of SSMS with ICP-MS data and a critical discussion of compiled values. *Fresenius' J. Anal. Chem.* **350**(4/5), 310.

Jones, B. T., Evans, E. H., Giglio, J. J., Castillano, T. M., and Caruso, J. A. (1998). Inductively coupled and microwave induced plasma sources for mass spectrometry. *Microchem. J.* **58**(2), 236.

Jones, V. D. (1998). Evaluation of nebulizer performance within the ICP-MS measurement system of analysis of SRS radiological waste tank simulated solutions. *ASTM Spec. Tech. Pub.*, p. 41.

Jorge, A. P. D. S., Enzweiler, J., Shibuya, E. K., Sarkis, J. E. S., and Figueiredo, A. M. G. (1998). Platinum-group elements and gold determination in NiS fire assay buttons by UV laser ablation ICP-MS. *Geostand. Newsl.* **22**(1), 47.

Kalamegham, R., and Ash, K. O. (1992). A simple ICP-MS procedure for the determination of total mercury in whole blood and urine. *J. Clin. Lab. Anal.* **6**(4), 190–193.

Kane, J. S., Beary, E. S., Murphy, K. E., and Paulsen, P. J. (1995). Impact of inductively coupled plasma mass spectrometry on certification programmes for geochemical reference materials. *Analyst (London)* **120**(5), 1505.

Kaneco, S., Nomizu, T., Tanaka, T., Mizutani, N., and Kawaguchi, H. (1995). Optimization of operating conditions in individual airborne particle analysis by inductively coupled plasma mass spectrometry. *Anal. Sci.: Int. J. Jpn. Soc. Anal. Chem.* **11**(5), 835.

Kantipuly, C. J. (1996). Inductively coupled plasma mass spectrometry for metals analysis in environmental samples. *Hazard. Ind. Waste: Proc. Proc. Mid-Atl. Ind. Waste Conf.*, p. 805.

Karandashev, V. K., Turanov, A. N., Kuss, H. M., Kumpmann, I., Zadnepruk, L. V., and Baulin, V. E. (1998). Extraction chromatographic separation of Y, REE, Bi, Th, and U from the matrix suitable for their determination in pure iron and low-alloyed steels by ICP-MS and ICP-AES. *Mikrochim. Acta* **130**(1–2), 47.

Karnanassios, V., Drouin, P., and Reynolds, G. G. (1995). Electrically heated wire-loop, in torch vaporization (ITV) sample introduction system for ICP-AES with photomultiplier tube detection and ICP-MS. *Spectrochim. Acta, Part B* **50**(4/7), 415.

Kato, K., Ito, M., and Watanabe, K. (2000). Determination of thorium and uranium in activated concrete by inductively coupled plasma mass spectrometry after anion- exchange separation. *Fresenius' J. Anal. Chem.* **366**(1), 54.

Katoh, T., Akiyama, M., Ohtsuka, H., and Nakamura, S. (1996). Determination of atmospheric trace metal concentrations by isotope dilution inductively coupled plasma mass spectrometry after separation from interfering elements by solvent extraction. *J. Anal. At. Spectrom.* **11**(1), 69.

Kawabata, K., Kishi, Y., Kawaguchi, O., Watanabe, Y., and Inoue, Y. (1991). Determination of rare-earth elements by inductively coupled plasma mass spectrometry with ion chromatography. *Anal. Chem.* **63**(19), 2137–2140.

Kawabata, K., Inoue, Y., Takahashi, H., and Endo, G. (1994). Determination of arsenic species by inductively coupled plasma mass spectrometry with ion chromatography. *Appl. Organomet. Chem.* **8**(3), 245.

Kawamoto, K., Sumino, T., Takada, J., Tanaka, Y., and Akaboshi, M. (1998). Applications of Radioanalytical methods to biological and clinical systems - Determination of rare earth and other elements in algae by ICP-MS and neutron activation analysis. *J. Radioanal. Nucl. Chem.* **236**(1–2), 119.

Kawamoto, K., Takada, J., Tanaka, Y., and Akaboshi, M. (1999). Short communications – Determination of rare earth and other elements in standard biological samples by ICP-MS and activation analysis. *J. Radioanal. Nucl. Chem.* **242**(2), 527.

Kawamura, H., Tagomori, H., Matsuoka, N., Takashima, Y., Tawaki, S., and Momoshima, N. (1999). Determination of stable lead isotope ratios in environmental samples: Combination of microwave digestion and ICP-MS. *J. Radioanal. Nucl. Chem.* **242**(3), 717.

Kawasaki, A., and Arai, S. (1996). Evaluation of digestion methods for multi-elemental analysis of organic wastes by inductively coupled plasma mass spectrometry. *Soil Sci. Plant Nutr.* **42**(2), 251.

Keil, O., Dahmen, J., and Volmer, D. A. (1999). Automated matrix separation and preconcentration for the trace level determination of metal impurities in ultrapure inorganic salts by high-resolution ICP-MS. *Fresenius' J. Anal. Chem.* **364**(8), 694.

Kershisnik, M. M., Kalamegham, R., Ash, K. O., Nixon, D. E., and Ashwood, E. R. (1992). Using ìsupï16Oìsupï35Cl to correct for chloride interference improves accuracy of urine arsenic determinations by inductively coupled plasma mass spectrometry. *Clin. Chem. (Winston-Salem, N.C.)* **38**(11), 2197.

Ketterer, M. E., and Biddle, D. A. (1992). Multivariate calibration in inductively coupled plasma mass spectrometry. 2. Effect of changes in abundances of interfering polyatomic ions. *Anal. Chem.* **64**(17), 1819.

Ketterer, M., and Khourey, C. J. (1998). High-precision determination of 020304U/020308U activity ratios in natural waters and carbonates by ICPMS. *ASTM Spec. Tech. Pub.*, p. 120.

Ketterer, M. E., Peters, M. J., and Tisdale, P. J. (1991). Verification of a correction procedure for measurement of lead isotope ratios by inductively coupled plasma mass spectrometry. *J. Anal. At. Spectrom.* **6**(6), 439.

Khalema, K. (1995). Electrochemical generation of arsenic and selenium hydrides with detection via inductively coupled plasma mass spectrometry. Unpublished Ph.D. Thesis, University of Cincinnati, Cincinnati, OH.

Kim, Y.-J., Kim, C.-K., Kim, C.-S., Yun, J.-Y., and Rho, B.-H. (1999). End of the proceedings—Determination of 226Ra in environmental samples using high-resolution inductively coupled plasma mass spectrometry. *J. Radioanal. Nucl. Chem.* **240**(2), 613.

Kin, F. D., Prudencio, M. I., Gouveia, M. A., and Magnusson, E. (1999). Determination of rare earth elements in geological reference materials: A comparative study by INM and ICP-MS. *Geostand. Newsl.* **23**(1), 47.

Kinzer, J. A., Olesik, J. W., and Olesik, S. V. (1996). Effect of laminar flow in capillary electrophoresis: Model and experimental results on controlling analysis time and resolution with inductively coupled plasma mass spectrometry detection. *Anal. Chem.* **68**(18), 3250.

Kir'yanov, G. I., and Titov, V. V. (1994). Mass spectrometry, with an inductively coupled plasma, of inorganic trace impurities in ultrapure water. *At. Energ.* **77**(2), 623.

Kitagawa, S., Tanaka, H., Okamoto, K., Etoh, T., and Matsubara, M. (1996). Determination of trace levels of platinum in biological materials by inductively coupled plasma mass spectrometry. *Bunseki Kagaku* **45**(6), 511.

Klaue, B., and Blum, J. D. (1999). Trace analyses of arsenic in drinking water by inductively coupled plasma mass spectrometry: High resolution versus hydride generation. *Anal. Chem.* **71**(7), 1408.

Klemm, W., Bombach, G., and Becker, K. P. (1999). Investigations of trace elements in high salinity waters by ICP-MS. *Fresenius' J. Anal. Chem.* **364**(5), 429.

Klinkenberg, H., Zicheng, P., Beeren, T., and Flach, K. (1992). Modifications of the roughing pump system of an Elan 500 ICP-MS for analysing complex matrices. *Spectrochim. Acta, Part B* **47B**(4), 585–590.

Klinkenberg, H., Beeren, T., Van Borm, W., Mevissen, B., and Van Dongen, C. (1993). Instrumental improvement of the XYZ-translation system for a Perkin-Elmer Sciex Elan 500 ICP-MS. *Spectrochim. Acta, Part B* **48B**(3), 475.

Klinkenberg, H., Beeren, T., Van Borm, W., van der Linden, F., and Raets, M. (1993). The use of an enriched isotope as an on-line internal standard in inductively coupled plasma mass spectrometry: A reference method for a proposed determination of tellurium in industrial waste water by means of graphite furnace atomic absorption spectrometry. *Spectrochim. Acta, Part B* **48B**(5), 649.

Klinkenberg, H., Beeren, T., and Van Borm, W. (1994). Multielement analysis using flow injection inductively coupled plasma mass spectrometry: Analytical aspects of multielement determinations in highly concentrated solutions of phosphoric acid, sodium phosphate and sodium nitrate. *Spectrochim. Acta, Part B* **49B**(2), 171.

Klinkenberg, H., Van Borm, W., and Souren, F. (1996). A theoretical adaptation of the classical isotope dilution technique for practical routine analytical determinations by means of inductively coupled plasma mass spectrometry. *Spectrochim. Acta, Part B* **51**(1), 139.

Klinkenberg, H., Van Borm, W., and Kip, B. J. (1997). The continuous on-line monitoring of the trace elemental composition of industrial ethene gas by means of inductively coupled plasma mass spectrometry. *Spectrochim. Acta, Part B* **52**(11), 1695.

Klinkenberg, H., van der Wal, S., de Koster, C. and Bart, J. (1998). On the use of inductively coupled plasma mass spectrometry as an element specific detector for liquid chromatography: Optimization of an industrial tellurium removal process. *J. Chromatogr.* **794**(1–2), 219.

Klinkhammer, G., German, C. R., Elderfield, H., Greaves, M. J., and Mitra, A. (1994). Rare earth elements in hydrothermal fluids and plume particulates by inductively coupled plasma mass spectrometry. *Mar. Chem.* **45**(3), 179–186.

Klueppel, D., Jakubowski, N., Messerschmidt, J., and Stuewer, D. (1998). Speciation of platinum metabolites in plants by size-exclusion chromatography and inductively coupled plasma mass spectrometry. *J. Anal. At. Spectrom.* **13**(4), 255.

Ko, F. H., and Yang, M. H. (1996). On-line removal of interferences via anion-exchange column separation for the determination of germanium, arsenic and selenium in biological samples by inductively coupled plasma mass spectrometry. *J. Anal. At. Spectrom.* **11**(6), 413.

Kogan, V. V., and Hinds, M. W. (1994). Assay of gold by laser ablation inductively coupled plasma mass spectrometry. *Precious Met.*, p. 415.

Kogan, V. V., Hinds, M. W., and Ramendik, G. I. (1994). The direct determination of trace metals in gold and silver materials by laser ablation inductively coupled plasma mass spectrometry without matrix matched standards. *Spectrochim. Acta, Part B* **49B**(4), 333.

Kokelj, F., Daris, F., Lutmann, A., and Braida, N. (1994). Nickel, chromate and cobalt in toilet soaps analysed by inductively coupled plasma mass spectrometry. *Contact Dermatitis* **31**(4), 270.

Komamura, M., and Tsumura, A. (1994). The transfer factors of long-lived radionuclides from soil to polished rice measured by ICP-MS. *Radioisotopes* **43**(1), 1.

Komoda, M., Chiba, K., and Uchida, H. (1996). Determination of trace impurities on silicon-wafer surface by isotope dilution analysis using electrothermal vaporization/inductively coupled plasma mass spectrometry. *Anal. Sci.: Int. J. Jpn. Soc. Anal. Chem.* **12**(1), 21.

Kontak, D. J., and Jackson, S. (1995). Laser-ablation ICP-MS micro-analysis of calcite cement from a Mississippi-Valley-type Zn-Pb deposit, Nova Scotia: Dramatic variability in REE content. *Can. Mineral.* **33**(2), 445.

Koons, R. D. (1998). Analysis of gunshot primer residue collection swabs by inductively coupled plasma-mass spectrometry. *J. Forensic Sci.* **43**(4), 748.

Kopajtic, Z., Roellin, S., Wernli, B., and Hochstrasser, C. (1995). Determination of trace element impurities in nuclear materials by inductively coupled plasma mass spectrometry in a glove-box. *J. Anal. At. Spectrom.* **10**(11), 947.

Koplik, R., Curdova, E. E., and Suchanek, M. (1998). Trace element analysis in CRM of plant origin by inductively coupled plasma mass spectrometry. *Fresenius' J. Anal. Chem.* **360**(3), 449.

Koplik, R., Mestek, O., Fingerova, H., and Suchanek, M. (1999). Validation protocol for the determination of copper in plant samples by isotope dilution inductively coupled plasma mass spectrometry. *J. Anal. At. Spectrom.* **14**(2), 241.

Kotrebai, M. a. (2000). Characterization of selenoamino acids and other related selenium compounds of medical and nutritional importance. Unpublished Ph.D. Thesis, University of Massachusetts, Amherst.

Kotrebai, M., Bird, S. M., Tyson, J. F., and Block, E. (1999). Characterization of selenium species in biological extracts by enhanced ion-pair liquid chromatography with inductively coupled plasma-mass spectrometry and by referenced electrospray ionization-mass spectrometry. *Spectrochim. Acta, Part B* **54**(11), 1573.

Koyama, H., Omura, K., Ejima, A., Kasanuma, Y., Watanabe, C., and Satoh, H. (1999). Separation of selenium-containing proteins in human and mouse plasma using tandem high-performance liquid chromatography columns coupled with inductively coupled plasma-mass spectrometry. *Anal. Biochem.* **267**(1), 84.

Kozerski, G. E., Fiorentino, M. A., and Ketterer, M. E. (1997). Determination of aqueous Fe III/II electron self-exchange rates using enriched stable isotope labels, ion chromatography, and inductively coupled plasma mass spectrometry. *Anal. Chem.* **69**(4), 783.

Kozono, S., Itoh, T., Yoshinaga, A., and Ohkawa, S. (1994). Trace analysis of a beryllium window for a solid state detector system by inductively coupled plasma mass spectrometry. *Anal. Sci.: Int. J. Jpn. Soc. Anal. Chem.* **10**(3), 477.

Kozono, S., Sakamoto, H., Takashi, R., and Haraguchi, H. (1998). Multielement determination of ultratrace impurities in high purity tantalum metals by flow injection/inductively coupled plasma mass spectrometry. *Anal. Sci.: Int. J. Jpn. Soc. Anal. Chem.* **14**(4), 757.

Kozono, S., Takashi, R., and Haraguchi, H. (2000). Determination of ultratrace impurities in high purity tantalum materials by on-line anion exchange matrix separation and inductively coupled plasma mass spectrometry. *Anal. Sci. Int. J. Jpn. Soc. Anal. Chem.* **16**(1), 69.

Kozuka, S., Yokote, Y., Abe, K., Hayashi, M., and Matsunaga, H. (1995). Determination of impurities in strontium titanate ceramics by inductively coupled plasma mass spectrometry. *Fresenius' J. Anal. Chem.* **351**(8), 801.

Kozuka, S., Yamada, Y., Takenaka, M., Hayashi, M., and Matsunaga, H. (1997). Determination of uranium and thorium in aluminum by inductively coupled plasma mass spectrometry. *Anal. Sci.: Int. J. Jpn. Soc. Anal. Chem.* **13**(6), 101.

Krachler, M., Radner, H., and Irgolic, K. J. (1996). Microwave digestion methods for the determination of trace elements in brain and liver samples by inductively coupled plasma mass spectrometry. *Fresenius' J. Anal. Chem.* **355**(2), 120.

Krachler, M., Rossipal, E., and Irgolic, K. J. (1998). Trace elements in formulas based on cow and soy milk and in Austrian cow milk determined by inductively coupled plasma mass spectrometry. *Biol. Trace Elem. Res.* **65**(1), 53.

Krachler, M., Alimonti, A., Petrucci, F., Irgolic, K. J., Forastiere, F., and Caroli, S. (1998). Analytical problems in the determination of platinum-group metals in urine by quadrupole and magnetic sector field inductively coupled plasma mass spectrometry. *Anal. Chimi. Acta* **363**(1), 1.

Krachler, M., Alimonti, A., Petrucci, F., and Forastiere, F. (1998). Influence of sample pretreatment on the determination of trace elements in urine by quadrupole and magnetic sector field inductively coupled plasma mass spectrometry. *J. Anal. At. Spectrom.* **13**(8), 701.

Krause, P., Kriews, M., Dannecker, W., Garbe-Schonberg, C. D., and Kersten, M. (1993). Determination of 206/207 Pb isotope ratios by ICP-MS in particulate matter from the North Sea environment. *Fresenius' J. Anal. Chem.* **347**(8–9), 324–329.

Krivan, V., and Theimer, K. H. (1997). Trace characterization of high-purity molybdenum and tungsten by electrothermal atomic absorption spectrometry, inductively coupled plasma atomic emission spectrometry, inductively coupled plasma mass spectrometry and total reflection X-ray fluorescence spectrometry involving analyte-matrix separation. *Spectrochim. Acta, Part B* **52**(14), 2061.

Krupp, E. M., Gruemping, R., Furchtbar, U. R. R., and Hirner, A. V. (1996). Speciation of metals and metalloids in sediments with LTGC/ICP-MS. *Fresenius' J. Anal. Chem.* **354**(5–6), 546.

Krushevska, A., Kotrebai, M., Lasztity, A., Barnes, R. M., and Amarasiriwardena, D. (1996). Application of tertiary amines for arsenic and selenium signal enhancement and polyatomic interference reduction in ICP-MS analysis of biological samples. *Fresenius' J. Anal. Chem.* **355**(7–8), 793.

Krushevska, A., Waheed, S., Nobrega, J., Amarisiriwardena, D., and Barnes, R. M. (1998). Reducing polyatomic interferences in the ICP-MS determination of chromium and vanadium in biofluids and tissues. *Appl. Spectrosc.* **52**(2), 205.

Kumagai, H., Yamanaka, M., Sakai, T., and Yokoyama, T. (1998). Determination of trace metals in sea-water by inductively coupled plasma mass spectrometry interfaced with an ion chromatographic separation system: Effectiveness of nitrilotriacetate chelating resin as the column stationary phase for preconcentration and elimination of matrix effects. *J. Anal. At. Spectrom.* **13**(6), 579.

Kumamaru, T., Yamamoto, M., Nakata, F., and Tsubota, H. (1994). Direct measurement of isotope ratio of antimony in seawater by hydride generation/inductively coupled plasma mass spectrometry. *Anal. Sci.: Int. J. Jpn. Soc. Anal. Chem.* **10**(4), 651.

Kumar, S. J., and Gangadharan, S. (1999). Determination of trace elements in naphtha by inductively coupled plasma mass spectrometry using water-in-oil emulsions. *J. Anal. At. Spectrom.* **14**(6), 967.

Kumar, S. J., Meeravali, N. N., and Arunachalam, J. (1998). Determination of trace impurities in high purity gallium by inductively coupled plasma mass spectrometry and cross validation of results by transverse heated graphite furnace atomic absorption spectrometry. *Anal. Chim. Acta* **371**(2), 305.

Kumar, U., Dorsey, J. G., Caruso, J. A., and Evans, E. H. (1994). Metalloporphyrin speciation by liquid chromatography and inductively coupled plasma-mass spectrometry. *J. Chromatogr. Sci.* **32**(7), 282.

Kunze, J., Koelling, S., Reich, M., and Wimmer, M. A. (1998). ICP-MS Determination of titanium and zirconium in human serum using an ultrasonic nebulizer with desolvator membrane. *At. Spectrosc.* **19**(5), 164.

Kuss, H.-M., and Muller, M. (1995). Spectral interference of polyatomic ions of Cr, Fe, Mn, Mo, Ni, and W in steel using inductively coupled plasma mass spectrometry. *Arch. Eisenhuettenwes.* **66**(12), 516.

Kuss, H. M., Bossman, D., and Mueller, M. (1994). Silicon determination in steel by ICP-MS. *At. Spectrosc.* **15**(4), 148.

Laborda, F., Baxter, M. J., Crews, H. M., and Dennis, J. (1994). Reduction of polyatomic interferences in inductively coupled plasma mass spectrometry by selection of instrumental parameters and using an argon-nitrogen plasma: Effect on multi-element analyses. *J. Anal. At. Spectrom.* **9**(6), 727.

Laborda, F., Medrano, J., and Castillo, J. R. (2000). Data acquisition of transient signals in inductively coupled plasma mass spectrometry. *Anal. Chim. Acta* **407**(1), 301.

Lachas, H., Richaud, R., Jarvis, K. E., and Herod, A. A. (1999). Determination of 17 trace elements in coal and ash reference materials by ICP-MS applied to milligram sample sizes. *Analyst. (London)* **124**(2), 177.

Lachas, H., Richaud, R., Herod, A. A., Dugwell, D. R., and Kandiyoti, R. (2000). Determination of trace elements by inductively coupled plasma mass spectrometry in biomass and fuel oil reference materials using milligram sample sizes. *Rapid Commun. Mass Spectrom.*, pp. 335–343.

Lahaye, Y., Lambert, D., and Walters, S. (1997). Ultraviolet laser sampling and high resolution inductively coupled plasma-mass spectrometry of NIST and BCR-2G glass reference materials. *Geostand. Newsl.* **21**(2), 205.

Laly, S., Nakagawa, K., Arimura, T., and Kimijima, T. (1996). Optimization of electrothermal vaporization, inductively coupled plasma mass spectrometry conditions for the determination of iron, copper, nickel and zinc in semiconductor grade acids. *Spectrochim. Acta, Part B* **51**(11), 1393.

Lam, J. W., and McLaren, J. W. (1990). Use of aerosol processing and nitrogen - argon plasmas for reduction of oxide interference in inductively coupled plasma mass spectrometry. *J. Anal. At. Spectrom.* **5**(6), 419.

Lam, J. W. H., McLaren, J. W., and Methven, B. A. J. (1995). Determination of chromium in biological tissues by inductively coupled plasma mass spectrometry. *J. Anal. At. Spectrom.* **10**(8), 551.

Lamothe, P. J., and Wilson, S. A. (1999). The determination of forty four elements in aqueous samples by inductively coupled plasma-mass spectrometry. *Geo. Sur. Open-File Rep. (U.S.)* **99–151**.

Langer, D., and Holcombe, J. A. (1999). Simple transient extension chamber to permit full mass scans with electrothermal vaporization inductively coupled plasma mass spectrometry. *Appl. Spectrosc.* **53**(10), 1244.

Langer, D., and Holcombe, J. A. (1999). A method for the direct analysis of new and used lubricating oils using electrothermal vaporization inductively coupled plasma mass spectrometry (ETV-ICPMS). *Appl. Spectrosc.* **44**(3), 274.

Lanza, F., and Trincherini, P. R. (2000). The determination of cadmium, lead and vanadium by high resolution ICP-MS in antartic snow samples. *Ann. Chim. (Rome)* **90**(1), 61.

Lapitajs, G., Greb, U., Dunemann, L., Begerow, J., Moens, L., and Verrept, P. (1995). ICP-MS in the determination of trace and ultratrace elements in the human body. *Int. Lab. Eur. Ed.* **25**(4), 21.

Larsen, E. H., Pritzl, G., and Hansen, S. H. (1993). Arsenic speciation in seafood samples with emphasis on minor constituents: An investigation using high-performance liquid chromatography with detection by inductively coupled plasma mass spectrometry. *J. Anal. At. Spectrom.* **8**(8), 1075–1084.

Larsen, E. H., Knuthsen, P., and Hansen, M. (1999). Seasonal and regional variations of iodine in Danish dairy products determined by inductively coupled plasma mass spectrometry. *J. Anal. At. Spectrom.* **14**(1), 41.

Larsen, F. H., and Ludwigsen, M. B. (1997). Determination of iodine in food-related certified reference materials using wet ashing and detection by inductively coupled plasma mass spectrometry. *J. Anal. At. Spectrom.* **12**(4), 435.

Lasenby, D. C., and Veinott, G. (1998). Ingestion of trace metals by the opossum shrimp, Mysis relicta, determined by laser ablation sampling-inductively coupled plasma mass spectrometry. *Can. J. Zool.* **76**(8), 1605.

Lasztity, A., Krushevska, A., Kotrebal, M., and Barnes, R. M. (1995). Arsenic determination in environmental, biological and food samples by inductively coupled plasma mass spectrometry. *J. Anal. At. Spectrom.* **10**(7), 505.

Latkoczy, C., Prohaska, T., Stingeder, G., and Teschler-Nicola, M. (1998). Strontium isotope ratio measurements in prehistoric human bone samples by means of high-resolution inductively coupled plasma mass spectrometry (HR-ICP-MS). *J. Anal. At. Spectrom.* **13**(6), 561.

Le, X.-C., Cullen, W. R., and Reimer, K. J. (1994). Speciation of arsenic compounds by HPLC with hydride generation atomic absorption spectrometry and inductively coupled plasma mass spectrometry detection. *Talanta* **41**(4), 495.

Le, X.-C., Li, X. F., Lai, V., and Ma, M. (1998). Simultaneous speciation of selenium and arsenic using elevated temperature liquid chromatography separation with inductively coupled plasma mass spectrometry detection. *Spectrochim. Acta, Part B* **53**(6–8), 899.

Leach, J. J. (1998). Calibration of laser ablation inductively coupled plasma mass spectrometry using dried solution aerosols for the quantitative analysis of solids. Unpublished M.s. Thesis, Iowa State University, Ames.

Leach, J. J., Allen, L. A., Aeschliman, D. B., and Houk, R. S. (1999). Calibration of laser ablation inductively coupled plasma mass spectrometry using standard additions with dried solution aerosols. *Anal. Chem.* **71**(2), 440.

Le Cornec, F., and Correge, T. (1997). Determination of uranium to calcium and strontium to calcium ratios in corals by inductively coupled plasma mass spectrometry. *J. Anal. At. Spectrom.* **12**(9), 969.

Lee, K. H., Jiang, S. J., and Liu, H. W. (1998). Determination of mercury in urine by electrothermal vaporization isotope dilution inductively coupled plasma mass spectrometry. *J. Anal. At. Spectrom.* **13**(11), 1227.

Lee, K. H., Liu, S. H., and Jiang, S. J. (1998). Determination of cadmium and lead in urine by electrothermal vaporization isotope dilution inductively coupled plasma mass spectrometry. *Analyst (London)* **123**(7), 1557.

Lee, K. H., Oshima, M., Takayanagi, T., and Motomizu, S. (1999). Determination of trace elements in water samples by inductively coupled plasma-mass spectrometry connected with air flow-and water flow-injection techniques. *J. Flow Inject. Anal.*, pp. 255–264.

Lee, K. H., Oshima, M., Takayanagi, T., and Motomizu, S. (2000). Simultaneous determination of lanthanoids and yttrium in rock reference samples by inductively coupled plasma-mass spectrometry coupled with cation exchange pretreatment. *Bull. Chem. Soc. Jpn.* **73**(3), 615.

Lee, K. M., Appleton, J., Cooke, M., Keenan, F., and Sawicka-Kapusta, K. (1999). Determination of nickel in blood, urine and faeces - Inductively coupled plasma mass spectrometry. *Acta Derm.-Venereolo. Suppl.* **395**(1), 10.

Lee, K. M., Appleton, J., Cooke, M., Sawicka-Kapusta, K., and Damek, M. (1999). Development of a method for the determination of heavy metals in calcified tissues by inductively coupled plasma-mass spectrometry. *Fresenius' J. Anal. Chem.* **364**(3), 245.

Leiterer, M., and Muench, U. (1994). Determination of heavy metals in groundwater samples—ICP-MS analysis and evaluation. *Fresenius' J. Anal. Chem.* **350**(4/5), 204.

Leiterer, M., Einax, J. W., Loeser, C., and Vetter, A. (1997). Trace analysis of metals in plant samples with inductively coupled plasma-mass spectrometry. *Fresenius' J. Anal. Chem.* **359**(4–5), 423.

Leloup, C., Marty, P., Dall'Ava, D., and Perdereau, M. (1997). Quantitative analysis for impurities in uranium by laser ablation inductively coupled plasma mass spectrometry: Improvements in the experimental setup. *J. Anal. At. Spectrom.* **12**(9), 945.

Leopold, I., Gunther, D., and Neumann, D. (1998). Application of high performance liquid chromatography—Inductively coupled plasma mass spectrometry to the investigation of phytochelatin complexes and their role in heavy metal detoxification in plants. *Analusis* **26**(6), M28.

Lewen, N., Schenkenberger, M., Larkin, T., and Conder, S. (1995). The determination of palladium in fosinopril sodium (monopril) by ICP-MS. *J. Pharm. Biomed. Anal.* **13**(7), 879.

Lewis, L. A. (1998). 99Tc bioassay by inductively coupled plasma mass spectrometry (ICP-MS). Unpublished Ph.D. Thesis, University of Tennessee, Knoxville.

Lewis, L. A., and Schweitzer, G. K. (1998). 99Tc bioassay: A direct comparison of liquid scintillation radiation detection and ICP-MS mass detection of the 99Tc isotope. *ASTM Spec. Tech. Pub.*, p. 99.

Li, B., Zhang, Y., and Yin, M. (1997). Determination of trace amounts of rare earth elements in high-purity cerium oxide by inductively coupled plasma mass spectrometry after separation by solvent extraction. *Analyst (London)* **122**(6), 543.

Li, B., Sun, Y., and Yin, M. (1999). Determination of cerium, neodymium and samarium in biological materials at low levels by isotope dilution inductively coupled plasma mass spectrometry. *J. Anal. At. Spectrom.* **14**(12), 1843–1848.

Li, C., Chai, C., Li, X., and Mao, X. (1998). Determination of platinum-group elements and gold in two Russian candidate reference materials SCHS-1 and SLg-1 by ICP-MS after fire assay preconcentration. *Geostand. Newsl.* **22**(2), 195.

Li, C., Chai, C., Mao, X., and Ouyang, H. (1998). Chemical speciation study of platinum group elements in geological samples by stepwise dissolution and inductively coupled plasma mass spectrometry. *Anal. Chim. Acta* **374**(1), 93.

Li, F., Goessler, W., and Irgolic, K. J. (1999). Determination of trimethylselenonium iodide, selenomethionine, selenious acid, and selenic acid using high-performance liquid chromatography with on-line detection by inductively coupled plasma mass spectrometry or flame atomic absorption spectrometry. *J. Chromatogr.* **830**(2), 337.

Li, H., Keohane, B. M., Sun, H., and Sadler, P. J. (1997). Determination of bismuth in serum and urine by direct injection nebulization inductively coupled plasma mass spectrometry. *J. Anal. At. Spectrom.* **12**(10), 1111.

Li, Y.-C., and Jiang, S.-J. (1998). Determination of Cu, Zn, Cd and Pb in fish samples by slurry sampling electrothermal vaporization inductively coupled plasma mass spectrometry. *Anal. Chim. Acta* **359**(1–2), 205.

Liang, Q., Jing, H., and Grégoire, D. C. (2000). Determination of trace elements in granites by inductively coupled plasma mass spectrometry. *Talanta* **51**(3), 507–513.

Liao, H. C., and Jiang, S. J. (1999). EDTA as the modifier for the determination of Cd, Hg and Pb in fish by slurry sampling electrothermal vaporization inductively coupled plasma mass spectrometry. *J. Anal. At. Spectrom.* **14**(10), 1583.

Liaw, M. J., and Jiang, S. J. (1996). Determination of copper, cadmium and lead in sediment samples by slurry sampling electrothermal vaporization inductively coupled plasma mass spectrometry. *J. Anal. At. Spectrom.* **11**(8), 555.

Liaw, M. J., Jiang, S. J., and Li, Y. C. (1997). Determination of mercury in fish samples by slurry sampling electrothermal vaporization inductively coupled plasma mass spectrometry. *Spectrochim. Acta, Part B* **52**(6), 779.

Lichte, F. E. (1995). Determination of elemental content of rocks by laser ablation inductively coupled plasma mass spectrometry. *Anal. Chem.* **67**(14), 2479.

Liezers, M., Tye, C. T., Mennie, D., and Koller, D. (1995). Ultra-low level (pg/L) actinide determinations and superior isotope ratio precisions by quadrupole ICP-MS. *ASTM Spec. Tech. Pub.*, p. 61.

Lindemann, T., Prange, A., Dannecker, W., and Neidhart, B. (1999). Simultaneous determination of arsenic, selenium and antimony species using HPLC/ICP-MS. *Fresenius' J. Anal. Chem.* **364**(5), 462.

Liu, H., Montaser, A., and Dolan, S. P. (1996). Evaluation of a low sample consumption, high-efficiency nebulizer for elemental analysis of biological samples using inductively coupled plasma mass spectrometry. *J. Anal. At. Spectrom.* **11**(4), 307.

Llano, J. J. M. de, Andreu, E. J., and Knecht, E. (1996). Use of inductively coupled plasma-mass spectrometry for the quantitation of the binding and uptake of colloidal gold-low-density lipoprotein conjugates by cultured cells. *Anal. Biochem.* **243**(2), 210.

Llorente, I., Gomez, M., and Camara, C. (1997). Improvement of selenium determination in water by inductively coupled plasma mass spectrometry through use of organic compounds as matrix modifiers. *Spectrochim. Acta, Part B* **52**(12), 1825.

Lofthouse, S. D., Greenway, G. M., and Stephen, S. C. (1997). Microconcentric nebuliser for the analysis of small sample volumes by inductively coupled plasma mass spectrometry. *J. Anal. At. Spectrom.* **12**(12), 1373.

Lofthouse, S. D., Greenway, G. M., and Stephen, S. C. (1998). Comparison of inductively coupled plasma mass spectrometry with a microconcentric nebuliser and total reflection X-ray spectrometry for the analysis of small liquid volume samples. *J. Anal. At. Spectrom.* **13**(12), 1333.

Lofthouse, S. D., Greenway, G. M., and Stephen, S. C. (1999). Miniaturisation of a matrix separation/preconcentration procedure for inductively coupled plasma mass spectrometry using 8-hydroxyquinoline immobilised on a microporous silica frit. *J. Anal. At. Spectrom.* **14**(12), 1839–1842.

Longbottom, J. E., Martin, T. D., Edgell, K. W., and Long, S. E. (1994). Determination of trace elements in water by inductively coupled plasma-mass spectrometry: Collaborative study. *J. Assoc. Off. Anal. Chem. Int.* **77**(4), 1004.

Longerich, H. P. (1993). Oxychlorine ions in inductively coupled plasma mass spectrometry: Effect of chlorine speciation as Cl- and ClO4. *J. Anal. At. Spectrom.* **8**(3), 439–444.

Longerich, H. P., Friel, J. K., Fraser, C., Jackson, S. E., and Fryer, B. J. (1990). Analysis of the drinking water of mothers of neutral tube defect infants and of normal infants for 14 selected trace elements by inductively coupled plasma-mass spectrometry (ICP-MS). *Can. J. Spectrosc.* **36**(1), 15.

Longerich, H. P., Friel, J. K., Fraser, C., Jackson, S. E., and Fryer, B. J. (1991). Analysis of the drinking water of mothers of neutral tube defect infants and of normal infants for 14 selected trace elements by inductively coupled plasma-mass spectrometry (ICP-MS). *Can. J. Spectrosc.* **36**(1), 15.

Longerich, H. P., Gunther, D., and Jackson, S. E. (1996). Elemental fractionation in laser ablation inductively coupled plasma mass spectrometry. *Fresenius' J. Anal. Chem.* **355**(5–6), 538.

Lorber, A., Karpas, Z., and Halicz, L. (1996). Flow injection method for determination of uranium in urine and serum by inductively coupled plasma mass spectrometry. *Anal. Chim. Acta* **334**(3), 295.

Lord, C. J. (1994). Determination of lead and lead isotope ratios in gasoline by inductively coupled plasma mass spectrometry. *J. Anal. At. Spectrom.* **9**(5), 599.

Los Alamos National Laboratory. (1997). "Determination of Actinides in WIPP Brines by Inductively Coupled Plasma Mass Spectrometry." Los Alamos National Laboratory, Los Alamos, N.M.

Louie, H., and Soo, S.Y.-P. (1992). Use of nitrogen and hydrogen in inductively coupled plasma mass spectrometry. *J. Anal. At. Spectrom.* **7**(3), 557.

Louise Armstrong, H. E., Corns, W.T., Stockwell, P. B., O'Connor, G., Ebdon, L., and Hywel Evans, E. (1999). Comparison of AFS and ICP-MS detection coupled with gas chromatography for the determination of methylmercury in marine samples. *Anal. Chim. Acta* **390**(1), 245.

Lowe, D. S., and Stahl, R. G. (1992). Determination of trace elements in organic solvents by inductively coupled plasma mass spectrometry. *Anal. Proc.* **29**(7), 277.

Lu, P. L. Huang, K. S., and Jiang, S. J. (1993). Determination of traces of copper, cadmium and lead in biological and environmental samples by flow-injection isotope dilution inductively coupled plasma mass spectrometry. *Anal. Chim. Acta* **284**(1), 181.

Lu, Q. (1997). Speciation and quantification of zinc in biological fluids by capillary electrophoresis and inductively coupled plasma mass spectrometry. Unpublished Ph.D. Thesis, University of Massachusetts, Amherst.

Lu, Q., and Barnes, R. M. (1996). Evaluation of an ultrasonic nebulizer interface for capillary electrophoresis and inductively coupled plasma mass spectrometry. *Microchem. J.* **54**(2), 129.

Luedke, C., Hoffmann, E., Skole, J., and Kriews, M. (1999). Determination of trace metals in size fractionated particles from arctic air by electrothermal vaporization inductively coupled plasma mass spectrometry. *J. Anal. At. Spectrom.* **14**(11), 1685.

Luong, E. T. (1999). Inductively coupled plasma mass spectrometry for stable isotope metabolic tracer studies of living systems. Unpublished Ph.D. Thesis, Iowa State University, Ames.

Luong, E. T., Houk, R. S., and Serfass, R. E. (1997). Chromatographic isolation of molybdenum from human blood plasma and determination by inductively coupled plasma mass spectrometry with isotope dilution. *J. Anal. At. Spectrom.* **12**(7), 703.

Lustig, S., Zang, S., Michalke, B., Schramel, P., and Beck, W. (1997). Platinum determination in nutrient plants by inductively coupled plasma mass spectrometry with special respect to the hafnium oxide interference. *Fresenius' J. Anal. Chem.* **357**(8), 1157.

Lustig, S., Lampaert, D., De Cremer, K., and De Kimpe, J. (1999). Capability of flatbed electrophoresis (IEF and native PAGE) combined with sector field ICP-MS and autoradiography for the speciation of Cr, Ga, In, Pt and V in incubated serum samples. *J. Anal. At. Spectrom.* **14**(9), 1357.

Lyon, T. D. B., and Fell, G. S. (1990). Isotopic composition of copper in serum by inductively coupled plasma mass spectrometry. *J. Anal. At. Spectrom.* **5**(2), 135.

Lyon, T. D. B., Fell, G. S., McKay, K., and Scott, R. D. (1991). Accuracy of multi-element analysis of human tissue obtained at autopsy using inductively coupled plasma mass spectrometry. *J. Anal. At. Spectrom.* **6**, 559.

Machado, N., and Gauthier, G. (1996). Determination of 207Pb/206Pb ages on zircon and monazite by laser-ablation ICPMS and application to a study of sedimentary provenance and metamorphism in southeastern Brazil. *Geochim. Cosmochim. Acta* **60**(24), 5063.

Madeddu, B., and Rivoldini, A. (1996). Analysis of plant tissues by ICP-OES and ICP-MS using an improved microwave oven acid digestion. *At. Spectrosc.* **17**(4), 148.

Magnuson, M. L., Creed, J. T., and Brockhoff, C. A. (1996). Speciation of arsenic compounds by Ion chromatography with inductively coupled plasma mass spectrometry detection utilizing hydride generation with a membrane separator. *J. Anal. At. Spectrom.* **11**(9), 893.

Magnuson, M. L., Creed, J. T., and Brockhoff, C. A. (1996). Speciation of arsenic compounds in drinking water by capillary electrophoresis with hydrodynamically modified electroosmotic flow detected through hydride generation inductively coupled plasma mass spectrometry with a membrane gas-liquid separator. *J. Anal. At. Spectrom.* **12**(7), 689.

Magyar, B., Cousin, H., and Aeschlimann, B. (1992). Combined use of x-ray fluorescence spectrometry and laser ablation inductively coupled plasma mass spectrometry in the analysis of soils. *Anal. Proc.* **29**(7), 282.

Mahalingam, T. R., Vijayalakshmi, S., Prabhu, R. K., Thiruvengadasami, A., Mathews, C. K., and Shanmugasundaram, K. R. (1997). Studies on some trace and minor elements in blood: A survey of the kalpakkam (India) population. Part I: Standardization of analytical methods using ICP-MS and AAS. *Bio. Trace Elem. Res.* **57**(3), 191.

Mahoney, P. P., Ray, S. J., Li, G., and Hieftje, G. M. (1999). Preliminary investigation of electrothermal vaporization sample introduction for inductively coupled plasma time-of-flight mass spectrometry. *Anal. Chem.* **71**(7), 1378.

Maibusch, R., Kuss, H. M., Coedo, A. G., and Dorado, T. (1999). Spark ablation inductively coupled plasma mass spectrometry analysis of minor and trace elements in low and high alloy steels using single calibration curves. *J. Anal. At. Spectrom.* **14**(8), 1155.

Makarov, A., and Szpunar, J. (1998). The coupling of size-exclusion HPLC with ICP-MS in bioinorganic analysis. *Analusis* **26**(6), M44.

Makarov, A., and Szpunar, J. (1999). Species-selective determination of cobalamin analogues by reversed-phase HPLC with ICP-MS detection. *J. Anal. At. Spectrom.* **14**(9), 1323.

Makinson, P. R. (1995). The comparison of sample preparation techniques for the determination of technetium-99 in pure uranium compounds and subsequent analysis by inductively coupled plasma–mass spectrometry (ICP-MS). *ASTM Spec. Tech. Publ.*, p. 7.

Makishima, A., and Nakamura, E. (2000). Determination of titanium at microgram levels in milligram amounts of silicate materials by isotope dilution high resolution inductively coupled plasma mass spectrometry with flow injection. *J. Anal. At. Spectrom.* **15**, 263–268.

Makishima, A., Nakamura, E., and Nakano, T. (1997). Determination of boron in silicate samples by direct aspiration of sample HF solutions into ICPMS. *Anal. Chem.* **69**(18), 3754.

Mallory-Greenough, L. M., Greenough, J. D., and Owen, J. V. (1998). New data for old pots: Trace-element characterization of ancient Egyptian pottery using ICP-MS. *J. Archaeol. Sci.* **25**(1), 85.

Mank, A. J. G., and Mason, P. R. D. (1999). A critical assessment of laser ablation ICP-MS as an analytical tool for depth analysis in silica-based glass samples. *J. Anal. At. Spectrom.* **14**(8), 1143.

Manninen, P. K. G. (1994). Determination of extractable organic chlorine by electrothermal vaporization inductively coupled plasma mass spectrometry. *J. Anal. At. Spectrom.* **9**(3), 209–212.

Mannio, J., Jaervinen, O., Tuominen, R., and Verta, M. (1995). Survey of trace elements in lake waters of Finnish Lapland using the ICP-MS technique. *Sci. Total Environ.* **160**(161), 433.

Marabini, A. M., Passariello, B., and Barbaro, M. (1992). Determination of rate-earth elements in minerals and ores by inductively coupled plasma-mass spectrometry (ICP-MS). *Mater. Chem. Phys.* **31**(1–2), 101–106.

Marabini, A. M., Passariello, B., and Barbaro, M. (1992). Inductively coupled plasma-mass spectrometry: Capabilities and applications. *Microchem. J.* **46**(3), 302–312.

Marawi, I. (1994). Hydride preconcentration on palladium and subsequent determination by inductively coupled plasma mass spectrometry. Unpublished Ph.D. Thesis, University of Cincinnati, Cincinnati, OH.

Marawi, I., Wang, J., and Caruso, J. A. (1994). Graphite furnace hydride preconcentration and subsequent detection by inductively coupled plasma mass spectrometry. *Anal. Chim. Acta* **291**(1–2), 127.

Marawi, I., Olson, L. K., Wang, J., and Caruso, J. A. (1995). Utilization of metallic platforms in electrothermal vaporization inductively coupled plasma mass spectrometry. *J. Anal. At. Spectrom.* **10**(1), 7.

Marin, B., Valladon, M., Polve, M., and Monaco, A. (1997). Reproducibility testing of a sequential scheme for the determination of trace metal speciation in a marine reference sediment by inductively coupled plasma-mass spectrometry. *Anal. Chim. Acta* **342**(2–3), 91.

Marshall, J., and Franks, J. (1990). Multielement analysis and reduction of spectral interferences using electrothermal vaporization inductively coupled plasma-mass spectrometry. *At. Spectrosc.* **11**(5), 177.

Marshall, J., and Franks, J. (1991). Matrix interferences from methacrylic Acid solutions in inductively coupled plasma mass spectrometry. *J. Anal. Spectrom.* **6**(8), 591–600.

Marshall, J., Franks, J., Abell, I., and Tye, C. (1991). Determination of trace elements in solid plastic materials by laser ablation inductively coupled plasma mass spectrometry. *J. Anal. At. Spectrom.* **6**(2), 145–151.

Martin, G. J., Fournier, J. B., Allain, P., Mauras, Y., and Aguile, L. (1997). Optimization of analytical methods for origin assessment of orange juices. II. ICP-MS determination of trace and ultra-trace elements. *Analusis* **25**(1), 7.

Martin, M., and Volmer, D. A. (1999). Examining the analytical capabilities of a desolvating microconcentric nebulizer for inductively coupled plasma-mass spectrometry. *Rapid Commun. Mass Spectrom.* **13**(1), 84.

Martin-Esteban, A., Fernandez, P., Perez-Conde, C., and Gutierrez, A. (1995). On-line preconcentration of aluminium with immobilized Chromotrope 2B for the determination by flame atomic absorption spectrometry and inductively coupled plasma mass spectrometry. *Anal. Chim. Acta* **304**(1), 121.

Mason, A. Z., and Storms, S. D. (1993). Applications of directly coupled SE-HPLC/ICP-MS in environmental toxicology studies: A study of metal-ligand interactions in cytoplasmic samples. *Mar. Environ. Res.* **35**(1–2), 19.

Mason, P. R. D., Kaspers, K., and Van Bergen, M. J. (1999). Determination of sulfur isotope ratios and concentrations in water samples using ICP-MS incorporating hexapole ion optics. *J. Anal. At. Spectrom.* **14**(7), 1067.

Matschat, R., and Czerwensky, M. (1997). Trace analysis of selected high-purity metals using high-resolution inductively coupled plasma mass spectrometry and inductively coupled plasma optical emission spectrometry. *Phys. Status Solidi A* **161**(1), 567.

Matschat, R., Czerwensky, M., Hamester, M., and Pattberg, S. (1997). Investigations concerning the analysis of high-purity metals (Cd, Cu, Ga and Zn) by high resolution inductively coupled plasma mass spectrometry. *Fresenius' J. Anal. Chemi.* **359**(4–5), 418.

Matsunaga, T., Ishii, T., and Watanabe, H. (1996). Speciation of water-soluble boron compounds in radish roots by size exclusion HPLC/ICP-MS. *Anal. Sci.: Int. J. Jpn. Soc. Anal. Chem.* **12**(4), 673.

Matsunaga, T., Ishii, T., and Watanabe-Oda, H. (1997). HPLC/ICP-MS study of metals bound to borate-rhamnogalacturonan- II from plant cell walls. *Dev. Plant Soil Sci.*, p. 81.

Matsuno, K., Kawamoto, T., Kayama, F., and Kodama, Y. (1995). Determination of selenium in blood by ICP-MS. *Nippon Eiseigaku Zasshi* **50**(1), 1–25.

Matsuoka, N., Kawamura, H., Saeki, K., Koike, M., Momoshima, N., and Okabe, H. (1996). Application of ICP-MS to the analysis of lead isotope ratios in an ancient bronze mirror. *Bunseki Kagaku* **45**(2), 201.

Mattusch, J., and Wennrich, R. (1998). Determination of anionic, neutral, and cationic species of arsenic by ion chromatography with ICPMS detection in environmental samples. *Anal. Chem.* **70**(17), 3649.

Mauras, Y., Premel-Cabic, A., Berre, S., and Allain, P. (1993). Simultaneous determination of lead, bismuth and thallium in plasma and urine by inductively coupled plasma mass spectrometry. *Clin. Chim. Acta* **218**(2), 201.

May, T. W., and Wiedmeyer, R. H. (1998). The CETAC ADX-500 autodiluter system: A study of dilution performance with the ELAN 6000 ICP-MS and ELAN software. *At. Spectrosc.* **19**(5), 143.

May, T. W., and Wiedmeyer, R. H. (1998). A table of polyatomic interferences in ICP-MS. *At. Spectrosc.* **19**(5), 150.

May, T. W., Wiedmeyer, R. H., Brumbaugh, W. G., and Schmidt, C. J. (1997). The determination of metals in sediment pore waters and in 1N HCl-extracted sediments by ICP-MS. *At. Spectrosc.* **18**(5), 133.

May, T. W., Wiedmeyer, R. H., Brown, L. D., and Casteel, S. W. (1999). A lead isotope distribution study in swine tissue using ICP-MS. *At. Spectrosc.* **20**(6), 199.

McCandless, T. E., Baker, M. E., and Ruiz, J. (1997). Trace element analysis of natural gold by laser ablation ICP-MS: A combined external/internal standardisation approach. *Geostand. Newsl.* **21**(2), 271.

McCandless, T. E., Lajack, D. J., Ruiz, J., and Ghazi, A. M. (1997). Trace element determination of single fluid inclusions in quartz by laser ablation ICP-MS. *Geostand. Newsl.* **21**(2), 279.

McCartney, M., Rajendran, K., Olive, V., Busby, R. G., and McDonald, P. (1999). Development of a novel method for the determination of 0909Tc in environmental samples by ICP-MS. *J. Anal. At. Spectrom.* **14**(12), 1849–1852.

McDonald, I., Harris, J. W., and Vaughan, D. J. (1996). Determination of noble metals in sulphide inclusions from diamonds using inductively coupled plasma-mass spectrometry. *Anal. Chim. Acta* **333**(1–2), 41.

McElroy, F., Mennito, A., Debrah, E., and Thomas, R. (1998). Uses and applications of inductively coupled plasma mass spectrometry in the petrochemical industry. *Spectroscopy* **13**(2), 42.

McGuire, J. S., and Hite, D. A. (1998). Determination of micronutrients in feed products by inductively coupled plasma mass spectrometry. *J. Assoc. Off. Anal. Chem. Int.* **81**(5), 923.

McIntyre, R. S. C., Gringoire, D. C., and Chakrabarti, C. L. (1997). Vaporization of radium and other alkaline earth elements in electrothermal vaporization inductively coupled plasma mass spectrometry. *J. Anal. At. Spectrom.* **12**(5), 547.

McKay, K. (1993). New techniques in the pharmacokinetic analysis of cancer drugs. II. The ultratrace determination of platinum in biological samples by inductively coupled plasma-mass spectrometry. *Cancer Surv.*, pp. 407–414.

McKelvey, B. A., and Orians, K. J. (1998). The determination of dissolved zirconium and hafnium from seawater using isotope dilution inductively coupled plasma mass spectrometry. *Mar. Chem.* **60**(3), 245.

McLaren, J. W. (1993). An ICP-MS applications bibliography update. *At. Spectrosc.* **14**(6), 191–204.

McLaren, J. W., Lam, J. W., and Gustavsson, A. (1990). Evaluation of a membrane interface sample introduction system for inductively coupled plasma mass spectrometry. *Spectrochim. Acta, Part B* **45**(9), 1091.

McLaren, J. W., Lam, J. W. H., Berman, S. S., Akatsuka, K., and Azeredo, M. A. (1993). On-line method for the analysis of sea-water for trace elements by inductively coupled plasma mass spectrometry. Plenary lecture. *J. Anal. At. Spectrom.* **8**(2), 279–286.

McLaren, J. W., Methven, B. A. J., Lam, J. W. H., and Berman, S. S. (1995). The use of inductively coupled plasma mass spectrometry in the production of environmental certified reference materials. *Mikrochim. Acta* **119**(3–4), 287.

McLean, J. A., Zhang, H., and Montaser, A. (1998). A direct injection high-efficiency nebulizer for inductively coupled plasma mass spectrometry. *Anal. Chem.* **70**(5), 1012.

McSheehy, S., and Szpunar, J. (2000). Speciation of arsenic in edible algae by bi-dimensional size-exclusion anion exchange HPLC with dual ICP-MS and electrospray MS/MS detection. *J. Anal. At. Spectrom.* 79–88.

Menegario, A. A., and Gine, M. F. (1997). On-line removal of anions for plant analysis by inductively coupled plasma mass spectrometry. *J. Anal. At. Spectrom.* **12**(6), 671.

Menegario, A. A., and Gine, M. F. (2000). Rapid sequential determination of arsenic and selenium in waters and plant digests by hydride generation inductively coupled plasma-mass spectrometry. *Spectrochim. Acta, Part B* **55**(4), 355–362.

Menegario, A. A., Gine, M. F., Bendassolli, J. A., and Bellalo, A. C. S. (1998). Sulfur isotope ratio (34S:32S) measurements in plant material by inductively coupled plasma mass spectrometry. *J. Anal. At. Spectrom.* **13**(9), 1065.

Mestek, O. (1998). Determination of arsenic in high-saline mineral waters by inductively coupled plasma mass spectrometry inorganic and organometallic chemistry. *Collect. Czech. Chem. Commun.* **63**(3), 347.

Mestek, O., Curdova, E., Koplik, R., and Zima, T. (1997). Direct determination of Cu and Zn in whole human blood by ICP-MS. *Chem. Listy* **91**(12), 1059.

Mestek, O., Koplik, R., Fingerova, H., and Suchanek, M. (2000). Determination of thallium in environmental samples by inductively coupled plasma mass spectrometry: Comparison and validation of isotope dilution and external calibration methods. *J. Anal. At. Spectrom.*, 403–408.

Michalke, B. (1999). Potential and limitations of capillary electrophoresis inductively coupled plasma mass spectrometry. *J. Anal. At. Spectrom.* **14**(9), 1297.

Michalke, B. (2000). Advantages and improvements in selenium speciation—Improvements in coupling capillary electrophoresis on-line to inductively coupled plasma mass spectrometry are described in the field of selenium speciation. *Spectroscopy* **15**(4), 30.

Michalke, B., and Schramel, P. (1997). Coupling of capillary electrophoresis with ICP-MS for speciation investigations. *Fresenius' J. Anal. Chem.* **357**(6), 594.

Michalke, B., and Schramel, P. (1997). Hyphenation of capillary electrophoresis to inductively coupled plasma mass spectrometry as an element-specific detection method for metal specification. *J. Chromatogr.* **750**(1–2), 51.

Michalke, B., and Schramel, P. (1998). The coupling of capillary electrophoresis to ICP-MS. *Analusis* **26**(6), M51.

Michalke, B., and Schramel, P. (1998). Selenium speciation by interfacing capillary electrophoresis with inductively coupled plasma-mass spectrometry. *Electrophoresis* **19**(2), 270.

Michalke, B., and Schramel, P. (1999). Detection—Iodine speciation in biological samples by capillary electrophoresis—Inductively coupled plasma mass spectrometry. *Electrophoresis* **20**(12), 2547.

Michalke, B., Schramel, O., and Kettrup, A. (1999). Capillary electrophoresis coupled to inductively coupled plasma mass spectrometry (CE/ICP-MS) and to electrospray ionization mass spectrometry (CE/ESI-MS): An approach for maximum species information in speciation of selenium. *Fresenius' J. Anal. Chem.* **363**(5), 456.

Miekeley, N., and Anlato, M. O. (1997). Fast Hg determination in biological samples by ICP-MS using minitube furnace catalytic combustion (MFCC). *At. Spectrosc.* **18**(6), 186.

Milgram, K. E. (1997). Abatement of spectral interferences in elemental mass spectrometry: design and construction of inductively coupled plasma ion sources for Fourier transform ion cyclotron resonance instrumentation. Unpublished Ph.D. Thesis, University of Florida, Talahassee.

Milgram, K. E., White, F. M., Goodner, K. L., Watson, C. H., Koppenaal, D. W., Barinaga, C. J., Smith, B. H., Winefordner, J. D., Marshall, A. G., Houk, R. S., and Eyler, J. R. (1997). High-resolution inductively coupled plasma Fourier transform ion cyclotron resonance mass spectrometry. *Anal. Chem.* **69**(18), 3714.

Milton, D. A., and Chenery, S. R. (1998). The effect of otolith storage methods on the concentrations of elements detected by laser-ablation ICPMS. *J. Fish Biol.* **53**(4), 785.

Ming, Y., and Bing, L. (1998). Determination of rare earth elements in human hair and wheat flour reference materials by inductively coupled plasma mass spectrometry with dry ashing and microwave digestion. *Spectrochim. Acta, Part B* **53**(10), 1447.

Ming-Tsai, W., and Shiuh-Jen, J. (1999). Determination of mercury in urine and seawater by flow injection vapor generation isotope dilution inductively coupled plasma mass spectrometry. *J. Chin. Chem. Soc.*, pp. 871–878.

Minnich, M. G. (1996). Unique applications of solvent removal in inductively coupled plasma mass spectrometry. Unpublished M.S. Thesis, Iowa State University, Ames.

Minnich, M. G., and Houk, R. S. (1998). Comparison of cryogenic and membrane desolvation for attenuation of oxide, hydride and hydroxide ions and ions containing chlorine in inductively coupled plasma mass spectrometry. *J. Anal. At. Spectrom.* **13**(3), 167.

Minnich, M. G., Houk, R. S., Woodin, M. A., and Christiani, D. C. (1997). Method to screen urine samples for vanadium by inductively coupled plasma mass spectrometry with cryogenic desolvation. *J. Anal. At. Spectrom.* **12**(12), 1345.

Mitterrand, B., Leprovost, P., Delaunay, J., and Vian, A. M. (1998). Determination of technetium-99, neptunium-237 and isotopes of thorium in uranyl nitrate solutions from a reprocessing plant, using double-focusing ICP-MS. *ASTM Spec. Tech. Pub.*, p. 64.

Miyatani, T., Suzuki, H., and Yoshimoto, O. (1994). Determination of trace amounts of boron in high purity graphite by ICP-MS. *Bunseki Kagaku* **43**(6), 495.

Miyazaki, A., and Reimer, R. A. (1993). Determination of lead isotope ratios and concentrations in sea-water by inductively coupled plasma mass spectrometry after preconcentration using Chelex-100. *J. Anal. At. Spectrom.* **8**(3), 449–452.

Moberg, L., Pettersson, K., Gustavsson, I., and Karlberg, B. (1999). Determination of cadmium in fly ash and metal alloy reference materials by inductively coupled plasma mass spectrometry and chemometrics. *J. Anal. At. Spectrom.* **14**(7), 1055.

Mochizuki, T., Sakashita, A., Iwata, H., Ishibashi, Y., and Gunji, N. (1991). Application of slurry nebulization to trace elemental analysis of some biological samples by inductively coupled plasma mass spectrometry. *Fresenius' J. Anal. Chem.* **339**(12), 889.

Moenke-Blankenburg, L., Schumann, T., Gunther, D., Kuss, H.-M., and Paul, M. (1992). Quantitative analysis of glass using inductively coupled plasma atomic emission and mass spectrometry, laser micro-analysis inductively coupled plasma atomic emission spectrometry and laser ablation inductively coupled plasma mass spectrometry. *J. Anal. At. Spectrom.* **7**(2), 251–254.

Moens, L., and Dams, R. (1995). NAA and ICP-MS: A comparison between two methods for trace and ultra-trace element analysis. *J. Radioanal. Nucl. Chem.* **192**(1), 29.

Moens, L., and Jakubowski, N. (1998). Double-focusing mass spectrometers in ICPMS— From its very first days, the Achilles' heel of inductively coupled plasma MS has been the number of spectroscopic and nonspectroscopic interferences. The only general method to overcome this limitation is to go to high mass resolution with a double-focusing instrument. *Anal. Chem.* **70**(7), 251A.

Moens, L., Vanhoe, H., Vanhaecke, F., Goossens, J., Campbell, M., and Dams, R. (1994). Application of inductively coupled plasma mass spectrometry to the certification of reference materials from the community bureau of reference. *J. Anal. At. Spectrom.* **9**(3), 187–192.

Moens, L., Verrept, P., Dams, R., and Greb, U. (1994). New high-resolution inductively coupled plasma mass spectrometry technology applied for the determination of V, Fe, Cu, Zn and Ag in human serum. *J. Anal. At. Spectrom.* **9**(9), 1075.

Moens, L., Verrept, P., Boonen, S., and Vanhaecke, F. (1995). Solid sampling electrothermal vaporization for sample introduction to inductively coupled plasma atomic emission spectrometry and inductively coupled plasma mass spectrometry. *Spectrochim. Acta, Part B* **50**(4/7), 463.

Moens, L., Smaele, T. D., Dams, R., Van den Broeck, P., and Sandra, P. (1997). Sensitive, simultaneous determination of organomercury, -lead, and -tin compounds with headspace solid phase microextraction capillary gas chromatography combined with inductively plasma mass spectrometry. *Anal. Chem.* **69**(8), 1604.

Mohamad Ghazi, A., McCandless, T. E., Vanko, D. A., and Ruiz, J. (1996). New quantitative approach in trace elemental analysis of single fluid inclusions: Applications of laser ablation inductively coupled plasma mass spectrometry (LA-ICP-MS). *J. Anal. At. Spectrom.* **11**(9), 667.

Moller, P., Dulski, P., and Luck, J. (1992). Determination of rare earth elements in sea-water by inductively coupled plasma-mass spectrometry. *Spectrochim. Acta, Part B* **47B**(12), 1379.

Momoshima, N., and Shimata, S. (1996). Production of 95Tc tracer for 0909Tc analysis by ICP-MS. *Radioisotopes* **45**(8), 511.

Momoshima, N., Sayad, M., and Takashima, Y. (1993). Analytical procedure for technetium-99 in seawater by ICP-MS. *Radiochim. Acta*, pp. 73–78.

Momoshima, N., Kakiuchi, H., Maeda, Y., Hirai, E., and Ono, T. (1997). Identification of the contamination source of plutonium in environmental samples with isotopic ratios determined by inductively coupled plasma mass spectrometry and alpha- spectrometry. *J. Radioanal. Nucl. Chem.* **222**(1–2), 213.

Montaser, A., ed. (1998). "Inductively Coupled Plasma Mass Spectrometry." Wiley, New York.

Montaser, A., Hsiaoming, T., Ishii, I., Sang-Ho, N., and Mingxiang, C. (1991). Argon inductively coupled plasma mass spectrometry with thermospray, ultrasonic, and pneumatic nebulization. *Anal. Chem.* **63**(22), 2660–2664.

Montes Bayon, M., Garcia Alonso, J. I., and Sanz-Medel, A. (1998). Semiquantitative elemental analysis of water samples using double focusing inductively coupled plasma mass spectrometry. *J. Anal. At. Spectrom.* **13**(9), 1027.

Moor, C., Boll, P., and Wiget, S. (1997). Determination of impurities in micro-amounts of silver alloys by electrothermal vaporization inductively coupled plasma mass spectrometry (ETV-ICP-MS) after in-situ digestion in the graphite furnace. *Fresenius' J. Anal. Chem.* **359**(4–5), 404.

Moor, C., Devos, W., Guecheva, M., and Kobler, J. (2000). Inductively coupled plasma mass spectrometry: A versatile tool for a variety of different tasks. *Fresenius' J. Anal. Chem.* **366**(2), 159.

Mora, J., Canals, A., Hernandis, V., and Van Veen, E. H. (1998). Evaluation of a microwave desolvation system in inductively coupled plasma mass spectrometry with low acid concentration solutions. *J. Anal. At. Spectrom.* **13**(3), 175.

Mora, J., Gras, L., van Veen, E. H., and de Loos-Vollebregt, M. T. C. (1999). Electrothermal vaporization of mineral acid solutions in inductively coupled plasma mass spectrometry: Comparison with sample nebulization. *Spectrochim. Acta, Part B* **54**(6), 959.

Moreton, J. A., and Delves, H. T. (1998). Simple direct method for the determination of total mercury levels in blood and urine and nitric acid digests of fish by inductively coupled plasma mass spectrometry. *J. Anal. At. Spectrom.* **13**(7), 659.

Moreton, J. A., and Delves, H. T. (1999). Use of Virkon as a disinfectant for clinical samples carrying a high risk of infection in inductively coupled plasma mass spectrometry. *J. Anal. At. Spectrom.* **14**(5), 893.

Moreton, J. A., and Delves, H. T. (1999). Measurement of total boron and 10B concentration and the detection and measurement of elevated 10B levels in biological samples by inductively coupled plasma mass spectrometry using the determination of 10B:11B ratios. *J. Anal. At. Spectrom.* **14**(10), 1545.

Morikawa, H. (1994). Studies on the analysis of ceramics by inductively coupled plasma atomic emission spectrometry and secondary ion mass spectrometry. *Bunseki Kagaku* **43**(5), 431.

Morita, H., Kita, T., Umeno, M., and Morita, M. (1994). Analysis of serum elements and the contaminations from devices used for serum preparation by inductively coupled plasma mass spectrometry. *Sci. Total Environ.* **151**(1), 9.

Morita, M., Ito, H., Lincheid, M., and Otsuka, K. (1994). Resolution of interelement spectral overlaps by high-resolution inductively coupled plasma mass spectrometry. *Anal. Chem.* **66**(9), 1588.

Morita, S., Kim, C. K., Takaku, Y., Seki, R., and Ikeda, N. (1991). Determination of technetium-99 in environmental samples by inductively coupled plasma mass spectrometry. *Int. J. Radia. Appl. Instrum., Part A* **42**(6), 531.

Morita, S., Tobita, K., and Kurabayashi, M. (1993). Determination of technetium-99 in environmental samples by inductively coupled plasma mass spectrometry. *Radiochim. Acta*, pp. 63–68.

Morrison, C. A., Lambert, D. D., Morrison, R. J. S., Ahlers, W. W., and Nicholls, I. A. (1995). Laser ablation—inductively coupled plasma—mass spectrometry: an investigation of elemental responses and matrix effects in the analysis of geostandard materials. *Chem. Geo.* **119**(1–4), 13.

Morse, C. E. (1994). The use of mixed gas plasma for trace elemental analysis of metal alloys using inductively coupled plasma mass spectrometry. Unpublished M.S. Thesis, Marshall University, Huntington, WV.

Mortlock, R. A., and Froelich, P. N. (1996). Determination of germanium by isotope dilution-hydride generation inductively coupled plasma mass spectrometry. *Anal. Chim. Acta* **332**(2–3), 277.

Motelica-Heino, M., Le Coustumer, P., Thomassin, J. H., and Gauthier, A. (1998). Macro and microchemistry of trace metals in vitrified domestic wastes by laser ablation ICP-MS and scanning electron microprobe X-ray energy dispersive spectroscopy. *Talanta* **46**(3), 407.

Mukai, H., Ambe, Y., and Morita, M. (1990). Flow injection inductively coupled plasma mass spectrometry for the determination of platinum in airborne particulate matter. *J. Anal. At. Spectrom.* **5**(1), 75.

Mulligan, K. J., Davidson, T. M., and Caruso, J. A. (1990). Feasibility of the direct analysis of urine by inductively coupled argon plasma mass spectrometry for biological monitoring of exposure to metals. *J. Anal. At. Spectrom.* **5**(4), 301.

Munker, C. (1998). Nb/Ta fractionation in a Cambrian arc/back arc system, New Zealand: Source constraints and application of refined ICPMS techniques. *Chem. Geol.* **144**(1–2), 23.

Munksgaard, N. C., and Parry, D. L. (1998). Lead isotope ratios determined by ICP-MS: Monitoring of mining-derived metal particulates in atmospheric fallout, Northern Territory, Australia. *Sci. Total Environ.* **217**(1), 113.

Munksgaard, N. C., Batterham, G. J., and Parry, D. L. (1998). Lead isotope ratios determined by ICP-MS: Investigation of anthropogenic lead in seawater and sediment from the Gulf of Carpentaria, Australia. *Mar. Pollu. Bull.* **36**(7), 527.

Muramatsu, Y., and Yoshda, S. (1995). Determination of 1291 and 1271 in environmental samples by neutron activation analysis (NAA) and inductively coupled plasma mass spectrometry (ICP-MS). *J. Radioanal. Nucl. Chem.* **197**(1), 149.

Muramatsu, Y., and Yoshida, S. (1999). Application of ICP-MS to the analyses of Pu, U and Th in environmental samples. *Radioisotopes* **48**(7), 472.

Muramatsu, Y., Uchida, S., Tagami, K., and Yoshida, S. (1999). Determination of plutonium concentration and its isotopic ratio in environmental materials by ICP-MS after separation using ion-exchange and extraction chromatography. *J. Anal. At. Spectrom.* **14**(5), 859.

Murillo, M., Carrion, N., and Chirinos, J. (1993). Determination of sulfur in crude oils and related materials with a parr bomb digestion method and inductively coupled plasma mass spectrometry. *J. Anal. At. Spectrom.* **8**(3), 493–496.

Murphy, D. M., Garbarino, J. R., Taylor, H. E., Hart, B. T. and Beckett, R. (1993) Determination of size and element composition distributions of complex colloids by sedimentation field-flow fractionation-inductively coupled plasma mass spectrometry. *J. Chromatogr.* **642**, 459.

Murphy, G. M., Mawhinney, H., and Parker, W. (1991). Some practical considerations in the design and construction of an ICP-MS facility. *At. Spectrosc.*, pp. 225–227.

Murphy, K. E., and Paulsen, P. J. (1995). The determination of lead in blood using isotope dilution inductively coupled plasma mass spectrometry. *Fresenius' J. Anal. Chem.* **352**(1–2), 203.

MuThiz, C. S., Marchante-Gayon, J. M., Alonso, J. I. G., and Sanz-Medel, A. (1998). Comparison of electrothermal atomic absorption spectrometry, quadrupole inductively coupled plasma mass spectrometry and double-focusing sector field inductively coupled plasma mass spectrometry for the determination of aluminium in human serum. *J. Anal. At. Spectrom.* **13**(4), 283.

MuThiz, C. S., Marchante-Gayon, J. M., Alonso, J. I. G., and Sanz-Medel, A. (1999). Multi-elemental trace analysis of human serum by double-focusing ICP-MS. *J. Anal. At. Spectrom.* **14**(2), 193.

MuThoz Olivas, R., Quetel, C. R., and Donard, O. F. X. (1995). Sensitive determination of selenium by inductively coupled plasma mass spectrometry with flow injection and hydride generation in the presence of organic solvents. *J. Anal. At. Spectrom.* **10**(10), 865.

Muto, H., Abe, T., Takizawa, Y., Kawabata, K., and Yamaguchi, K. (1994). Simultaneous multi-elemental analysis of daily food samples by inductively coupled plasma mass spectrometry. *Sci. Total Environ.* **144**(1–3), 231.

Muto, T. (1995). Determination of some long-lived radionuclide by inductively coupled plasma mass spectrometry. *Radioisotopes* **44**(7), 497.

Nageotte, S. M., and Day, J. P. (1998). Lead concentrations and isotope ratios in street dust determined by electrothermal atomic absorption spectrometry and inductively coupled plasma mass spectrometry. *Analyst (London)* **123**(1), 59.

Naka, H., and Grégoire, D. C. (1996). Determination of trace amounts of sulfur in steel by electrothermal vaporization—Inductively coupled plasma mass spectrometry. *J. Anal. At. Spectrom.* **11**(5), 359.

Naka, H., and Kurayasu, H. (1993). Determination of trace impurities in high-purity quartz by inductively coupled plasma mass spectrometry. *ISIJ Int.* **33**(12), 1252.

Naka, H., and Kurayasu, H. (1996). Determination of trace impurities in silicon carbide by ICP-MS. *Bunseki Kagaku* **45**(12), 1139.

Nakamoto, Y., and Tomiyama, T. (1997). Highly sensitive determination of antimony and selenium in crude oils by oxygen-bomb combustion/ICP-MS. *Anal. Sci.: Int. J. Jpn. Soc. Anal. Chem.* **13**(4), 665.

Nam, S.-H. (1995). Elemental analysis with helium inductively coupled plasma mass spectrometry. Unpublished Ph.D. Thesis, George Washington University, Washington, DC.

Nam, S.-H., Masamaba, W. R. L., and Montaser, A. (1993). Investigation of helium inductively coupled plasma mass spectrometry for the detection of metals and nonmetals in aqueous solutions. *Anal. Chem.* **65**(20), 2784.

Nam, S.-H., Lim, J. S., and Montaser, A. (1994). High-efficiency nebulizer for argon inductively coupled plasma mass spectrometry. *J. Anal. At. Spectrom.* **9**(12), 1357.

Nam, S.-H., Masamba, W. R. L., and Montaser, A. (1994). Helium inductively coupled plasma-mass spectrometry: Studies of matrix effects and the determination of arsenic and selenium in urine. *Spectrochim. Acta, Part B* **49**(12//14), 1325.

Nam, S. H., Zhang, H., Cai, M., Lim, J. S., and Montaser, A. (1996). A status report on helium inductively coupled plasma mass spectrometry. *Fresenius' J. Anal. Chem.* **355**(5–6), 510.

Naohara, J., and Yamashita, E. (1996). Determination of selenium in the commercially bottled drinking water by ICP-MS. *Nippon Shokuhin Kogyo Gakkaishi* **43**(10), 1133.

Narasaki, H., and Cao, J. Y. (1996). Determination of arsenic and selenium in river water by hydride generation inductively coupled plasma-mass spectrometry with high resolution. *Anal. Sci.: Int. J. Jpn. Soc. Anal. Chem.* **12**(4), 623.

Narine, S. S., Hughes, R., and Slavin, A. J. (1999). The use of inductively coupled plasma mass spectrometry to provide an absolute measurement of surface coverage, and comparison with the quartz crystal microbalance. *Appl. Surf. Sci.* **137**(1), 204.

Narita, Y., Tanaka, S., and Santosa, S. J. (1999). A study on the concentration, distribution, and behavior of metals in atmospheric particulate matter over the North Pacific Ocean by using inductively coupled plasma mass spectrometry equipped with laser ablation. *J. Geophys. Res.* **104**(D21), 26.

Narusawa, Y., and Matsubara, I. (1994). Simultaneous determination of germanium, arsenic and some other trace elements in biological reference materials by inductively coupled plasma mass spectrometry. *Nippon Kagaku Kaishi*, p. 195.

Navarro, S. X., Dziewatkoski, M. P., and Enyedi, A. J. (1999). Isolation of cadmium excluding mutants of *Arabidopsis thaliana* using a vertical Mesh transfer system and ICP-MS. *J. Environ. Sci. Health, Part A* **34**(9), 1797.

Neilsen, J. L., Abildtrup, A., Christensen, J., and Watson, P. (1998). Laser ablation inductively coupled plasma-mass spectrometry in combination with gel electrophoresis: A new strategy for speciation of metal binding serum proteins. *Spectrochim. Acta, Part B* **53**(2), 339.

Nelms, S. M., Greenway, G. M., and Hutton, R. C. (1995). Application of multi-element time-resolved analysis to a rapid on-line matrix separation system for inductively coupled plasma mass spectrometry. *J. Anal. At. Spectrom.* **10**(11), 929.

Nelms, S. M., Greenway, G. M., and Koller, D. (1996). Evaluation of controlled-pore glass immobilized iminodiacetate as a reagent for automated on-line matrix separation for inductively coupled plasma mass spectrometry. *J. Anal. At. Spectrom.* **11**(10), 907.

Nesbitt, R. W., Hirata, T., Butler, I. B., and Milton, J. A. (1997). UV laser ablation ICP-MS: Some applications in the Earth sciences. *Geostand. Newsl.* **21**(2), 231.

Neubauer, K., and Vollkopf, U. (1999). The benefits of a dynamic reaction cell to remove carbon- and chloride-based spectral interferences by ICP-MS. *At. Spectrosc.* **20**(2), 64.

Nguyen, S. N., Miller, P. E., Wild, J. F., and Hickman, D. P. (1996). Simultaneous determination of 237NP, 232Th and U isotopes in urine samples using extraction chromatography, ICP-MS and gamma-ray spectroscopy. *Radioact. Radiochem.* **7**(3), 16.

Nguyen, T. H., Bowman, J., and Leermakers, M. (1998). EDXRF and ICP-MS Analysis of environmental samples. *X-Ray Spectrom.* **27**(4), 265.

Nham, T. T. (1998). American Spectroscopy Laboratory: Typical detection limits for an ICP-MS. *Am. Lab.* **30**(16), 17A.

Nicholson, J. K., Lindon, J. C., Scarfe, G., Wilson, I. D., Abou-Shakra, F., Castro-Perez, J., Eaton, A., and Preece, S. (2000). High-performance liquid chromatography and inductively coupled plasma mass spectrometry (HPLC-ICP-MS) for the analysis of xenobiotic metabolites in rat urine: Application to the metabolites of 4-bromoaniline. *Analyst (London)* **125**, 235–236.

Nicholson, S., Sanders, T. W., and Blaine, L. M. (1993). The determination of low levels of 99Tc in environmental samples by inductively coupled plasma-mass spectrometry. *Sci. Total Environ.* **138**, 275.

Nicola, M., Rosin, C., Tousset, N., and Nicolai, Y. (1999). Trace metals analysis in estuarine and seawater by ICP-MS using on line preconcentration and matrix elimination with chelating resin. *Talanta* **50**(2), 433.

Nirel, P. M. V., and Lutz, T. M. (1991). Plasma source mass spectrometry (ICP-MS) application to multi-element analysis in environmental samples: A practical evaluation. *J. Trace Microprobe Tech.*, p. 95.

Niu, H. (1994). Fundamental studies of the plasma extraction and ion beam formation processes in inductively coupled plasma mass spectrometry. Unpublished Ph.D. Thesis, Iowa State University, Ames.

Niu, H., and Houk, R. S. (1994). Langmuir probe measurements of the ion extraction process in inductively coupled plasma mass spectrometry - I. Spatially resolved determination of electron density and electron temperature. *Spectrochim. Acta, Part B* **49**(12//14), 1283.

Niu, H., and Houk, R. S. (1996). Fundamental aspects of ion extraction in inductively coupled plasma mass spectrometry. *Spectrochim. Acta, Part B* **51**(8), 779.

Nixon, D. E., and Moyer, T. P. (1996). Routine clinical determination of lead, arsenic, cadmium, and thallium in urine and whole blood by inductively coupled plasma mass spectrometry. *Spectrochim. Acta, Part B* **51**(1), 13.

Nixon, D. E., Moyer, T. P., and Burritt, M. F. (1999). The determination of selenium in serum and urine by inductively coupled plasma mass spectrometry: comparison with Zeeman graphite furnace atomic absorption spectrometry. *Spectrochim. Acta, Part B* **54**(6), 931.

Nobrega, J. A., Gelinas, Y., Krushevska, A., and Barnes, R. M. (1997). Determination of elements in biological and botanical materials by inductively coupled plasma atomic emission and mass spectrometry after extraction with a tertiary amine reagent. *J. Anal. At. Spectrom.* **12**(10), 1239.

Nobrega, J. A., Gelinas, Y., Krushevska, A., and Barnes, R. M. (1997). Direct determination of major and trace elements in milk by inductively coupled plasma atomic emission and mass spectrometry. *J. Anal. At. Spectrom.* **12**(10), 1243.

Noltner, T., Maisenbacher, P., and Puchelt, H. (1990). Microwave acid digestion of geological and biological standard reference materials for trace element analysis by inductively coupled plasma-mass spectrometry. *Spectrosc. Int.* **2**(7), 36.

Nonose, N., and Kubota, M. (1998). Determination of metal impurities in sulfamic acid by isotope dilution electrothermal vaporization inductively coupled plasma mass spectrometry. *J. Anal. At. Spectrom.* **13**(2), 151.

Nonose, N. S., Matsuda, N., Fudagawa, N., and Kubota, M. (1994). Some characteristics of polyatomic spectra in inductively coupled plasma mass spectrometry. *Spectrochim. Acta, Part B* **49**(10), 955.

Norman, M. D. (1997). Melting and metasomatism in the continental lithosphere: laser ablation ICPMS analysis of minerals in spinel lherzolites from eastern Australia. *Contrib. Mineral. Petrol.* **130**(3–4), 240.

Norman, M. D., Griffin, W. L., Pearson, N. J., and Garcia, M. O. (1998). Quantitative analysis of trace element abundances in glasses and minerals: a comparison of laser ablation inductively coupled plasma mass spectrometry, solution inductively coupled plasma mass spectrometry, proton microprobe and electron microprobe data. *J. Anal. At. Spectrom.* **13**(5), 477.

Nowinski, P., and Hodge, V. (1994). Evaluation of ICP-MS/Microwave oven preparation for the rapid analysis of ore samples for gold and the platinum-group metals. *At. Spectrosc.* **15**(3), 109.

Nozaki, Y., Alibo, D. S., Amakawa, H., and Gamo, T. (1999). Dissolved rare earth elements and hydrography in the Sulu Sea - ICP-MS determinations in the East Caroline, Coral Sea, and South Fiji basins of the western Pacific. *Geochim. Cosmochim. Acta* **63**(15), 2171.

Nuttall, K. L., Gordon, W. H., and Ash, K. O. (1995). Inductively coupled plasma mass spectrometry for trace element analysis in the clinical laboratory. *Ann. Clin. Lab. Sci.* **25**(3), 264.

Nwogu, V. I. (1991). Electrothermal vaporization sample introduction for inductively coupled plasma atomic emission and inductively coupled plasma mass spectrometry. Unpublished Ph.D. Thesis, Georgia Institute of Technology, Atlanta.

Nyomora, A. M. S., Sah, R. N., Brown, P. H., and Miller, R. O. (1997). Boron determination in biological materials by inductively coupled plasma atomic emission and mass spectrometry: Effects of sample dissolution methods. *Fresenius' J. Anal. Chem.* **357**(8), 1185.

O'Connor, G., Ebdon, L., and Evans, E. H. (1997). Low pressure inductively coupled plasma ion source for molecular and atomic mass spectrometry: The effect of reagent gases. *J. Anal. At. Spectrom.* **12**(11), 1263.

O'Connor, G., Ebdon, L., and Evans, E. H. (1999). Qualitative and quantitative determination of tetraethyllead in fuel using low pressure ICP-MS. *J. Anal. At. Spectrom.* **14**(9), 1303.

Ochsenkuhn-Petropoulou, M., Ochsenkuhn, K., and Luck, J. (1991). Comparison of inductively coupled plasma mass spectrometry with inductively coupled plasma atomic emission spectrometry and instrumental neutron activation analysis for the determination of rare earth elements in Greek bauxites. *Spectrochim. Acta, Part B* **46**(1), 51.

Oedegaard, K., and Lund, W. (1997). Multi-element speciation of tea infusion using cation-exchange separation and size-exclusion chromatography in combination with inductively coupled plasma mass spectrometry. *J. Anal. At. Spectrom.* **12**(4), 403.

Oenning, G., and Bergdahl, I. A. (1999). Fractionation of soluble selenium compounds from fish using size-exclusion chramatography with on-line detection by inductively coupled plasma mass spectrometry. *Analyst (London)* **124**(10), 1435.

Oernemark, U., Taylor, P. D. P., and De Bievre, P. (1997). Certification of the rubidium concentration in water materials for the international measurement evaluation programme (IMEP) using isotope dilution inductively coupled plasma mass spectrometry. *J. Anal. At. Spectrom.* **12**(5), 567.

Oguri, K., Shimoda, G., and Tatsumi, Y. (1999). Quantitative determination of gold and the platinum-group elements in geological samples using improved NiS fire-assay and tellurium coprecipitation with inductively coupled plasma- mass spectrometry (ICP-MS). *Chem. Geo.* **157**(3), 189.

Okamoto, Y. (1999). Alkylating vaporisation of antimony using tungsten boat furnace-sample cuvette technique for inductively coupled plasma mass spectrometry. *J. Anal. At. Spectrom.* **14**(10), 1631.

Okamoto, Y. (2000). Generation of a methylbismuth species and its electrothermal vaporization for the determination of bismuth by inductively coupled plasma mass spectrometry. *Fresenius' J. Anal. Chem.* **366**(3), 309.

Okamoto, Y. (2000). Direct determination of lead in biological samples by electrothermal vaporization inductively coupled plasma mass spectrometry (ETV-ICP-MS) after furnace-fusion in the sample cuvette-tungsten boat furnace. *Fresenius' J. Anal. Chem.* **367**(3), 300.

Okino, A., Ishizuka, H., Hotta, E., and Shimada, R. (1996). Development of helium ICP-MS using an enhanced vortex flow torch. *Bunseki kagaku* **45**(6), 473.

Oladipo, M. O. A., Ajayi, O. O., Elegba, S. B., Alonge, S. O., and Adeleye, S. A. (1993). The determination of minor and trace elements in some Nigerian cigarettes and raw tobacco using inductively coupled plasma mass spectrometry (ICP-MS). *J. Environ. Sci. Health, Part A* **A28**(4), 839–858.

Olesik, J. W. (2000). Inductively coupled plasma mass spectrometry. *Prac. Spectrosc.* pp. 67–158.

Olney, T. N., Chen, W., and Douglas, D. J. (1999). Gas dynamics of the ICP-MS interface: Impact pressure probe measurements of gas flow profiles. *J. Anal. At. Spectrom.* **14**(1), 9.

Olsen, S. D., Filby, R. H., Brekke, T., and Isaksen, G. H. (1995). Determination of trace elements in petroleum exploration samples by inductively coupled plasma mass spectrometry and instrumental neutron activation analysis. *Analyst (London)* **120**(5), 1379.

Olsen, S. D., Westerlund, S., and Visser, R. G. (1997). Analysis of metals in condensates and naphtha by inductively coupled plasma mass spectrometry. *Analyst (London)* **122**(11), 1229.

Olson, L. K., Vela, N. P., and Caruso, J. A. (1995). Hydride generation, electrothermal vaporization and liquid chromatography as sample introduction techniques for inductively coupled plasma mass spectrometry. *Spectrochim. Acta, Part B* **50**(4/7), 355.

Onda, N. (1996). A new approach to extending the dynamic range in inductively coupled plasma-mass spectrometry. *Tōhoku J. Exp. Med.* **178**(1), 91.

Opitz, M., and Wunsch, G. (1997). Determination of trace impurities in tantalum pentoxide by ICP-AES and ICP-MS after trace-matrix-separation. *J. Prakt. Chem.* **339**(1), 44.

Orians, K. J., and Boyle, E. A. (1993). Determination of picomolar concentrations of titanium, gallium and indium in sea water by inductively coupled plasma mass spectrometry following an 8-hydroxyquinoline chelating resin preconcentration. *Anal. Chim. Acta* **282**(1), 63.

Outridge, P. M. (1996). Atomic spectroscopy perspectives: Potential applications of laser ablation ICP-MS in forensic biology and exploration geochemistry. *Spectroscopy* **11**(4), 21.

Outridge, P. M., Hughes, R. J., and Evans, R. D. (1996). Determination of trace metals in teeth and bones by solution nebulization ICP-MS. *At. Spectrosc.* **17**(1), 1.

Outridge, P. M., Doherty, W., and Grégoire, D. C. (1997). Determination of trace elemental signatures in placer gold by laser ablation-inductively coupled plasma-mass spectrometry as a potential aid for gold. *J. Geochem. Explor.* **60**(3), 229.

Owen, L. M. W., Rauscher, A. M., Fairweather-Tait, S. J., and Crews, H. M. (1996). Use of HPLC with inductively coupled plasma mass spectrometry (ICP-MS) for trace element speciation studies in biological materials. *Transactions* **24**(3), 947.

Owen, L. M. W., Crews, H. M., Hutton, R. C., and Walsh, A. (1990). Preliminary study of metals in proteins by high-performance liquid chromatography—Inductively coupled plasma mass spectrometry using multi-element time-resolved analysis. *Analyst (London)* **117**(3), 649–656.

Owen, L. M. W., Crews, H. M., Massey, R. C., and Bishop, N. J. (1995). Determination of copper, zinc and aluminium from dietary sources in the femur, brain and kidney of guinea pigs and a study of some elements in in vivo intestinal digesta by size-exclusion chromatography inductively coupled plasma mass spectrometry. *Analyst (London)* **120**(3), 705.

Packer, A. P., Gine, M. F., Miranda, C. E. S., and Dos Reis, B. F. (1997). Automated on-line preconcentration for trace metals determination in water samples by inductively coupled plasma mass spectrometry. *J. Anal. At. Spectrom.* **12**(5), 563.

Panayi, A. E., Spyrou, N. M., Ubertalli, L. C., White, M. A., and Part, P. (1999). Determination of trace elements in porcine brain by inductively coupled plasma-mass spectrometry, electrothermal atomic absorption spectrometry, and instrumental neutron activation analysis. *Bio. Trace Elem. Res.*, p. 529.

Panday, V. K., Becker, J. S., and Dietze, H. J. (1995). Semiconductor industry benefits from ICP-MS. Trace impurities in zircaloys by ICP-MS after removal of the matrix by liquid liquid extraction. *Res./Dev.* **37**(7), 51.

Panday, V. K., Becker, J. S., and Dietze, H. J. (1995). Trace analysis of rare earth elements and other impurities in high purity scandium by inductively coupled plasma mass spectrometry after liquid-liquid extraction of the matrix. *Fresenius' J. Anal. Chem.* **352**(3–4), 327.

Panday, V. K. Hoppstock, K., Becker, J. S., and Dietze, H. J. (1996). Determination of rare earth elements in environmental materials by ICP-MS after liquid-liquid extraction. *At. Spectrosc.* **17**(2), 98.

Panday, V. K., Becker, J. S., and Dietze, H. J. (1996). Determination of trace impurities in tantalum by inductively coupled plasma mass spectrometry after removal of the matrix by liquid-liquid extraction. *Anal. Chim. Acta* **329**(1–2), 153.

Papadakis, I., Taylor, P. D. P., and Bievre, P. D. (1997). SI-traceable values for cadmium and lead concentration in the candidate ref rence material, MURST-ISS Al Antarctic sediment, by combination of ICP-MS. *Anal. Chim. Acta* **346**(1), 17.

Parent, M., Vanhoe, H., Moens, L., and Dams, R. (1996). Determination of low amounts of platinum in environmental and biological materials using thermospray nebulization inductively coupled plasma–mass spectrometry. *Fresenius' J. Anal. Chem.* **354**(5–6), 664.

Parent, M., Vanhoe, H., Moens, L., and Dams, R. (1997). Investigation of HfO+ interference in the determination of platinum in a catalytic converter (cordierite) by inductively coupled plasma mass spectrometry. *Talanta* **44**(2), 221.

Park, C. J., and Suh, J. K. (1997). Determination of trace elements in rice flour by isotope dilution inductively coupled plasma mass spectrometry. *J. Anal. At. Spectrom.* **12**(5), 573.

Park, C. J., and Yim, S. A. (1999). Determination of nickel in water samples by isotope dilution inductively coupled plasma mass spectrometry with sample introduction by carbonyl vapor generation. *J. Anal. At. Spectrom.* **14**(7), 1061.

Park, C. J., Park, S. R., Yang, S. R., Han, M. S., and Lee, K. W. (1992). Determination of trace impurities in pure copper by isotope dilution inductively coupled plasma mass spectrometry. *J. Anal. At. Spectrom.* **7**(4), 641–646.

Park, C. J., Oh, P. J., Kim, H. Y., and Lee, D. S. (1999). Determination of 020206Ra in mineral waters by high-resolution inductively coupled plasma mass spectrometry after sample preparation by cation exchange. *J. Anal. At. Spectrom.* **14**(2), 223.

Park, C. J., Kim, K. J., Cha, M. J., and Lee, D. S. (2000). Determination of boron in uniformly-doped silicon thin films by isotope dilution inductively coupled plasma mass spectrometry. *Analyst (London)* 493–498.

Parouchais, T., Warner, I. M., Palmer, L. T., and Kobus, H. (1996). The analysis of small glass fragments using inductively coupled plasma mass spectrometry. *J. Forensic Sci.* **41**(3), 351.

Paschal, D. C., Caldwell, K. L., and Ting, B. G. (1995). Determination of lead in whole blood using inductively coupled argon plasma mass spectrometry with isotope dilution. *J. Anal. At. Spectrom.* **10**(5), 367.

Passariello, B., Barbaro, M., Quaresima, S., Casciello, A., and Marabini, A. (1996). Determination of mercury by inductively coupled plasma—mass spectrometry. *Microchem. J.* **54**(4), 348.

Patriarca, M. (1996). The contribution of inductively coupled plasma mass spectrometry to biomedical research. *Microchem. J.* **54**(3), 262.

Patriarca, M., Lyon, T. D. B., McGaw, B., and Fell, G. S. (1996). Determination of selected nickel isotopes in biological samples by inductively coupled plasma mass spectrometry with isotope dilution. *J. Anal. At. Spectrom.* **11**(4), 297.

Patriarca, M., Kratochwil, N. A., and Sadler, P. J. (1999). Simultaneous determination of Pt and I by ICP-MS for studies of the mechanism of reaction of diiodoplatinum anticancer complexes. *J. Anal. At. Spectrom.* **14**(4), 633.

Pattberg, S., and Matschat, R. (1999). Determination of trace impurities in high purity copper using sector-field ICP-MS: Continuous nebulization, flow injection analysis and laser ablation. *Fresenius' J. Anal. Chem.* **364**(5), 410.

Patterson, H. M., Thorrold, S. R., and Shenker, J. M. (1999). Analysis of otolith chemistry in Nassau grouper *(Epinephelus striatus)* from the Bahamas and Belize using solution-based ICP-MS. *Coral Reefs* **18**(2), 171.

Patterson, K. Y. (1997). Optimization in the analysis of nutritionally important stable isotope tracers using inductively coupled plasma-mass spectrometry. Unpublished Ph.D. Thesis, University of Maryland, College Park.

Patterson, K. Y., Veillon, C., Moser-Veillon, P. B., and Wallace, G. F. (1992). Determination of zinc stable isotopes in biological materials using isotope dilution inductively coupled plasma mass spectrometry. *Anal. Chim. Acta* **258**(2), 317–324.

Patterson, K. Y., Veillon, C., Hill, A. D., and Moser-Veillon, P. B. (1999). Measurement of calcium stable isotope tracers using cool plasma ICP-MS. *J. Anal. At. Spectrom.* **14**(11), 1673.

Paul, M. (1994). Analysis of solid samples by laser sampling ICP-MS. *At. Spectrosc.* **15**(1), 21–26.

Paulsen, S. C., and List, E. J. (1997). A study of transport and mixing in natural waters using ICP-MS: Water-particle interactions. *Water, Air, Soil Pollut.* **99**(1–4), 149.

Paya-Perez, A., Sala, J., and Mousty, F. (1993). Comparison of ICP-AES and ICP-MS for the analysis of trace elements in soil extracts. *Int. J. Environ. Anal. Chem.* **51**(1–4), 223–230.

Pearce, F. M. (1991). The use of ICP-MS for the analysis of natural waters and an evaluation of sampling techniques. *Environ. Geochem. Health* **13**(2), 50.

Pearce, N., and Schettler, G. (1994). Trace metal uptake by mussels in a recently deceased community, Lake Breitling, Germany: A laser ablation ICP-MS study. *Environ. Geochem. Health* **16**(2), 79.

Pearce, N. J. G., Perkins, W. T., Abell, I., Duller, G. A. T., and Fuge, R. (1992). Mineral microanalysis by laser ablation inductively coupled plasma mass spectrometry. *J. Anal. At. Spectrom.* **7**(1), 53–58.

Pearce, N. J. G., Perkins, W. T., and Fuge, R. (1992). Developments in the quantitative and semiquantitative determination of trace elements in carbonates by laser ablation inductively coupled plasma mass spectrometry. *J. Anal. At. Spectrom.* **7**(4), 595–598.

Pearce, N. J. G., Westgate, J. A., and Perkins, W. T. (1996). Developments in the analysis of volcanic glass shards by laser ablation ICP-MS: Quantitative and single internal standard-multi-element methods. *Quaternary Int.*, p. 213.

Pearce, N. J. G., Westgate, J. A., Perkins, W. T., Eastwood, W. J., and Shane, P. (1999). The application of laser ablation ICP-MS to the analysis of volcanic glass shards from tephra deposits: Bulk glass and single shard analysis. *Global Planet. Change* **21**(1), 151.

Pearson, D. G., and Woodland, S. J. (2000). Solvent extraction/anion exchange separation and determination of PGEs (Os, Ir, Pt, Pd, Ru) and Re-Os isotopes in geological samples by isotope dilution ICP-MS. *Chem. Geol.* **165**(1), 87.

Pecheyran, C. (1998). Cold plasma extends trace metal detection capability—ICP-MS capabilities are extended to measure calcium, iron and potassium to acceptable levels. ICP-AES and ICP-MS: Trends at Pittcon'98 (in French). *Semicond. Int.* **21**(8), 261.

Pecheyran, C., Quetel, C. R., Martin Lecuyer, F. M., and Donard, O. F. X. (1998). Simultaneous determination of volatile metal (Pb, Hg, Sn, In, Ga) and nonmetal species (Se, P, As) in different atmospheres by cryofocusing and detection by ICPMS. *Anal. Chem.* **70**(13), 2639.

Pedersen, G. A., and Larsen, E. H. (1997). Speciation of four selenium compounds using high performance liquid chromatography with on-line detection by inductively coupled plasma mass spectrometry or flame atomic absorption spectrometry. *Fresenius' J. Anal. Chem.* **358**(5), 591.

Pepelnik, R., Prange, A., and Niedergesass, R. (1994). Comparative study of multi-element determination using inductively coupled plasma mass spectrometry, total reflection x-ray fluorescence spectrometry and neutron activation analysis. *J. Anal. At. Spectrom.* **9**(9), 1071.

Pepelnik, R., Prange, A., Jantzen, E., Krause, P., and Wuempling, V. (1997). Development and application of ICP-MS in Elbe river research. *Fresenius' J. Anal. Chem.* **359**(4–5), 346.

Perez-Jordan, M. Y., Soldevila, J., Salvador, A., and Pastor, A. (1999). Inductively coupled plasma mass spectrometry analysis of wines. *J. Anal. At. Spectrom.* **14**(1), 33.

Pergantis, S. A., Heithmar, E. M., and Hinners, T. A. (1995). Microscale flow injection and microbore high-performance liquid chromatography coupled with inductively coupled plasma mass spectrometry via a high-efficiency nebulizer. *Anal. Chem.* **67**(24), 4530.

Pergantis, S. A., Heithmar, E. M., and Hinners, T. A. (1997). Speciation of arsenic animal feed additives by microbore high-performance liquid chromatography with inductively coupled plasma mass spectrometry. *Analyst (London)* **122**(10), 1063.

Perkins, W. (1992). Alan Date Memorial Award: Role of inductively coupled plasma mass spectrometry in natural environment research. *J. Anal. At. Spectrom.* **7**(4), 25N–33N.

Perkins, W. T., Fuge, R., and Pearce, N. J. G. (1991). Quantitative analysis of trace elements in carbonates using laser ablation inductively coupled plasma mass spectrometry. *J. Anal. At. Spectrom.* **6**(6), 445.

Perkins, W. T., Pearce, N. J. G., and Fuge, R. (1992). Analysis of zircon by laser ablation and solution inductively coupled plasma mass spectrometry. *J. Anal. At. Spectrom.* **7**(4), 611–616.

Perkins, W. T., Pearce, N. J. G., and Jeffries, T. E. (1993). Laser ablation inductively coupled plasma mass spectrometry: A new technique for the determination of trace and ultra-trace elements in silicates. *Geochim. Cosmochim. Acta* **57**(2), 475.

Perkins, W. T., Pearce, N. J. G., and Westgate, J. A. (1997). The development of laser ablation ICP-MS and calibration strategies: Examples from the analysis of trace elements in volcanic glass shards and sulfide minerals. *Geostand. Newsl.* **21**(2), 175.

Perry, B. J., and Balazs, R. E. (1994). ICP-MS method for the determination of platinum in suspensions of cells exposed to cisplatin. *Anal. Proc.* **31**(9), 269.

Perry, B. J., Barefoot, R. R., and Loon, J. C. V. (1995). Inductively coupled plasma mass spectrometry for the determination of platinum group elements and gold. *Trends Anal. Chem.* **14**(8), 388.

Persaud, A. T., Beauchemin, D., Jamieson, H. E., and McLean, R. J. (1999). Partial leaching as an aid to slurry nebulization for the analysis of soils by ICP-MS with flow injection and mixed-gas plasmas. *Can. J. Chem.* **77**(4), 409.

Peters, G. R., and Beauchemin, D. (1992). Versatile interface for gas chromatographic detection or solution nebulization analysis by inductively coupled plasma mass spectrometry: Preliminary results. *J. Anal. At. Spectrom.* **7**(6), 965–970.

Peters, G. R., and Beauchemin, D. (1993). Characterization of an interface allowing either nebulization or gas chromatography as the sample introduction system in ICPMS. *Anal. Chem.* **65**(2), 97.

Peters, G. R., and Beauchemin, D. (1993). Effect of pre-evaporating the solvent on the analytical performance of inductively coupled plasma mass spectrometry. *Spectrochim. Acta, Part B* **48B**(12), 1481.

Pickford, C. J., Haines, J., Hearn, R., and McAughey, J. (1998). Determination of radionuclides in biological fluids using a high resolution ICP-MS in low resolution mode. *ASTM Spec. Tech. Pub.*, p. 111.

Pilger, C., Leis, F., Tschoepel, P., and Broekaert, J. A. C. (1995). An evaluation of the place of ICP-OES and ICP-MS in the environmental laboratory. Analysis of silicon carbide powders with ICP-MS subsequent to sample dissolution without and with matrix removal. *Chem. N. Z.* **59**(1), 32.

Pilon, F., Lorthioir, S., Birolleau, J. C., and Lafontan, S. (1996). Determination of trace elements in radioactive and toxic materials by inductively coupled plasma mass spectrometry. *J. Anal. At. Spectrom.* **11**(9), 759.

Pin, C., and Joannon, S. (1997). Low level analysis of lanthanides in eleven silicate rock standards by ICP-MS after group separation using cation exchange chromotography. *Geostand. Newsl.* **21**(1), 43.

Pin, C., Lacombe, S., Telouk, P., and Imbert, J. L. (1992). Isotope dilution inductively coupled plasma mass spectrometry: A straightforward method for rapid and accurate determination of uranium and thorium in silicate rocks. *Anal. Chim. Acta* **256**(1), 153–162.

Pin, C., Telouk, P., and Imbert, J. L. (1995). Direct determination of the samarium: Neodymium ratio in geological materials by inductively coupled plasma quadrupole mass spectrometry with cryogenic desolvation. Comparison with isotope dilution thermal ionization mass spectrometry. *J. Anal. At. Spectrom.* **10**(2), 93.

Pingitore, N. E., Leach, J. D., Villalobos, J., and Peterson, J. A. (1997). Provenance determination from ICP-MS elemental and isotopic compositions of El Paso area ceramics. *Mate. Res. Soc. Sym. Proc.*, p. 59.

Pingitore, N. E., Hill, D., Villalobos, J., and Leach, J. (1997). ICP-MS Isotopic signatures of lead ceramic glazes, Rio Grande Valley, New Mexico, 1315–1700. *Mate. Res. Soc. Sym. Pro.*, p. 217.

Platzner, I., Sala, J. V., Mousty, F., and Trincherini, P. R. (1994). Signal enhancement and reduction of interferences in inductively coupled plasma mass spectrometry with an argon- trifluoromethane mixed aerosol carrier gas. *J. Anal. At. Spectrom.* **9**(6), 719.

Plessen, H.-G., and Erzinger, J. (1998). Determination of the platinum-group elements and gold in twenty rock reference materials by inductively coupled plasma-mass spectrometry (ICP-MS) after pre-concentration by nickel sulfide fire assay. *Geostand. Newsl.* **22**(2), 187.

Poitrasson, F., and Siv Hjorth, D. (1999). Direct isotope ratio measurement of ultra-trace lead in waters by double focusing inductively coupled plasma mass spectrometry with an ultrasonic nebuliser and a desolvation unit. *J. Anal. At. Spectrom.* **14**(10), 1573.

Poitrasson, F., Pin, C., Telouk, P., and Imbert, J. L. (1993). Assessment of a simple method for the determination of Nb and Ta at the sub-mu g/g level in silicate rocks by ICP-MS. *Geostand. Newsl.* **17**(2), 209–216.

Policke, T. A., Bolin, R. N., and Harris, T. L. (1998). Uranium isotope measurements by quadrupole ICP-MS for process monitoring of enrichment *ASTM Spe. Tech. Pub.*, p. 3.

Pollmann, D., Pilger, C., Hergenroeder, R., and Leis, F. (1994). Noise power spectra of inductively coupled plasma mass spectrometry using a cooled spray chamber. *Spectrochim. Acta, Part B* **49**(7), 683.

Pollmann, D., Leis, F., Toelg, G., and Tschoepel, P. (1994). Multielement trace determinations in Al2O3 ceramic powders by inductively coupled plasma mass spectrometry with special reference to on-line trace preconcentration. *Spectrochim. Acta, Part B* **49**(12//14), 1251.

Potter, D. (1994). An ICP-MS instrument for the modern laboratory. *Am. Lab.* **26**(11), 35.

Poussel, E., Mermet, J.-M., and Deruaz, D. (1994). Dissociation of analyte oxide ions in inductively coupled plasma mass spectrometry. *J. Anal. At. Spectrom.* **9**(2), 61–66.

Powell, M. J., Quan, E. S. K., Boomer, D. W., and Wiederin, D. R. (1992). Inductively coupled plasma mass spectrometry with direct injection nebulization for mercury analysis of drinking water. *Anal. Chem.* **64**(19), 2253.

Powell, M. J., Boomer, D. W., and Wiederin, D. R. (1995). Determination of chromium species in environmental samples using high-pressure liquid chromatography direct injection nebulization and inductively coupled plasma mass spectrometry. *Anal. Chem.* **67**(14), 2474.

Pozebon, D., Dressler, V. L., and Curtius, A. J. (1998). Determination of Mo, U and B in waters by electrothermal vaporization: Inductively coupled plasma mass spectrometry. *Talanta* **47**(4), 849.

Pozebon, D., Dressler, V. L., and Curtius, A. J. (1998). Determination of arsenic, selenium and lead by electrothermal vaporization inductively coupled plasma mass spectrometry using iridium-coated graphite tubes. *J. Anal. At. Spectrom.* **13**(1), 7.

Pozebon, D., Dressier, V. L., and Curtius, A. J. (1998). Determination of copper, cadmium, lead, bismuth and selenium(IV) in sea-water by electrothermal vaporization inductively coupled plasma mass spectrometry after on-line separation. *J. Anal. At. Spectrom.* **13**(5), 363.

Prabhu, R. K., Vijayalakshmi, S., Mahalingam, T. R., Viswanathan, K. S., and Mathews, C. K. (1993). Laser vaporization inductively coupled plasma mass spectrometry: A technique for the analysis of small volumes of solutions. *J. Anal. At. Spectrom.* **8**(4), 565–570.

Prange, A., and Jantzen, E. (1995). Determination of organometallic species by gas chromatography inductively coupled plasma mass spectrometry. *J. Anal. At. Spectrom.* **10**(2), 105.

Pretorius, W., Foulkes, M., Ebdon, L., and Rowland, S. (1993). HPLC coupled with ICPMS for the determination of metalloporphyrins in coal extracts. *J. High Resolut. Chromatogr.* **16**(3), 157.

Pretty, J. R., Evans, E. H., Blubaugh, E. A., Shen, W.-L., Caruso, J. A., and Davidson, T. M. (1990). Minimisation of sample matrix effects and signal enhancement for trace analytes using anodic stripping voltammetry with detection by inductively coupled plasma atomic emission spectrometry and inductively coupled plasma mass spectrometry. *J. Anal. At. Spectrom.* **5**(8), 710.

Pretty, J. R., Blubaugh, E. A., and Caruso, J. A. (1993). Detemination of arsenic (III) and selenium (IV) using an on-line anodic stripping voltammetry flow cell with detection by inductively coupled plasma atomic emission spectrometry and inductively coupled plasma mass spectrometry. *Anal. Chem.* **65**(23), 3396.

Pretty, J. R., Blubaugh, E. A., Caruso, J. A., and Davidson, T. M. (1994). Determination of chromium (VI) and vanadium(V) using an on-line anodic stripping voltammetry flow cell with detection by inductively coupled plasma mass spectrometry. *Anal. Chem.* **66**(9), 1540.

Pretty, J. R., Duckworth, D. C., and Berkel, G. J. V. (1997). Anodic stripping voltammetry coupled on-line with inductively coupled plasma mass spectrometry: Optimization of a thin-layer flow cell system for analyte signal enhancement. *Anal. Chem.* **69**(17), 3544.

Pretty, J. R., Duckworth, D. C., and Berkel, G. J. V. (1998). Electrochemical sample pretreatment coupled on-line with ICP-MS: Analysis of uranium using an anodically conditioned glassy carbon working electrode. *Anal. Chem.* **70**(6), 1141.

Price, G. D., and Pearce, N. J. G. (1997). Biomonitoring of pollution by Cerastoderma edule from the British Isles: A laser ablation ICP-MS study. *Mar. Pollu. Bull.* **34**(12), 1025.

Probst, T. U., Zeh, P., and Kim, J. I. (1995). Multielement determinations in ground water ultrafiltrates using inductively coupled plasma mass spectrometry and monostandard neutron activation analysis. *Fresenius' J. Anal. Chem.* **351**(8), 745.

Probst, T. U. (1996). Studies on the long-term stabilities of the background of radionuclides in inductively coupled plasma mass spectrometry (ICP-MS). *Fresenius' J. Anal. Chem.* **354**(7–8), 782.

Probst, T. U., Berryman, N. G., Lemmen, P., and Weissfloch, L. (1997). Comparison of inductively coupled plasma atomic emission spectrometry and inductively coupled plasma mass spectrometry with quantitative neutron capture radiography for the determination of boron in biological samples from cancer therapy. *J. Anal. At. Spectrom.* **12**(10), 1115.

Prohaska, T., Pfeffer, M., Tulipan, M., Stingeder, G., Mentler, A., and Wenzel, W. W. (1999). Speciation of arsenic of liquid and gaseous emissions from soil in a microcosmos experiment by liquid and gas chromatography with inductively coupled plasma mass spectrometer (ICP-MS) detection. *Fresenius' J. Anal. Chem.* **364**(5), 467.

Prohaska, T., Hann, S., Latkoczy, C., and Stingeder, G. (1999). Determination of rare earth elements U and Th in environmental samples by inductively coupled plasma double focusing sectorfield mass spectrometry (ICP-SMS). *J. Anal. At. Spectrom.* **14**(1), 1.

Prohaska, T., Latkoczy, C., and Stingeder, G. (1999). Precise sulfur isotope ratio measurements in trace concentration of sulfur by inductively coupled plasma double focusing sector field mass spectrometry. *J. Anal. At. Spectrom.* **14**(9), 1501.

Prohaska, T., Kollensperger, G., Krachler, M., De Winne, K., Stingeder, G., and Moens, L. (2000). Determination of trace elements in human milk by inductively coupled plasma sector field mass spectrometry (ICP-SFMS). *J. Anal. At. Spectrom.*, 335–340.

Prohaska, T., Watkins, M., Latkoczy, C., Wenzel, W. W., and Stingeder, G. (2000). Lead isotope ratio analysis by inductively coupled plasma sector field mass spectrometry (ICP-SMS) in soil digests of a depth profile. *J. Anal. At. Spectrom.* **15**, 365–370.

Prudnikov, E. D., and Barnes, R. M. (1998). Estimation of detection limits in inductively coupled plasma mass spectrometry. *Fresenius' J. Anal. Chem.* **362**(5), 465.

Prudnikov, E. D., and Barnes, R. M. (1999). Theoretical calculation of the standard deviation in inductively coupled plasma mass spectrometry. *J. Anal. At. Spectrom.* **14**(1), 27.

Pruszkowski, E., Neubauer, K., and Thomas, R. (1998). An overview of clinical applications by ICP-MS. *At. Spectrosc.* **19**(4), 111.

Puchyr, R. F., Bass, D. A., Gajewski, R., Calvin, M., Marquardt, W., Urek, K., Druyan, M. E., and Quig, D. (1998). Preparation of hair for measurement of elements by inductively coupled plasma-mass spectrometry (ICP-MS). *Bio. Trace Elem. Res.* **62**(3), 167.

Pupyshev, A. A., Muzgin, V. N., and Lutsak, A. K. (1999). Thermochemical processes and ion transport in inductively coupled plasma mass spectrometry: Theoretical description and experimental confirmation. *J. Anal. At. Spectrom.* **14**(9), 1485.

Quetel, C. R., Thomas, B., Donard, O. F. X., and Grousset, F. E. (1997). Factorial optimization of data acquisition factors for lead isotope ratio determination by inductively coupled plasma mass spectrometry. *Spectrochim. Acta, Part B* **52**(2), 177.

Quetel, C. R., Prohaska, T., Hamester, M., Kerl, W., and Taylor, P. D. (2000). Examination of the performance exhibited by a single detector double focusing magnetic sector ICP-MS instrument for uranium isotope abundance ratio measurements over almost three orders of magnitude and down to pg g0–01 concentration levels. *J. Anal. At. Spectrom.*, 353–358.

Quijano, M. A., Gutierrez, A. M., Perez-Conde, M. C., and Camara, C. (1996). Determination of selenocystine, selenomethionine, selenite and selenate by high-performance liquid chromatography coupled to inductively coupled plasma mass spectrometry. *J. Anal. At. Spectrom.* **11**(6), 407.

Quijano, M. A., Gutierrez, A. M., Perez-Conde, M. C., and Camara, C. (1999). Determination of selenium species in human urine by high performance liquid chromatography and inductively coupled plasma mass spectrometry. *Talanta* **50**(1), 165.

Rabb, S. A. (2000). The investigation of concomitant species and acid matrix effects in inductively coupled plasma optical emission spectroscopy and mass spectrometry and preliminary studies of direct solids analysis. Unpublished Ph.D. Thesis, Ohio State University, Columbus.

Rädlinger, G., and Heumann, K. G. (1997). Determination of halogen species of humic substances using HPLC/ICP-MS coupling. *Fresenius' J. Anal. Chem.* **359**(4–5), 430.

Rädlinger, G., and Heumann, K. G. (1998). Iodine determination in food samples using inductively coupled plasma isotope dilution mass spectrometry. *Anal. Chem.* **70**(11), 2221.

Raith, A., and Hutton, R. C. (1994). Quantitation methods using laser ablation ICP-MS. Part 1: Analysis of powders. *Fresenius' J. Anal. Chem.* **350**(4/5), 242.

Raith, A., Godfrey, J., and Hutton, R. C. (1996). Quantitation methods using laser ablation ICP-MS. *Fresenius' J. Anal. Chem.* **354**(2), 163.

Ramanujam, V. M. S., Egger, N. G., Alcock, N. W., and Anderson, K. E. (1995). Measurement of La and Lu in metal-containing texaphyrins by ICP-MS. *At. Spectrosc.* **16**(5), 211.

Ramanujam, V. M. S., Yokoi, K., Egger, N. G., Dayal, H. H., Alcock, N. W., and Sandstead, H. H. (1999). Polyatomics in zinc isotope ratio analysis of plasma samples by inductively coupled plasma-mass spectrometry and applicability of nonextracted samples for zinc kinetics. *Bio. Trace Elem. Res.* **68**(2), 143.

Rameback, H., and Skalberg, M. (1998). Waste and fuel cycle applications—Separation of neptunium, plutonium, americium and curium from uranium with di-(2-ethylhexyl)-phosphoric acid (HDEHP) for radiometric and ICP-MS analysis. *J. Radioanal. Nucl. Chem.* **235**(1–2), 229.

Rameback, H., Albinsson, Y., Skalberg, M., and Eklund, U. B. (1998). Determination of 99Tc in bentonite clay samples using inductively coupled plasma mass spectrometry. *Fresenius' J. Anal. Chem.* **362**(4), 391.

Ramessur, R. T., Parry, S. J., and Jarvis, K. E. (1998). Characterization of some trace metals from the export processing zone and a coastal tourist area in Mauritius using inductively coupled plasma mass spectrometry. *Environ. Int.* **24**(7), 773.

Rattray, R., and Salin, E. D. (1995). Aerosol deposition direct sample insertion for ultra-trace elemental analysis by inductively coupled plasma mass spectrometry. *J. Anal. At. Spectrom.* **10**(10), 829.

Rauch, S., Motelica-Heino, M., Morrison, G. M., and Donard, O. F. (2000). Critical assessment of platinum group element determination in road and urban river sediments using ultrasonic nebulisation and high resolution ICP-MS. *J. Anal. At. Spectrom.* **15**, 329–334.

Rayman, M. P., Abou-Shakra, F. R., and Ward, N. I. (1996). Determination of selenium in blood serum by hydride generation inductively coupled plasma mass spectrometry. *J. Anal. At. Spectrom.* **11**(1), 61.

Razagui, I. B. A., and Haswell, S. J. (1997). The determination of mercury and selenium in maternal and neonatal scalp hair by inductively coupled plasma-mass spectrometry. *J. Anal. Toxicol.* **21**(2), 149.

Reed, N. M., Cairns, R. O., and Hutton, R. C. (1994). Characterization of polyatomic ion interferences in inductively coupled plasma mass spectrometry using a high resolution mass spectrometer. *J. Anal. At. Spectrom.* **9**(8), 881.

Reid, J. E., Horn, I., Longerich, H. P., Forsythe, L., and Jenner, G. A. (1999). Determination of Zr and Hf in a flux-free fusion of whole rock samples using laser ablation inductively coupled plasma-mass spectrometry (LA-ICP-MS) with isotope dilution calibration. *Geostand. Newsl.* **23**(2), 149.

Reimann, C., Hall, G. E. M., Siewers, U., Bjorvatn, K., Morland, G., Skarphagen, H., and Strand, T. (1997). Radon, fluoride and 62 elements as determined by ICP-MS in 145 Norwegian hard rock groundwater samples. *Sci. Total Environ.* **192**(1), 1.

Reimann, C., Siewers, U., Skarphagen, H., and Banks, D. (1999). Influence of filtration on concentrations of 62 elements analysed on crystalline bedrock groundwater samples by ICP-MS. *Sci. Total Environ.* **234**(1), 155.

Reimann, C., Siewers, U., Skarphagen, H., and Banks, D. (1999). Does bottle type and acid-washing influence trace element analyses by ICP-MS on water samples? A test covering 62 elements and four bottle types: High density polyethene (HDPE), polypropene (PP), fluorinated ethene propene copolymer (FEP) and perfluoroalkoxy polymer (PFA). *Sci. Total Environ.* **239**(1), 111.

Reinhard, K. J., and Ghazi, A. M. (1992). Evaluation of lead concentrations in 18th-century Omaha Indian skeletons using ICP-MS. *Am. J. Phys. Anthropol.* **89**(2), 183.

Remond, G., Batel, A., Roques-Carmes, C., Wehbi, D., Abell, I., and Seroussi, G. (1990). Scanning mechanical microscopy of laser ablated volumes related to inductively coupled plasma-mass spectrometry. *Scanning Microsc.* **4**(2), 249.

Ren, J. M., Rattray, R., Salin, E. D., and Grégoire, D. C. (1995). Assessment of direct solid sample analysis by graphite pellet electrothermal vaporization inductively coupled plasma mass spectrometry. *J. Anal. At. Spectrom.* **10**(11), 1027.

Resano, M., Verstraete, M., Vanhaecke, F., Moens, L., van Alphen, A., and Denoyer, E. R. (2000). Simultaneous determination of Co, Mn, P and Ti in PET samples by solid sampling electrothermal vaporization ICP-MS. *J. Anal. At. Spectrom.* **15**, 389–396.

Revel, G., and Ayrault, S. (2000). Environment—Comparative use of INAA and ICP-MS methods for environmental studies. *J. Radioanal. Nucl. Chem.* **244**(1), 73.

Richaud, R., Lachas, H., Lazaro, M. J., Clarke, L. J., Jarvis, K. E., Herod, A. A., Gibb, T. C., and Kandiyoti, R. (2000). Trace elements in coal derived liquids: Analysis by ICP-MS and Mossbauer spectroscopy. *Fuel* **79**(1), 57.

Richaud, R., Lachas, H., Healey, A. E., Reed, G. P., Haines, J., Jarvis, K. E., Herod, A. A., Dugwell, D. R., and Kandiyoti, R. (2000). Trace element analysis of gasification plant samples by i.c.p.-m.s.: Validation by comparison of results from two laboratories. *Fuel* **79**(9), 1077.

Richaud, R., Lazaro, M. J., Lachas, H., Miller, B. B., Herod, A. A., Dugwell, D. R., and Kandiyoti, R. (2000). Identification of organically associated trace elements in wood and coal by inductively coupled plasma mass spectrometry. *Rapid Commun. Mass Spectrom.*, pp. 317–328.

Richner, P., and Evans, D. (1993). Automated laser ablation sampling (ALAS) for ICP-MS. *At. Spectrosc.* **14**(6), 157–161.

Richner, P., and Wunderli, S. (1993). Differentiation between organic and inorganic chlorine by electrothermal vaporization inductively coupled plasma mass spectrometry: Application to the determination of polychlorinated biphenyls in waste oils. *J. Anal. At. Spectrom.* **8**(1), 45–50.

Richner, P., Evans, D., Wahrenberger, C., and Dietrich, V. (1994). Applications of laser ablation and electrothermal vaporization as sample introduction techniques for ICP-MS. *Fresenius' J. Anal. Chem.* **350**(4/5), 235.

Richner, R. C. (1995). Determination of Technetium-99 by inductively coupled plasma mass spectrometry. Unpublished Ph.D. Thesis, University of Missouri, Columbia.

Richter, R. C., Koirtyohann, S. R., and Jurisson, S. S. (1997). Determination of technetium-99 in aqueous solutions by inductively coupled plasma mass spectrometry: Effects of chemical form and memory. *J. Anal. At. Spectrom.* **12**(5), 557.

Richter, R. C., Swami, K., Chace, S., and Husain, L. (1998). Determination of arsenic, selenium, and antimony in cloud water by inductively coupled plasma mass spectrometry. *Fresenius' J. Anal. Chem.* **361**(2), 168.

Rietz, B., Heydorn, K., and Pritzl, G. (1997). Determination of aluminium in fish tissues by means of INM and ICP-MS. *J. Radioanal. Nucl. Chem.* **216**(1), 113.

Riglet, C., Provitina, O., Dautheribes, J.-L., and Revy, D. (1992). Determination of traces of neptunium-237 in enriched uranium solutions using inductively coupled plasma mass spectrometry. *J. Anal. At. Spectrom.* **7**(6), 923–928.

Riondato, J., Vanhaecke, F., Moens, L., and Dams, R. (2000). Fast and reliable determination of (ultra-)trace and/or spectrally interfered elements in water by sector field ICP-MS. *J. Anal. At. Spectrom.* **15**, 341–346.

Rivas, C., Ebdon, L., Evans, E. H., and Hill, S. J. (1996). An evaluation of reversed-phase and ion-exchange chromatography for use with inductively coupled plasma-mass spectrometry for the determination of organotin compounds. *Appl. Organomet. Chem.* **10**(1), 61.

Robb, P., Owen, L. M. W., and Crews, H. M. (1995). Stable isotope approach to fission product element studies of soil-to-plant transfer and in vitro modelling of ruminant digestion using inductively coupled plasma mass spectrometry. *J. Anal. At. Spectrom.* **10**(9), 625.

Roberts, N. B., Walsh, H. P. J., Klenerman, L., and Kelly, S. A. (1996). Determination of elements in human femoral bone using inductively coupled plasma atomic emission spectrometry and inductively coupled plasma mass spectrometry. *J. Anal. At. Spectrom.* **11**(2), 133.

Robinson, P., Townsend, A. T., Yu, Z., and Munker, C. (1999). Determination of scandium, yttrium and rare earth elements in rocks by high resolution inductively coupled plasma-mass spectrometry. *Geostand. Newsl.* **23**(1), 31.

Rocholl, A. B. E., Simon, K., Jochum, K. P., Bruhn, F., Gehann, R., Kramar, U., Luecke, W., Molzahn, M., Pernicka, E., Seufert, M., Spettel, B., and Stummeier, J. (1997). Chemical characterisation of NIST silicate glass reference material SRM 610 by ICP-MS, TIMS, LIMS, SSMS, INAA, AAS and PIXE. *Geostand. Newsl.* **21**(1), 101.

Rodushkin, I. (1998). Capabilities of high resolution inductively coupled plasma mass spectrometry for trace element determination in plant sample digests. *Fresenius' J. Anal. Chem.* **362**(6), 541.

Rodushkin, I., and Axelsson, M. D. (2000). Application of double focusing sector field ICP-MS for multielemental characterization of human hair and nails. Part I. Analytical methodology. *Sci. Total Environ.* **250**(1), 83.

Rodushkin, I., and Ruth, T. (1997). Determination of trace metals in estuarine and sea-water reference materials by high resolution inductively coupled plasma mass spectrometry. *J. Anal. At. Spectrom.* **12**(10), 1181.

Rodushkin, I., Ruth, T., and Klockare, D. (1998). Non-spectral interferences caused by a saline water matrix in quadrupole and high resolution inductively coupled plasma mass spectrometry. *J. Anal. At. Spectrom.* **13**(3), 159.

Rodushkin, I., Odman, F., and Appelblad, P. K. (1999). Multielement determination and lead isotope ratio measurement in alcoholic beverages by high-resolution inductively coupled plasma mass spectrometry. *J. Food Compos. Anal.* **12**(4), 243.

Rodushkin, I., Odman, F., and Branth, S. (1999). Multielement analysis of whole blood by high resolution inductively coupled plasma mass spectrometry. *Fresenius' J. Anal. Chem.* **364**(4), 338.

Roehl, R., and Alforque, M. M. (1993). Comparison of the determination of hexavalent chromium by ion chromatography coupled with ICP-MS or with colorimetry. *At. Spectrosc.* **11**(6), 210.

Roehl, R., Gomez, J., and Woodhouse, L. R. (1995). Correction of mass bias drift in inductively coupled plasma mass spectrometry measurements of zinc isotope ratios using gallium as an isotope ratio internal standard. *J. Anal. At. Spectrom.* **10**(1), 15.

Roltmann, L., and Heumann, K. G. (1994). Development of an on-line isotope dilution technique with HPLC/ICP-MS for the accurate determination of elemental species. *Fresenius' J. Anal. Chem.* **350**(4/5), 221.

Rosenberg, R. J., Zilliacus, R., and Manninen, P. K. G. (1994). Determination of transition metals in the primary water of pressurized water reactors by inductively coupled plasma mass spectrometry. *J. Anal. At. Spectrom.* **9**(6), 713.

Rosenberg, R. J., Zilliacus, R., Lakomaa, E. L., Rautiainen, A., and Mäkelä, A. (1996). Study of CdTe/CdS-thin films by isotope dilution, neutron activation analysis, inductively coupled plasma mass spectrometry and secondary ion mass spectrometry. *Fresenius' J. Anal. Chem.* **354**(1), 6.

Rosenkranz, B., O'Connor, G., and Evans, E. H. (2000). Low pressure inductively coupled plasma ion source for atomic and molecular mass spectrometry: Investigation of alternative reagent gases for organomercury speciation in tissue and sediment. *J. Anal. At. Spectrom.* **15**, 7–12.

Rosland, E., and Lund, W. (1998). Direct determination of trace metals in sea-water by inductively coupled plasma mass spectrometry. *J. Anal. At. Spectrom.* **13**(11), 1239.

Ross, B. S., and Hieftje, G. M. (1991). Alteration of the ion-optic lens configuration to eliminate mass-dependent matrix-interference effects in inductively coupled plasma-mass spectrometry. *Spectrochim. Acta, Part B* **46B**(9), 1263–1274.

Ross, B. S., and Hieftje, G. M. (1992). Selection of solvent load and first-stage pressure to reduce interference effects in inductively coupled plasma mass spectrometry. *J. Am. Soc. Mass Spectrom.* **3**(2), 128–138.

Ross, B. S., Chambers, D. M., Vickers, G. H., Yang, P., and Hieftje, G. M. (1990). Characterisation of a 9-mm torch for inductively coupled plasma mass spectrometry. *J. Anal. At. Spectrom.* **5**(5), 351.

Ross, B. S., Pengyuan, Y., Chambers, D. M., and Hieftje, G. M. (1991). Comparison of center-tapped and inverted load-coil geometries for inductively coupled plasma-mass spectrometry. *Spectrochim. Acta, Part B* **46B**(13), 1667.

Rupprecht, M., and Probst, T. (1997). Employing multivariate calibration for the determination of radionuclides by inductively coupled plasma-mass spectrometry. *Fresenius' J. Anal. Chem.* **359**(4–5), 442.

Rupprecht, M., and Probst, T. (1998). Development of a method for the systematic use of bilinear multivariate calibration methods for the correction of interferences in inductively coupled plasma-mass spectrometry. *Anal. Chim. Acta* **358**(3), 205.

Russell, R. A. (1997). Determination of metallo-organic and particulate wear metals in lubricating oils associated with hybrid ceramic bearings by inductively coupled plasma mass spectrometry. Unpublished Ph.D. Thesis, University of Florida, Talahassee.

Ruth, K., Schmidt, P., and Mori, E. J. (1992). ICP-MS applications in silicon semiconductor manufacturing. *Spectroscopy* **7**(9), 36.

Sabine Becker, J., and Dietze, H. J. (1997). Double-focusing sector field inductively coupled plasma mass spectrometry for highly sensitive multi-element and isotopic analysis. *J. Anal. At. Spectrom.* **12**(9), 881.

Sacher, F., Raue, B., and Brauch, H. J. (1999). Trace-level determination of bromate in drinking water by IC/ICP-MS. *ASTM Spec. Publ.*, p. 91.

Saito, T., and Saito, K. (1996). Simultaneous multi-determination of brain trace element concentrations by inductively coupled plasma mass spectrometry. *Tohoku J. Exp. Med.* **178**(1), 11.

Sakao, S., and Uchida, H. (1999). Determination of trace elements in shellfish tissue samples by inductively coupled plasma mass spectrometry. *Anal. Chim. Acta* **382**(1), 215.

Sakao, S., Ogawa, Y., and Uchida, H. (1997). Determination of trace elements in sea weed samples by inductively coupled plasma mass spectrometry. *Anal. Chim. Acta* **355**(2–3), 121.

Sakata, K., and Kawabata, K. (1994). Reduction of fundamental polyatomic ions in inductively coupled plasma mass spectrometry. *Spectrochim. Acta, Part B* **49**(10), 1027.

Salov, V. V., Yoshinaga, J., Shibata, Y., and Morita, M. (1992). Determination of inorganic halogen species by liquid chromatography with inductively coupled argon plasma mass spectrometry. *Anal. Chem.* **64**(20), 2425.

Salvato, N. (1995). Analysis of lead and cadmium in foodstuff. Comparison between two of the most advanced analytical methods: ETA-AAS Zemman and ICP-MS. *Ind. Alimen.* **34**(5, No. 337), 504.

Santosa, S. J., Tanaka, S., and Yamanaka, K. (1997). Sequential determination of trace metals in sea water by inductively coupled plasma mass spectrometry after electrothermal vaporization of their dithiocarmabamate complexes in methyl isobutyl ketone. *Environ. Moni. Assess.* **44**(1–3), 515.

Santosa, S. J., Tanaka, S., and Yamanaka, K. (1997). Inductively coupled plasma mass spectrometry for the sequential determination of trace metals in sea water after electrothermal vaporization. *Fresenius' J. Anal. Chem.* **357**(8), 1122.

Sariego MuThiz, C., Marchante Gayon, J. M., Garcia Alonso, J. I., and Sanz-Medel, A. (1999). Accurate determination of iron, copper and zinc in human serum by isotope dilution analysis using double focusing ICP-MS. *J. Anal. At. Spectrom.* **14**(9), 1505.

Sartoros, C., and Salin, E. D. (1999). Automatic selection of internal standards in inductively coupled plasma-mass spectrometry. *Spectrochim. Acta, Part B* **54**(11), 1557.

Sartoros, C., Goltz, D. M., and Salin, E. D. (1998). Program considerations for simplex optimization of ion lenses in ICP-MS. *Appl. Spectrosc.* **52**(5), 643.

Sato, K., Kohri, M., and Okouchi, H. (1995). The speciation of organotin compounds in seawater by miceller LC/ICP-MS. *Bunseki Kagaku* **44**(7), 561.

Sato, K., Kohri, M., and Okochi, H. (1996). The speciation of organotin compounds in seawater by hydride purge-and-trap/ICP-MS. *Bunseki Kagaku* **45**(6), 575.

Sawatari, H., Toda, T., Saizuka, T., Kimata, C., Itoh, A., and Haraguchi, H. (1995). Multielement determination of rare earth elements in coastal seawater by inductively coupled plasma mass spectrometry after preconcentration using chelating resin. *Bull. Chem. Soc. Jpn.* **68**(11), 3065.

Sawatari, H., Fujimori, E., and Haraguchi, H. (1995). Multi-element determination of trace elements in seawater by gallium coprecipitation and inductively coupled plasma mass spectrometry. *Anal. Sci.: Int. J. Jpn. Soc. Anal. Chem.* **11**(3), 369.

Sawatari, H., Hayashi, T., Fujimori, E., Hirose, A., and Haraguchi, H. (1996). Multi-element determination of trace elements in coastal seawater by ICP-MS and ICP-AES after aluminum coprecipitation associated with magnesium. *Bull. Chem. Soc. Jpn.* **69**(7), 1925.

Sayama, Y., Hayashibe, Y., and Fukuda, M. (1995). Determination of cadmium and bismuth in high-purity zinc by inductively coupled plasma mass spectrometry with on-line matrix separation. *Fresenius' J. Anal. Chem.* **353**(2), 162.

Schaumloffel, D., and Prange, A. (1999). A new interface for combining capillary electrophoresis with inductively coupled plasma-mass spectrometry. *Fresenius' J. Anal. Chem.* **364**(5), 452.

Schettler, G., and Pearce, N. J. G. (1996). Metal pollution recorded in extinct Dreissena polymorpha communities, Lake Breitling, Havel Lakes system, Germany: A laser ablation inductively coupled plasma mass spectrometry study. *Hydrobiologia* **317**(1), 1.

Schijf, J., and Byrne, R. H. (1999). Determination of stability constants for the mono- and difluoro-complexes of Y and the REE, using a cation-exchange resin and ICP-MS. *Polyhedron* **18**(22), 2839.

Schmit, J. P., Youla, M., and Gelinas, Y. (1991). Multi-element analysis of biological tissues by inductively coupled plasma mass spectrometry. *Anal. Chim. Acta* **249**(2), 495–502.

Schnetger, B., Muramatsu, Y., and Yoshida, S. (1998). Iodine (and other halogens) in twenty six geological reference materials by ICP-MS and Ion chromatography. *Geostan. News.* **22**(2), 181.

Schoeppenthau, J., and Dunemann, L. (1994). Hyphenated HPLC/ICP-MS and HPLC/ICP-OES techniques for the characterization of metal and non-metal species. *Fresenius' J. Anal. Chem.* **349**(12), 794.

Scholze, H., Hoffmann, E., Luedke, C., and Platalla, A. (1996). Analysis of leaves by using the laser ICP-MS with isotope dissolution method. *Fresenius' J. Anal. Chem.* **355**(7–8), 892.

Schramel, P., and Hasse, S. (1994). Iodine determination in biological materials by ICP-MS. *Mikrochim. Acta* **116**(4), 205.

Schramel, P., and Wendler, I. (1995). Molybdenum determination in human serum (plasma) by ICP-MS coupled to a graphite furnace. *Fresenius' J. Anal. Chem.* **351**(6), 567.

Schramel, P., Wendler, I., and Lustig, S. (1995). Capability of ICP-MS (pneumatic nebulization and ETV) for Pt-analysis in different matrices at ecologically relevant concentrations. *Fresenius' J. Anal. Chem.* **353**(2), 115.

Schramel, P., Wendler, I., and Angerer, J. (1997). The determination of metals (antimony, bismuth, lead, cadmium, mercury, palladium, platinum, tellurium, thallium, tin and tungsten) in urine samples by inductively coupled plasma–mass spectrometry. *Int. Arch. Occup. Environ. Health* **69**(3), 219.

Schroeder, E., Hamester, M., and Kaiser, M. (1998). Properties and characteristics of a laser ablation ICP-MS system for the quantitative elemental analysis of glasses. *Appl. Surf. Sci.*, p. 292.

Schutz, A., Bergdahl, I. A., Ekholm, A., and Skerfving, S. (1996). Measurement by ICP-MS of lead in plasma and whole blood of lead workers and controls. *Occup. Environ. Med.* **53**(11), 736.

Scott, D. J., and Gauthier, G. (1996). Comparison of TIMS (U–Pb) and laser ablation microprobe ICP-MS (Pb) techniques for age determination of detrital zircons from Paleoproterozoic metasedimentary rocks from northeastern Laurentia, Canada, with tectonic implications. *Chem. Geol.* **131**(1–4), 127.

Screnci, D., Galettis, P., Baguley, B. C., and McKeage, M. J. (1998). Optimization of an ICP-MS assay for the detection of trace levels of platinum in peripheral nerves. *At. Spectrosc.* **19**(5), 172.

Sekaly, A. L., Back, M. H., Chakrabarti, C. L., and Grégoire, D. C. (1998). Measurements and analysis of dissociation rate constants of metal-fulvic acid complexes in aqueous solutions. Part I: Simulation of decay curves obtained by inductively coupled plasma–mass spectrometry to evaluate a method to measure the distribution of first-order rate constants. *Spectrochim. Acta, Part B* **53**(6–8), 837.

Selby, M. (1994). Approaches to interference-free elemental analysis with ICP-MS. *At. Spectrosc.* **15**(1), 27–35.

Sen Gupta, J. G., and Bertrand, N. B. (1995). Direct ICP-MS determination of trace and ultratrace elements in geological materials after decomposition in a microwave oven. I. Quantitation of Y, Th, U and the lanthanides. *Talanta* **42**(11), 1595.

Sen Gupta, J. G., and Bertrand, N. B. (1995). Direct ICP-MS determination of trace and ultratrace elements in geological materials after decomposition in a microwave oven. Part II. Quantitation of Ba, Cs, Ga, Hf, In, Mo, Nb, Pb, Rb, Sn, Sr, Ta and Tl. *Talanta* **42**(12), 1947.

Seubert, A. (1994). On-line coupling between chromatography and inductively coupled plasma mass spectrometry—a current assessment. *Fresenius' J. Anal. Chem.* **350**(4/5), 210.

Seubert, A., Petzold, G., and McLaren, J. W. (1995). Synthesis and application of an inert type of 8- hydroxyquinoline-based chelating ion exchanger for seawater analysis using online inductively coupled plasma mass spectrometry detection. *J. Anal. At. Spectrom.* **10**(5), 371.

Sha, L. K., and Chappell, B. W. (1999). Apatite chemical composition, determined by electron microprobe and laser-ablation inductively coupled plasma mass spectrometry, as a probe into granite petrogenesis—Some mineralogical and petrological constraints. *Geochim. Cosmochim. Acta* **63**(22), 3861–3881.

Shabani, M. B., and Masuda, A. (1991). Sample introduction by on-line two-stage solvent extraction and back-extraction to eliminate matrix interference and to enhance sensitivity in the determination of rare-earth elements with inductively coupled plasma mass spectrometry. *Anal. Chem.* **63**(19), 2099–2104.

Shabani, M. B., and Masuda, A. (1992). Determination of trace rhenium in sea water by inductively coupled plasma mass spectrometry with on-line preconcentration. *Anal. Chim. Acta* **261**(1–2), 315–322.

Shabani, M. B., Akagi, T., and Masuda, A. (1992). Preconcentration of trace rare-earth elements in seawater by complexation with bis(2-ethylhexyl) hydrogen phosphate and 2-ethylhexyl dihydrogen phosphate adsorbed on a Csub 18 cartridge and determination by inductively coupled plasma mass spectrometry. *Anal. Chem.* **64**(7), 737–742.

Shabani, M. B., Sahoo, S. K., and Masuda, A. (1992). Determination of bismuth at ultra-trace levels in sea-water by inductively coupled plasma mass spectrometry after preconcentration with solvent extraction and Back-extraction. *Analyst (London)* **117**(9), 1477.

Shaw, P. (1994). Is inductively coupled plasma-mass spectrometry a specialist tool? *Int. Laboratory* **19**(5), 25.

Shen, W.-L., Caruso, J. A., Fricke, F. L., and Satzger, R. D. (1990). Electrothermal vaporisation interface for sample introduction in inductively coupled plasma mass spectrometry. *J. Anal. At. Spectrom.* **5**(6), 451.

Shepherd, T. J., and Chenery, S. R. (1995). Laser ablation ICP-MS elemental analysis of individual fluid inclusions: An evaluation study. *Geochim. Cosmochim. Acta* **59**(19), 3997.

Sheppard, B. S., Shen, W.-L., Caruso, J. A., Heitkemper, D. T., and Fricke, F. L. (1990). Elimination of the argon chloride interference on arsenic speciation in inductively coupled plasma mass spectrometry using ion chromatography. *J. Anal. At. Spectrom.* **5**(6), 431.

Sheppard, B. S., Shen, W.-L., Davidson, T. M., and Caruso, J. A. (1990). Helium-argon inductively coupled plasma for plasma source mass spectrometry. *J. Anal. At. Spectrom.* **5**(8), 697.

Sheppard, B. S., Shen, W. L., and Caruso, J. A. (1991). Investigation of matrix-induced interferences in mixed-gas helium-argon inductively coupled plasma mass spectrometry. *J. Am. Soc. Mass Spectrom.* **2**(5), 355.

Sheppard, B. S., Heitkemper, D. T., and Gaston, C. M. (1994). Microwave digestion for the determination of arsenic, cadmium and lead in seafood products by inductively coupled plasma atomic emission and mass spectrometry. *Analyst. (London)* **119**(8), 1683.

Shi, H. J., and Liu, H. S. (1999). Determination of trace cerium in doped crystal Ce:K3.0Li2.0Nb5.0O15.0 by ICP-MS. *J. Anal. At. Spectrom.* **14**(11), 1771.

Shibata, N., Fudagawa, N., and Kubota, M. (1992). Determination of trace metals in potassium hydrogenphthalate by inductively coupled plasma mass spectrometry. *Anal. Chim. Acta* **265**(1), 93.

Shibata, N., Fudagawa, N., and Kubota, M. (1993). Oxide formation in electrothermal vaporization inductively coupled plasma mass spectrometry. *Spectrochim. Acta, Part B* **48B**(9), 1127.

Shibata, Y. (1996). Inductively coupled plasma (ICP) atomic emission spectrometry and ICP mass spectrometry: Their biomedical and environmental applications. *Eisei Kagaku* **42**(5), 385.

Shibata, Y., Morita, M., and Fuwa, K. (1992). Selenium and arsenic in biology: Their chemical forms and biological functions. *Adv. Biophys.*, pp.31–80.

Shimamura, T., and Iwashita, M. (1997). Determination of major and trace elements for certified reference material of riverine water JAC 0031 and JAC 0032 by ICP-MS. *Anal. Sci.: Int. J. Jpn. Soc. Anal. Chem.* **13**(2), 177.

Shinotsuka, K., and Ebihara, M. (1997). Precise determination of rare earth elements, thorium and uranium in chondritic meteorites by inductively coupled plasma mass spectrometry—A comparative study with radiochemical neutron activation analysis. *Anal. Chim. Acta* **338**(3), 237.

Shinotsuka, K., Hidaka, H., and Ebihara, M. (1995). Detailed abundances of rare earth elements, thorium and uranium in chondritic meteorites: An ICP-MS study. *Meteoritics* **30**(6), 694.

Shinotsuka, K., Hidaka, H., Ebihara, M., and Nakahara, H. (1996). ICP-MS analysis of geological standard rocks for Yttrium, Lanthanoids, thorium and uranium. *Anal. Sci.: Int. J. Jpn. Soc. Anal. Chem.* **12**(6), 917.

Shiraishi, K. (1998). Multi-element analysis of 18 food groups using semi-quantitative ICP-MS. *J. Radioanal. Nucl. Chem.* **238**(1), 67.

Shiraishi, K., McInroy, J. F., and Igarashi, Y. (1990). Simultaneous multielement analysis of diet samples by inductively coupled plasma mass spectrometry and inductively coupled plasma atomic emission spectrometry. *J. Nutr. Sci. Vitaminol.* **36**(1), 81–86.

Shiraishi, K., Takaku, Y., Yoshimizu, K., Igarashi, Y., Masuda, K., McInroy, J. F., and Tanaka, G.-I. (1991). Determination of thorium and uranium in total diet samples by inductively coupled plasma mass spectrometry. *J. Anal. At. Spectrom.* **6**(4), 335.

Shiraishi, K., Nakajima, T., Takaku, Y., Tsumura, A., Yamasaki, S., Los, I. P., Kamarikov, I. Y., Buzinny, M. G., and Zelensky, A. V. (1993). Elemental analysis of freshwater samples collected in the former USSR by inductively coupled plasma mass spectrometry. *J. Radioanal. Nucl. Chem.* **173**(2), 313–322.

Shum, S. C.-K. (1993). Measurement of elemental speciation by liquid chromatography: inductively coupled plasma mass spectrometry (LC-ICP-MS) with the direct injection nebulizer (DIN). Unpublished Ph.D. Thesis, Iowa State University, Ames.

Shum, S. C. K., and Houk, R. S. (1993). Elemental speciation by anion exchange and size exclusion chromatography with detection by inductively coupled plasma mass spectrometry with direct injection nebulization. *Anal. Chem.* **65**(21), 2972.

Shum, S. C. K., Pang, H.-M., and Houk, R. S. (1992). Speciation of mercury and lead compounds by microbore column liquid chromatography—Inductively coupled plasma mass spectrometry with direct injection nebulization. *Anal. Chem.* **64**(20), 2444.

Shuqin, C., Hangting, C., and Xianjin, Z. (1999). Determination of mercury in biological samples using organic compounds as matrix modifiers by inductively coupled plasma mass spectrometry. *J. Anal. At. Spectrom.* **14**(8), 1183.

Shuttleworth, S., and Kremser, D. T. (1998). Assessment of laser ablation and sector field inductively coupled plasma mass spectrometry for elemental analysis of solid samples. *J. Anal. At. Spectrom.* **13**(8), 697.

Sieniawska, C. E., Mensikov, R., and Delves, H.T. (1999). Determination of total selenium in serum, whole blood and erythrocytes by ICP-MS. *J. Anal. At. Spectrom.* **14**(2), 109.

Siethoff, C., Feldmann, I., Jakubowski, N., and Linscheid, M. (1999). Quantitative determination of DNA adducts using liquid chromatography/electrospray ionization mass spectrometry and liquid chromatography/high resolution inductively coupled plasma mass spectrometry. *J. Mass Spectrom.* **34**(4), 421.

Silva, W. G. P. de, Campos, R. C., and Miekeley, N. (1998). A simple digestion procedure for the determination of copper, molybdenium, and vanadium in plants by graphite furnace atomic absorption spectrometry and mass inductively coupled plasma spectrometry. *Anal. Lett.* **31**(6), 1061.

Simon, K., Wiechert, U., Hoefs, J., and Grote, B. (1997). Microanalysis of minerals by laser ablation ICPMS and SIRMS. *Fresenius' J. Anal. Chem.* **359**(4–5), 458.

Sinclair, D. J., Knisley, L. P. J., and McCulloch, M.T. (1998). High resolution analysis of trace elements in corals by laser ablation ICP-MS. *Geochim. Cosmochim. Acta* **62**(11), 1889.

Singh, J., McLean, J. A., Pritchard, D. E., Montaser, A., and Patierno, S. R. (1998). Carcinogenicity-sensitive quantitation of chromium-DNA adducts by inductively coupled plasma mass spectrometry with a direct injection high-efficiency nebulizer. *Toxicol. Sci.* **46**(2), 260.

Skelly Frame, E. M., and Uzgiris, E. E. (1998). Gadolinium determination in tissue samples by inductively coupled plasma mass spectrometry and inductively coupled plasma atomic emission spectrometry in evaluation of the action of magnetic resonance imaging contrast agents. *Analyst (London)* **123**(4), 675.

Slaets, S., Adams, F., Pereiro, I. R., and Lobinski, R. (1999). Optimization of the coupling of multicapillary GC with ICP-MS for mercury speciation analysis in biological materials. *J. Anal. At. Spectrom.* **14**(5), 851.

Slavin, W. (1999). Inductively coupled plasma mass spectrometry. *Spectrochim. Acta, Part B* **54**(2), 247.

Smith, C. W. (1991). "Analytical Inductively Coupled Plasma Mass Spectrometry: Advantages, Limitations, Research Directions and Applications, a Literature Review," Div. rep., MSL 91-72 (LS). Mineral Sciences Laboratories, Ottawa.

Smith, F. G., and Houk, R. S. (1990). Alleviation of polyatomic ion interferences for determination of chlorine isotope ratios by inductively coupled plasma mass spectrometry. *J. Am. Soc. Mass Spectrom.* **1**(4), 284.

Smith, R. G. (1993). Determination of mercury in environmental samples by isotope dilution/ICPMS. *Anal. Chem.* **65**(18), 2485.

Sohrin, Y., Iwamoto, S.-i., Akiyama, S., Fujita, T., Kugii, T., Obata, H., Nakayama, E., Goda, S., Fujishima, Y., Hasegawa, H., Ueda, K., and Matsui, M. (1998). Determination of trace elements in seawater by fluorinated metal alkoxide glass-immobilized 8-hydroxyquinoline concentration and high-resolution inductively coupled plasma mass spectrometry detection. *Anal. Chim. Acta* **363**(1), 11.

Soldevila, J., El Himri, M., Pastor, A., and de la Guardia, M. (1998). Evaluation of operational parameters affecting semiquantitative multi-elemental analysis by inductively coupled plasma mass spectrometry. *J. Anal. At. Spectrom.* **13**(8), 803.

Sparks, C. M. (1996). Sample transport in electrothermal vaporization-inductively coupled plasma-mass spectrometry. Unpublished Ph.D.Thesis, University of Texas, Austin.

Specht, A. A., and Beauchemin, D. (1998). Automated on-line isotope dilution analysis with ICP-MS using sandwich flow injection. *Anal. Chem.* **70**(5), 1036.

St'astna, M., Nemcova, I., and Zyka, J. (1999). ICP-MS for the determination of trace elements in clinical samples. *Anal. Lett.* **32**(13), 2531.

Stefanova, V., Kmetov, V., and Futekov, L. (1997). Air segmented discrete introduction in inductively coupled plasma mass spectrometry. *J. Anal. At. Spectrom.* **12**(11), 1271.

Steiner, J. W., and Duncan, P. L. (1991). Use of ICP-MS for determining cashmere yield. *At. Spectrosc.*, pp. 235–238.

Stetzenbach, K. J., Amano, M., Kreamer, D. K., and Hodge, V. F. (1994). Testing the limits of ICP-MS: Determination of trace elements in ground water at the part-per-trillion level. *Ground Water* **32**(6), 976.

Stewart, B., and Darbouret, D. (1998). Advancements in the production of ultrapure water for ICP-MS metals analysis. *Am. Lab.* **30**(9), 36.

Stewart, I. I., and Olesik, J. W. (1998). Transient acid effects in inductively coupled plasma optical emission spectrometry and inductively coupled plasma mass spectrometry. *J. Anal. At. Spectrom.* **13**(9), 843.

Stewart, I. I., and Olesik, J. W. (1998). Steady state acid effects in ICP-MS. *J. Anal. At. Spectrom.* **13**(12), 1313.

Stewart, I. I., Hensman, C. E., and Olesik, J. W. (2000). Influence of gas sampling on analyte transport within the ICP and ion sampling for ICP-MS studied using individual, isolated sample droplets. *Semicond. Int.* **23**(3), 99.

Stewart, I. I., Hensman, C. E., and Olesik, J. W. (2000). Influence of gas sampling on analyte transport within the ICP and ion sampling for ICP-MS studied using individual, isolated sample droplets. *Appl. Spectrosc.* **54**(2), 164.

Stix, J., Gauthier, G., and Ludden, J. N. (1995). A critical look at quantitative laser-ablation ICP-MS analysis of natural and synthetic glasses. *Can. Mineral.* **33**(2), 435.

Story, W. C., Caruso, J. A., Heitkemper, D. T., and Perkins, L. (1992). Elimination of the chloride interference on the determination of arsenic using hydride generation inductively coupled plasma mass spectrometry. *J. Chromatogr. Sci.* **30**(11), 427.

Strekopytov, S. V., and Dubinin, A. V. (1996). Determination of Mo, W, and Th contents in standard samples SDO (oceanic sediments and ores) by ICP-MS technique. *Oceanology* **36**(6), 816.

Strekopytov, S. V., and Dubinin, A. V. (1997). Determination of Zr, Hf, Mo, W, and Th in standard reference samples of ocean sediments by inductively coupled plasma mass spectrometry. *J. Anal. Chem.* **52**(12), 1171.

Stroh, A. (1992). Analysis of rare earth elements in natural waters by ICP-MS. *At. Spectrosc.* **13**(3), 89–92.

Stroh, A. (1993). Determination of Pb and Cd in whole blood using isotope dilution ICP-MS. *At. Spectrosc.* **14**(5), 141–143.

Stroh, A., and Vollkopf, U. (1992). Analysis of difficult samples by flow injection inductively coupled plasma mass spectrometry. *Anal. Proc.* **29**(7), 274.

Stroh, A., and Vollkopf, U. (1993). Effects of Ca on instrument stability in the trace element determination of Ca-rich soils using ICP-MS. *At. Spectrosc.* **14**(3), 76–79.

Stroh, A., and Vollkopf, U. (1993). Optimization and use of flow injection vapour generation inductively coupled plasma mass spectrometry for the determination of arsenic, antimony and mercury in water and sea-water at ultratrace levels. *J. Anal. At. Spectrom.* **8**(1), 35–40.

Stroh, A., Bruckner, P., and Vollkopf, U. (1994). Multielement analysis of wine samples using ICP-MS. *At. Spectrosc.* **15**(2), 100–106.

Stroh, A., Bea, F., and Montero, P. G. (1995). Ultratrace-level determination of rare earth elements, thorium, and uranium in ultramafic rocks by ICP-MS. *At. Spectrosc.* **16**(1), 7.

Stuerup, S., and Buechert, A. (1996). Direct determination of copper and iodine in milk and milk powder in alkaline solution by flow injection inductively coupled plasma mass spectrometry. *Fresenius' J. Anal. Chem.* **354**(3), 323.

Stuerup, S., Hansen, M., and Moelgaard, C. (1997). Measurements of 44Ca:43Ca and 42Ca:43Ca isotopic ratios in urine using high resolution inductively coupled plasma mass spectrometry. *J. Anal. At. Spectrom.* **12**(9), 919.

Stuerup, S., Dahlgaard, H., and Nielsen, S. C. (1998). High resolution inductively coupled plasma mass spectrometry for the trace determination of plutonium isotopes and isotope ratios in environmental samples. *J. Anal. At. Spectrom.* **13**(12), 1321.

Stuewer, D., and Jakubowski, N. (1998). Elemental analysis by inductively coupled plasma mass spectrometry with sector field instruments: A progress report. *J. Mass Spectrom.* **33**(7), 579.

Stuhne-Sekalec, L., Xu, S. X., Parkes, J. G., Olivieri, N. F., and Templeton, D. M. (1992). Speciation of tissue and cellular iron with on-line detection by inductively coupled plasma-mass spectrometry. *Anal. Biochem.* **205**(2), 278.

Sturgeon, R. E., and Lam, J. W. (1999). The ETV as a thermochemical reactor for ICP-MS sample introduction. *J. Anal. At. Spectrom.* **14**(5), 785.

Sturgeon, R. E., Willie, S. N., Zheng, J., Kudo, A., and Grégoire, D. C. (1993). Determination of ultratrace levels of heavy metals in arctic snow by electrothermal vaporization inductively coupled plasma mass spectrometry. *J. Anal. At. Spectrom.* **8**(8), 1053–1058.

Su, W., Qi, L., Hu, R., and Zhang, G. (1998). Analysis of rare-earth elements in fluid inclusions by inductively coupled plasma-mass spectrometry (ICP-MS). *Kexue Tongbao* **43**(22), 1922.

Sumiya, S., and Morita, S. (1994). Determination of technetium-99 and neptunium-237 in environmental samples by inductively coupled plasma mass spectrometry. *J. Radioanal. Nucl. Chem.* **177**(1), 149–160.

Sung, Y., and Lim, H. B. (2000). Double membrane desolvator for direct analysis of isopropyl alcohol in inductively coupled plasma atomic emission spectrometry (ICP-AES) and inductively coupled plasma mass spectrometry (ICP-MS). *Microchem. J.* **64**(1), 51.

Sutton, K. L., Ponce de Leon, C. A., Ackley, K. L., Sutton, R. M., Stalcup, A. M., and Caruso, J. A. (2000). Development of chiral HPLC for selenoamino acids with ICP-MS detection: Application to selenium nutritional supplements. *Analyst (London)* **14**, 281–286.

Sutton, R. L. (1994). Analysis of liquid-phase tungsten hexafluoride residue by inductively coupled plasma mass spectrometry with ultrasonic nebulization. *J. Anal. At. Spectrom.* **9**(9), 1079.

Suzuki, K. T. (1996). Simultaneous speciation of endogenous and exogenous elements by HPLC/ICP-MS with enriched stable isotopes. *Tohoku J. Exp. Med.* **178**(1), 27.

Suzuki, K. T., Shiobara, Y., Ishiwata, K., and Ohmichi, M. (1997). Speciation by HPLC/ICP-MS with use of stable isotopes: Chemical reactions in the metabolism of selenium administered as selenite. *J. Inorg. Biochem.* **67**(1–4), 21.

Suzuki, Y., and Marumo, Y. (1996). Determination of trace impurities in lead shotgun pellets by ICP-MS. *Anal. Sci.: Int. J. Jpn. Soc. Anal. Chem.* **12**(1), 129.

Sylvester, P. J., and Eggins, S. M. (1997). Analysis of Re, Au, Pd, Pt and Rh in NIST glass certified reference materials and natural basalt glasses by laser ablation ICP-MS. *Geostand. Newsl.* **21**(2), 215.

Szpunar, J., Makarov, A., Pieper, T., Keppler, B. K., and Lobinski, R. (1999). Investigation of metallodrug-protein interactions by size- exclusion chromatography coupled with inductively coupled plasma mass spectrometry (ICP-MS). *Anal. Chim. Acta* **387**(2), 135.

Taddia, M., Bosi, M., and Poluzzi, V. (1993). Determination of tin in indium phosphide by electrothermal atomic absorption spectrometry and inductively coupled plasma mass spectrometry. *J. Anal. At. Spectrom.* **8**(5), 755–758.

Tagami, K., and Uchida, S. (1993). Separation procedure for the determination of technetium-99 in soil by ICP-MS. *Radiochim. Acta,* pp. 69–72.

Tagami, K., and Uchida, S. (1995). Fundamental studies using ICP-MS for the measurement of technetium-99 in a dried-up deposition sample. *J. Radioanal. Nucl. Chem.* **190**(1), 31.

Tagami, K., and Uchida, S. (1996). Analysis of technetium-99 in soil and deposition samples by inductively coupled plasma mass spectrometry. *Appl. Radiat. Isot.* **47**(9/10), 1057.

Takaku, Y., Masuda, K., Takahashi, T., and Shimamura, T. (1993). Determination of trace impurity rare earth elements in high-purity rare earth element samples using high-resolution inductively coupled plasma mass spectrometry. *J. Anal. At. Spectrom.* **8**(5), 687–690.

Takaku, Y., Masuda, K., Takahashi, T., and Shimamura, T. (1994). Determination of trace silicon in ultra-high-purity water by inductively coupled plasma mass spectrometry. *J. Anal. At. Spectrom.* **9**(12), 1385.

Takaku, Y., Shimamura, T., Masuda, K., and Igarashi, Y. (1995). Iodine determination in natural and tap water using inductively coupled plasma mass spectrometry. *Anal. Sci.: Int. J. Jpn. Soc. Anal. Chem.* **11**(5), 823.

Takeda, K., Yamaguchi, T., Akiyama, H., and Masuda, T. (1992). Determination of ultra-trace amounts of uranium and thorium in high-purity aluminium by inductively coupled plasma mass spectrometry. *Analyst (London)* **116**(5), 501.

Takeda, K., Watanabe, S., Naka, H., Okuzaki, J., and Fujimoto, T. (1998). Determination of ultra-trace impurities in semiconductor-grade water and chemicals by inductively coupled plasma mass spectrometry following a concentration step by boiling with mannitol. *Anal. Chim. Acta* **377**(1), 47.

Takenaka, M., Tomita, M., Kubota, A., and Tsuchiya, N. (1994). Depth profiling of ultra-trace chromium, iron, nickel, and copper in silicon wafers by electrothermal vaporization/ICP-MS. *Bunseki Kagaku* **43**(2), 173.

Tanaka, S., Yasushi, N., Sato, N., and Fukasawa, T. (1998). Rapid and simultaneous multi-element analysis of atmospheric particulate matter using inductively coupled plasma mass spectrometry with laser ablation sample introduction. *J. Anal. At. Spectrom.* **13**(2), 135.

Tanaka, T., Sakai, Y., and Kawaguchi, H. (1995). Effect of orifice diameter of torch injector tube in inductively coupled plasma mass spectrometry. *Anal. Sci.: Int. J. Jpn. Soc. Anal. Chem.* **11**(4), 673.

Tanaka, T., Yamamoto, K., Nomizu, T., and Kawaguchi, H. (1995). Laser ablation/inductively coupled plasma mass spectrometry with aerosol density normalization. *Anal. Sci.: Int. J. Jpn. Soc. Anal. Chem.* **11**(6), 967.

Tanaka, T., Hara, K., Tanimoto, A., Kasai, K., Kita, T., Tanaka, N., and Takayasu, T. (1996). Determination of arsenic in blood and stomach contents by inductively coupled plasma/mass spectrometry (ICP/MS). *Forensic Sci. Int.* **81**(1), 43.

Tanaka, T., Kobayashi, K., and Hiraide, M. (1999). Development of hydride generation low-pressure helium ICP-MS for the determination of selenium. *Nippon Kagaku Kaishi* **7**, 463.

Tang, Y. Q., Jarvis, K. E., and Williams, J. G. (1992). Determination of trace elements in 11 Chinese geological reference materials by ICP-MS. *Geostand. Newsl.* **16**(1), 61–70.

Tangen, A., and Lund, W. (1999). A multivariate study of the acid effect and the selection of internal standards for inductively coupled plasma mass spectrometry. *Spectrochim. Acta, Part B* **54**(13), 1831–1838.

Tangen, A., Trones, R., Greibrokk, T., and Lund, W. (1997). Microconcentric nebulizer for the coupling of micro liquid chromatography and capillary zone electrophoresis with inductively coupled plasma mass spectrometry. *J. Anal. At. Spectrom.* **12**(6), 667.

Tangen, A., Lund, W., Josefsson, B., and Borg, H. (1999). Interface for the coupling of capillary electrophoresis and inductively coupled plasma mass spectrometry. *J. Chromatogr.* **826**(1), 87.

Tanner, S. D. (1992). Space charge in ICP-MS: Calculation and implications. *Spectrochim. Acta, Part B* **47B**(6), 809–824.

Tanner, S. D. (1992). Space charge in inductively coupled plasma mass spectrometry: Calculation and implications. *Anal. Proc.* **29**(7), 281.

Tanner, S. D. (1995). Characterization of ionization and matrix suppression in inductively coupled cold plasma mass spectrometry. *J. Anal. At. Spectrom.* **10**(11), 905.

Tanner, S. D., and Baranov, V. I. (1999). Theory, design, and operation of a dynamic reaction cell for ICP-MS. *At. Spectrosc.* **20**(2), 45.

Tanner, S. D., and Baranov, V. I. (1999). A dynamic reaction cell for inductively coupled plasma mass spectrometry (ICP-DRC-MS). II. Reduction of interferences produced within the cell. *J. Am. Soc. Mass Spectrom.* **10**(11), 1083.

Tanner, S. D., Cousins, L. M., and Douglas, D. J. (1994). Reduction of space charge effects using a three-aperture gas dynamic vacuum interface for inductively coupled plasma-mass spectrometry. *Appl. Spectrosc.* **48**(11), 1367.

Tanner, S. D., Douglas, D. J., and French, J. B. (1994). Gas and ion dynamics of a three-aperture vacuum interface for inductively coupled plasma-mass spectrometry. *Appl. Spectrosc.* **48**(11), 1373.

Tanner, S. D., Paul, M., Beres, S. A., and Denoyer, E. R. (1995). The application of cold plasma conditions for the determination of trace levels of Fe, Ca, K, Na, and Li by ICP-MS. *At. Spectrosc.* **16**(1), 16.

Tao, G., Yamada, R., Fujikawa, Y., Kojima, R., Zheng, J., Fisher, D. A., Koerner, R. M., and Kudo, A. (2000). Determination of major metals in arctic snow by inductively coupled plasma mass spectrometry with cold plasma and microconcentric nebulization techniques. *Int. J. Environ. Anal. Chem.* **76**(2), 135.

Tao, H., and Miyazaki, A. (1995). Decrease of solvent water loading in inductively coupled plasma mass spectrometry by using a membrane separator. *J. Anal. At. Spectrom.* **10**(1), 1.

Tao, H., Lam, J. W. H., and McLaren, J. W. (1993). Determination of selenium in marine certified reference materials by hydride generation inductively coupled plasma mass spectrometry. *J. Anal. At. Spectrom.* **8**(8), 1067–1074.

Tao, H., Rajendran, R. B., Quetel, C. R., Nakazato, T., Tominaga, M., and Miyazaki, A. (1999). Tin speciation in the femtogram range in open ocean seawater by gas chromatography/inductively coupled plasma mass spectrometry using a shield torch at normal plasma conditions. *Anal. Chem.* **71**(19), 4208.

Taylor, D. B., Kingston, H. M., Nogay, D. J., and Koller, D. (1996). On-line solid-phase chelation for the determination of eight metals in environmental waters by inductively coupled plasma mass spectrometry. *J. Anal. At. Spectrom.* **11**(3), 187.

Taylor, H. E. (2000). Mass Spectrometry in the analysis of inorganic substances. *In* "The Encyclopedia of Analytical Chemistry" (R. A. Meyers, ed.). Wiley, West Sussex, England (in press).

Taylor, H. E., and Garbarino, J. R. (1991). The measurement of trace metals in water resource monitoring samples by inductively coupled plasma-mass spectrometry. *Spectrochim. Acta Rev.* **14**(1–2), 33–44.

Taylor, H. E. and Garbarino, J. R. (1992) Analytical applications of inductively coupled plasma-mass spectrometry. *In* "Inductively Coupled Plasmas in Analytical Atomic Spectrometry" (A. Montaser and D. W. Golightly, eds.), pp. 651–678. VCH, New York.

Taylor, H. E., and Shiller, A. M. (1997). The use of sedimentation field flow fractionation-inductively coupled plasma mass spectrometry for the chemical characterization of suspended particulate matter in environmental hydrologic systems. *Geo. Surv. Open-File Rep. (U. S.)* **97-496**, 1–100.

Taylor, H. E., Garbarino, J. R., Murphy, D. M., and Beckett, R. (1992). Inductively coupled plasma-mass spectrometry as an element-specific detector for field- flow fractionation particle separation. *Anal. Chem.* **64**(18), 2036.

Taylor, H. E., Huff, R. A., and Montaser, A. (1998). Novel applications of ICPMS. *In* "Inductively Coupled Plasma Mass Spectrometry" (A. Montaser, ed.), pp. 681–808. Wiley-VCH, New York.

Templeton, D. M., Xu, S. X., and Stuhne-Sekalec, L. (1994). Isotope-specific analysis of Ni by ICP-MS: Applications of stable isotope tracers to biokinetic studies. *Sci. Total Environ.* **148**(2–3), 253.

Thirlwall, M. F., and Walder, A. J. (1995). In situ hafnium isotope ratio analysis of zircon by inductively coupled plasma multiple collector mass spectrometry. *Chem. Geo.* **122**(1–4), 241.

Thomas, C., Jakubowski, N., Stuewer, D., and Broekaert, J. A. C. (1995). Thermospray device of improved design for application in ICP-MS. *J. Anal. At. Spectrom.* **10**(9), 583.

Thomas, C., Jakobowski, N., Stuewer, D., and Klockow, D. (1998). Speciation of organic selenium compounds by reversed-phase liquid chromatography and inductively coupled plasma mass spectrometry. Part I. Sector field instrument with low mass resolution. *J. Anal. At. Spectrom.* **13**(11), 1221.

Thomas, P., and Sniatecki, K. (1995). Inductively coupled plasma mass spectrometry: Application to the determination of arsenic species. *Fresenius' J. Anal. Chem.* **351**(4–5), 410.

Thomas, P., Finnie, J. K., and Williams, J. G. (1997). Feasibility of identification and monitoring of arsenic species in soil and sediment samples by coupled high-performance liquid chromatography inductively coupled plasma mass spectrometry. *J. Anal. At. Spectrom.* **12**(12), 1367.

Thomas, S., Morawska, L., Bofinger, N., and Selby, M. (1997). Investigation of the source of blank problems in the measurement of lead in sub-micrometre airborne particulates by inductively coupled plasma mass spectrometry. *J. Anal. At. Spectrom.* **12**(5), 553.

Tian, X., Emteborg, H., and Adams, F. C. (1999). Analytical performance of axial inductively coupled plasma time of flight mass spectrometry (ICP-TOFMS). *J. Anal. At. Spectrom.* **14**(12), 1807–1814.

Tian, X.-D., Zhuang, Z.-X., Chen, B., and Wang, X.-R. (1999). Determination of arsenic speciation by capillary electrophoresis and ICP-MS using a movable reduction bed hydride generation system. *At. Spectrosc.* **20**(4), 127.

Tilch, J., Luedke, C., and Hoffmann, E. (1996). Determination of metals in airborne particulates by LEAFS and ICP-MS after sampling on reusable graphite filters. *Fresenius' J. Anal. Chem.* **355**(7–8), 913.

Ting, B. G., Paschal, D. C., and Caldwell, K. L. (1996). Determination of thorium and uranium in urine with inductively coupled argon plasma mass spectrometry. *J. Anal. At. Spectrom.* **11**(5), 339.

Tirez, K., Seuntjens, P., and De Brucker, N. (1999). Full uncertainty calculation on quantitative determination of tracer (111Cd) cadmium and natural cadmium in soil column effluents with ICP-MS. *J. Anal. At. Spectrom.* **14**(9), 1475.

Tittes, W., Jakubowski, N., Stuewer, D., and Toeig, G. (1994). Reduction of some selected spectral interferences in inductively coupled plasma mass spectrometry. *J. Anal. At. Spectrom.* **9**(9), 1015.

Tomascak, P. B., Carlson, R. W., and Shirey, S. B. (1999). Accurate and precise determination of Li isotopic compositions by multi-collector sector ICP-MS. *Chem. Geo.* **158**(1), 145.

Tomascak, P. B., Tera, F., Helz, R. T., and Walker, R. J. (1999). The absence of lithium isotope fractionation during basalt differentiation: New measurements by multicollector sector ICP-MS—A new tool for investigating pegmatite petrogenesis. *Geochim. Cosmochim. Acta* **63**(6), 907.

Tomlinson, M. J., and Caruso, J. A. (1996). Speciation of chromium using thermospray nebulization as sample introduction into inductively coupled plasma mass spectrometry. *Anal. Chim. Acta* **322**(1–2), 1.

Tothill, P., Matheson, L. M., Smyth, J. F., and McKay, K. (1990). Inductively coupled plasma mass spectrometry for the determination of platinum in animal tissues and a comparison with atomic absorption spectrometry. *J. Anal. At. Spectrom.* **5**(7), 619.

Totland, M. M., Jarvis, I., and Jarvis, K. E. (1993). Determination of the platinum-group elements and gold in solid samples by slurry nebulisation ICP-MS. *Chem. Geo.* **104**(1–4), 175.

Totland, M. M., Jarvis, I., and Jarvis, K. E. (1995). Microwave digestion and alkali fusion procedures for the determination of the platinum-group elements and gold in geological materials by ICP-MS. *Chem. Geo.* **124**(1–2), 21.

Tovo, L. L., Clymire, J. W., Boyce, W. T., and Kinard, W. F. (1998). Actinide, elemental, and fission product measurements by at the Savannah River site. *ASTM Spec. Tech. Pub.*, p. 32.

Townsend, A. T. (1996). The development of ICP-MS continues—High resolution capability. *Chem. Aus.* **63**(10), 449.

Townsend, A. T. (1999). The determination of arsenic and selenium in standard reference materials using sector field ICP-MS in high resolution mode. *Fresenius' J. Anal. Chem.* **364**(6), 521.

Townsend, A. T. (2000). The accurate determination of the first row transition metals in water, urine, plant, tissue and rock samples by sector field ICP-MS. *J. Anal. At. Spectrom.* **15**, 307–314.

Townsend, A. T., and Edwards, R. (1998). Ultratrace analysis of Antarctic snow and ice samples using high resolution inductively coupled plasma mass spectrometry. *J. Anal. At. Spectrom.* **13**(5), 463.

Townsend, A. T., Yu, Z., McGoldrick, P., and Hutton, J. A. (1998). Precise lead isotope ratios in Australian galena samples by high resolution inductively coupled plasma mass spectrometry. *J. Anal. At. Spectrom.* **13**(8), 809.

Townsend, A. T., Miller, K. A., McLean, S., and Aldous, S. (1998). The determination of copper, zinc, cadmium and lead in urine by high resolution ICP-MS. *J. Anal. At. Spectrom.* **13**(11), 1213.

Trevor Delves, H., and Sieniawska, C. E. (1997). Simple method for the accurate determination of selenium in serum by using inductively coupled plasma mass spectrometry. *J. Anal. At. Spectrom.* **12**(3), 387.

Trones, R., Tangen, A., Lund, W., and Greibrokk, T. (1999). Packed capillary high-temperature liquid chromatography coupled to inductively coupled plasma mass spectrometry. *J. Chromatogr.* **835**(1), 105.

Tsoupras, G. (1996). ICP-MS application to semiconductor processing chemical materials analysis. *Analusis* **24**(9–10), 23.

Tsumura, A., Okamoto, R., Takaku, Y. I., and Yamasaki, S. I. (1995). Direct determination of uranium in rainwater by high resolution ICP-MS with an ultrasonic nebulizer. *Radioisotopes* **44**(2), 85.

Tsumura, A., Ichihashi, H., and Yamasaki, S. I. (1997). Extraction factors and distribution coefficients of trace elements as estimated by batch experiments using high resolution ICP-MS. *Radioisotopes* **46**(4), 230.

Tubaro, F., Barbangelo, F., Toniolo, R., Narda, F. D., and Bontempelli, G. (1999). Effect of the sample introduction system in ICP-MS on the formation of both oxygenated polyatomic and doubly charged ions. *An. Chim. (Rome)* **89**(11), 863.

Turner, I. L. (1996). An improved plasma RF system for ICP-MS. *Am. Lab.* **28**(4), 14.

Twiss, P., Watling, R. J., and Delev, D. (1994). Determination of thorium and uranium in faecal material from occupationally-exposed workers using ICP-MS. *At. Spectrosc.* **15**(1), 36–39.

Tykot, R. H., and Young, S. M. M. (1996). Archaeological applications of inductively coupled plasma-mass spectrometry. *ACS Symp. Ser.*, p. 116.

Tyson, J. F., Ge, H., and Denoyer, E. R. (1997). On-line dilution for ICP-MS with a flow injection recirculating loop manifold. *J. Anal. At. Spectrom.* **12**(10), 1163.

Uchida, H., and Ito, T. (1994). Comparative study of 27.12 and 40.68 MHz inductively coupled argon plasmas for mass spectrometry on the basis of analytical characteristic distributions. *J. Anal. At. Spectrom.* **9**(9), 1001.

Uchida, H., and Ito, T. (1995). Inductively coupled nitrogen plasma mass spectrometry assisted by adding argon to the outer gas. *J. Anal. At. Spectrom.* **10**(10), 843.

Uchida, H., and Ito, T. (1997). Evaluation of an inductively coupled air-argon plasma as an ion source for mass spectrometry. *J. Anal. At. Spectrom.* **12**(9), 913.

Uchida, S., and Tagami, K. (1998). Improvement of Tc separation procedure using a chromatographic resin for direct measurement by ICP-MS. *Anal. Chim. Acta* **357**(1–2), 1.

Uchida, S., and Tagami, K. (1999). A rapid separation method for determination of Tc-99 in environmental waters by ICP-MS. *Radioact. Radiochem.* **10**(2), 23.

Uchino, T., and Ebihara, M. (1995). Determination of rare earth elements in Precambrian sediments at Isua by inductively coupled plasma mass spectrometry. *J. Anal. At. Spectrom.* **10**(1), 25.

Uggerud, H. T., and Lund, W. (1997). Use of palladium and iridium as modifiers in the of arsenic and antimony by electrothermal vaporization inductively coupled plasma mass spectrometry, following in situ trapping of the hydrides. *J. Anal. At. Spectrom.* **12**(10), 1169.

Uggerud, H. T., and Lund, W. (1999). Modifier effects from palladium and iridium in the determination of arsenic and antimony using electrothermal vaporisation inductively coupled plasma mass spectrometry. *Spectrochim. Acta, Part B* **54**(11), 1625.

Ulens, K., Moens, L., Dams, R., and Van Winckel, S. (1994). Study of element distributions in weathered marble crusts using laser ablation inductively coupled plasma mass spectrometry. *J. Anal. At. Spectrom.* **9**(11), 1243.

Ulrich, A., Huchulski, C., Dannecker, W., and Vollkopf, U. (1992). Use of electrothermal vaporization inductively coupled plasma mass spectrometry for element determinations in complex matrices such as sandstone samples. *Anal. Proc.* **29**(7), 282.

Ulrich, A., Dannecker, W., Meiners, S., and Vollkopf, U. (1992). Use of electrothermal vaporization inductively coupled plasma mass spectrometry for single-element and multi-element determinations. *Anal. Proc.* **29**(7), 284.

Ulrich, N. (1998). Study of ion chromatographic behavior of inorganic and organic antimony species by using inductively coupled plasma mass spectrometric (ICP-MS) detection. *Fresenius' J. Anal. Chem.* **360**(7), 797.

U.S. Environmental Protection Agency (USEPA). (1990). "Inductively Coupled Plasma-Mass Spectrometry," Method 6020 CLP-M Version 3.1. USEPA, Office of Research and Development, Environmental Monitoring Systems Laboratory, Las Vegas, NV.

U.S. Environmental Protection Agency (USEPA). (1990). "Methods for the Determination of Inorganic Compounds in Drinking Water," Methods 300.0 and 200.8. USEPA, Office of Research and Development, Environmental Monitoring Systems Laboratory, Cincinnati, OH.

U.S. Environmental Protection Agency (USEPA). (1996). "Determination of Trace Elements in Ambient Waters by Inductively Coupled Plasma-Mass Spectrometry, Method 1638. USEPA, Office of Water Engineering and Analysis Division, Washington, DC/Springfield, VA.

U.S. Environmental Protection Agency (USEPA). (1996). "Determination of Trace Elements in Ambient Waters by On-line Chelation Preconcentration and Inductively Coupled Plasma-Mass Spectrometry," Method 1640. USEPA, Office of Water Engineering and Analysis Division, Washington, D C./Springfield, VA.

Vacchina, V., Polec, K., and Szpunar, J. (1999). Speciation of cadmium in plant tissues by size-exclusion chromatography with ICP-MS detection. *J. Anal. At. Spectrom.* **14**(10), 1557.

Valles Mota, J. P., Fernandez de la Campa, M. R., Garcia Alonso, J. I., and Sanz-Medel, A. (1999). Determination of cadmium in biological and environmental materials by isotope dilution inductively coupled plasma mass spectrometry: Effect of flow sample introduction methods. *J. Anal. At. Spectrom.* **14**(2), 113.

Valles Mota, J. P., Ruiz Encinar, J., Fernandez de la Campa, M. R., and Garcia Alonso, J. I. (1999). Determination of cadmium in environmental and biological reference materials using isotope dilution analysis with a double focusing ICP-MS: A comparison with quadrupole ICP-MS. *J. Anal. At. Spectrom.* **14**(9), 1467.

Vandecasteele, C., Vanhoe, H., Dams, R., and Versieck, J. (1990). Determination of trace elements in human serum by inductively coupled plasma-mass spectrometry: Comparison with nuclear analytical techniques. *Bio. Trace Elemt. Res.*, p. 553.

Vandecasteele, C., Vanhoe, H., Dams, R., Vanballenberghe, L., Wittoek, A., and Versieck, J. (1991). Geological and nuclear applications of inductively coupled plasma mass spectrometry Determination of strontium in human serum by inductively coupled plasma mass spectrometry and neutron activation analysis: A comparison. *Anal. Proc.* **37**(8), 382.

Vandecasteele, C., Van den Broeck, K., and Dutr, V. (1997). ICPMS, hydride-generation ICP-MS and CZE for the study of solidification/stabilisation of industrial waste containing Arsenic. *Stud. Environ. Sci.*, p. 469.

Vandecasteele, C., Van den Broeck, K., and Dutre, V. (2000). ICP-MS, hydride generation-ICP-MS, and CZE for the study (analysis and speciation) of solidification/stabilisation of industrial waste containing arsenic. *Waste Manag.* **20**(2), 211.

Van den, Broeck, K., and Vandecasteele, C. (1998). Elimination of interferences in the determination of arsenic and determination of arsenic in percolate waters from an industrial landfill by inductively coupled plasma mass spectrometry. *Anal. Lett.* **31**(11), 1891.

Van den Broeck, K., Vandecasteele, C., and Genus, J. M. C. (1997). Determination of arsenic by inductively coupled plasma mass spectrometry in mung bean seedlings for use as a bio-indicator of arsenic contamination. *J. Anal. At. Spectrom.* **12**(9), 987.

Van den Broeck, K., Helsen, L., Vandecasteele, C., and Van den Bulck, E. (1997). Determination and characterisation of copper, chromium and arsenic in chromated copper arsenate (CCA) treated wood and its pyrolysis residues by inductively coupled plasma mass spectrometry. *Analyst (London)* **122**(7), 695.

Van der Hoeff, A. A., Foster, R., and Sotera, J. (1995). Analyzing real-world samples: A review of ICP-MS design. *Int. Lab. Eur. Ed.* **25**(7), 13.

Van der Hoeff, A. A., Foster, R., and Sotera, J. (1995). Analyzing real-world samples: A review of ICP-MS design. *Am. Lab.* **27**(4), 48DD.

Vanderpool, R. A., and Buckley, W. T. (1999). Liquid–liquid extraction of cadmium by sodium diethyldithiocarbamate from biological matrixes for isotope dilution inductively coupled plasma mass spectrometry. *Anal. Chem.* **71**(3), 652.

Vander Putten, E., Dehairs, F., Andre, L., and Baeyens, W. (1999). Quantitative in situ microanalysis of minor and trace elements in biogenic calcite using infrared laser ablation—inductively coupled plasma mass spectrometry—a critical evaluation. *Anal. Chim. Acta* **378**(1), 263.

Van der Velde-Koerts, T., and De Boer, J. L. M. (1994). Minimization of spectral interferences in inductively coupled plasma mass spectrometry by simplex optimization and nitrogen addition to the aerosol carrier for multi-element environmental analysis. *J. Anal. At. Spectrom.* **9**(10), 1093.

van de Weijer, P., Vullings, P. J. M. G., Baeten, W. L. M., and de Laat, W. J. M. (1991). Determination of uranium and thorium in aluminium with flow injection and laser ablation inductively coupled plasma mass spectrometry. *J. Anal. At. Spectrom.* **6**(8), 609–614.

van de Weijer, P., Baeten, W. L. M., Bekkers, M. H. J., and Vullings, P. J. M. G. (1992). Fast semiquantitative survey analysis of solids by laser ablation inductively coupled plasma mass spectrometry. *J. Anal. At. Spectrom.* **7**(4), 599–604.

van Heuzen, A. A. (1991). Analysis of solids by laser ablation inductively coupled plasma mass spectrometry (LA ICP MS). I. Matching with a glass matrix. *Spectrochim. Acta, Part B* **46B**(14), 1803.

van Heuzen, A. A., and Morsink, J. B. W. (1991). Analysis of solids by laser ablation inductively coupled plasma mass spectrometry (LA ICP MS). II. Matching with a pressed pellet. *Spectrochim. Acta, Part B* **46B**(14), 1819.

van Huezen, A. A., and Nibbering, N. M. M. (1993). Elemental composition and origin of (poly-atomic) ions in inductively coupled plasma mass spectrometry disclosed by means of isotope exchange experiments. *Spectrochim. Acta, Part B* **48B**(8), 1013.

Van Holderbeke, M., Zhao, Y., Vanhaecke, F., and Moens, L. (1999). Speciation of six arsenic compounds using capillary electrophoresis—inductively coupled plasma mass spectrometry. *J. Anal. At. Spectrom.* **14**(2), 229.

Van Hoven, R. L. (1992). Determination of Pt, Pd, Rh, and Ir in geological materials by direct solid sampling of fire assay beads using spark ablation—inductively coupled plasma mass spectrometry. Unpublished Ph.D. Thesis, George Washington University, Washington, DC.

Van Hoven, R. L., Nam, S. H., Montaser, A., and Doughten, M. W. (1995). Direct solid sampling of fire assay beads by spark ablation inductively coupled plasma mass spectrometry. *Spectrochim. Acta, Part B* **50**(4/7), 549.

Van Veen, E. H., Bosch, S., and De Loos-Vollebregt, M. T. C. (1994). Spectral interpretation and interference correction in inductively coupled plasma mass spectrometry. *Spectrochim. Acta, Part B* **49**(12//14), 1347.

Van Veen, E. H., Bosch, S., and De Loos-Vollebregt, M. T. C. (1996). Precision-based optimization of multicomponent analysis in inductively coupled plasma mass spectrometry. *Spectrochim. Acta, Part B* **51**(6), 591.

Vanhaecke, F., and Moens, L. (1999). Recent trends in trace element determination and speciation using inductively coupled plasma mass spectrometry. *Fresenius' J. Anal. Chem.* **364**(5), 440.

Vanhaecke, F., Vanhoe, H., Dams, R., and Vandecasteele, C. (1992). The use of internal standards in ICP-MS. *Talanta* **39**(7), 737–742.

Vanhaecke, F., Vandecasteele, C., and Dams, R. (1992). Inductively coupled plasma mass spectrometry for the determination of aluminium, calcium, chlorine, iron, magnesium, manganese and sodium in fresh water. *Anal. Lett.* **25**(5), 919–936.

Vanhaecke, F., Dams, R., and Vandecasteele, C. (1993). 'Zone model' as an explanation for signal behaviour and non-spectral interferences in inductively coupled plasma mass spectrometry. *J. Anal. At. Spectrom.* **8**(3), 433–438.

Vanhaecke, F., Goossens, J., Dams, R., and Vandecasteele, C. (1993). The determination of molybdenum in a sea water candidate reference material by inductively coupled plasma-mass spectrometry. *Talanta* **40**(7), 975.

Vanhaecke, F., Boonen, S., Moens, L., and Dams, R. (1995). Solid sampling electrothermal vaporization inductively coupled plasma mass spectrometry for the determination of arsenic in standard reference materials of plant origin. *J. Anal. At. Spectrom.* **10**(2), 81.

Vanhaecke, F., Galbacs, G., Boonen, S., and Moens, L. (1995). Use of the Ar20+ signal as a diagnostic tool in solid sampling electrothermal vaporization inductively coupled plasma mass spectrometry. *J. Anal. At. Spectrom.* **10**(12), 1047.

Vanhaecke, F., Van Holderbeke, M., Moens, L., and Dams, R. (1996). Evaluation of a commercially available microconcentric nebulizer for inductively coupled plasma mass spectrometry. *J. Anal. At. Spectrom.* **11**(8), 543.

Vanhaecke, F., Boonen, S., Moens, L., and Dams, R. (1997). Isotope dilution as a calibration method for solid sampling electrothermal vaporization inductively coupled plasma mass spectrometry. *J. Anal. At. Spectrom.* **12**(2), 125.

Vanhaecke, F., De Wannemacker, G., Moens, L., and Dams, R. (1998). Dependence of detector dead time on analyte mass number in inductively coupled plasma mass spectrometry. *J. Anal. At. Spectrom.* **13**(6), 567.

Vanhaecke, F., Moens, L., and Dams, R. (1998). The accurate determination of copper in two groundwater candidate reference materials by means of high resolution inductively coupled plasma mass spectrometry using isotope dilution for calibration. *J. Anal. At. Spectrom.* **13**(10), 1189.

Vanhaecke, F., Gelaude, I., Moens, L., and Dams, R. (1999). Solid sampling electrothermal vaporization inductively coupled plasma mass spectrometry for the direct determination of Hg in sludge samples. *Anal. Chim. Acta* **383**(3), 253.

Vanhoe, H. (1993). A review of the capabilities of ICP-MS for trace element analysis in body fluids and tissues. *J. Trace Elem. Electrolytes Health Dis.* **7**(3), 131–140.

Vanhoe, H., Dams, R., Vandecasteele, C., and Versieck, J. (1993). Determination of boron in human serum by inductively coupled plasma mass spectrometry after a simple dilution of the sample. *Anal. Chim. Acta* **281**(2), 401.

Vanhoe, H., Van Allemeersch, F., Versieck, J., and Dams, R. (1993). Effect solvent type on the determination of total iodine in milk powder and human serum by inductively coupled plasma mass spectrometry. *Analyst (London)* **118**(8), 1015.

Vanhoe, H., Dams, R., and Versieck, J. (1994). Use of inductively coupled plasma mass spectrometry for the determination of ultra-trace elements in human serum. *J. Anal. At. Spectrom.* **9**(1), 23–32.

Vanhoe, H., Goossens, J., Moens, L., and Dams, R. (1994). Spectral interferences encountered in the analysis of biological materials by inductively coupled plasma mass spectrometry. *J. Anal. At. Spectrom.* **9**(3), 177–186.

Vanhoe, H., Moens, L., and Dams, R. (1994). Thermospray nebulization as sample introduction for inductively coupled plasma mass spectrometry. *J. Anal. At. Spectrom.* **9**(8), 815.

Vasilyeva, I. E., Shabanova, E. V., Sokolnikova, Y. V., and Proydakova, O. A. (1999). Selection of internal standard for determination of boron and phosphorus by ICP-MS in silicon photovoltaic materials. *J. Anal. At. Spectrom.* **14**(9), 1519.

Vaughan, M.-A., and Horlick, G. (1990). Effect of sampler and skimmer orifice size on analyte and analyte oxide signals in inductively coupled plasma-mass spectrometry. *Spectrochim. Acta, Part B* **45**(12), 1289.

Vaughan, M.-A., Baines, A. D., and Templeton, D. M. (1991). Multielement analysis of biological samples by inductively coupled plasma-mass spectrometry. II. Rapid survey method for profiling trace elements in body fluids. *Clin. Chem. (Winston-Salem, N.C.)* **37**(2), 210.

Veinott, G., Northcote, T., Rosenau, M., and Evans, R. D. (1999). Concentrations of strontium in the pectoral fin rays of the white sturgeon (Acipenser transmontanus) by laser ablation sampling—inductively coupled plasma—mass spectrometry as an indicator of marine migrations. *Can. J. Fish. Aqua. Sci.* **56**(11), 1981.

Vela, N. P., Olson, L. K., and Caruso, J. A. (1993). Elemental speciation with plasma MS. *Anal. Chem.* **65**(13), 585 A.

Venth, K., Danzer, K., Kundermann, G., and Blaufuss, K. H. (1996). Multisignal evaluation in ICP-MS. *Fresenius' J. Anal. Chem.* **354**(7–8), 811.

Vicente, O., Pelfort, E., Martinez, L., Olsina, R., and Marchevsky, E. (1998). Determination of rare earth elements in urinary calculi by ICP-MS. *At. Spectrosc.* **19**(5), 168.

Vicente, O., Padro, A., Martinez, L., and Olsina, R. (1998). Determination of some rare earth elements in seawater by inductively coupled plasma mass spectrometry using flow injection preconcentration. *Spectrochim. Acta, Part B* **53**(9), 1281.

Viczian, M., Lasztity, A., Wang, X., and Barnes, R. M. (1990). On-line isotope dilution and sample dilution by flow injection and inductively coupled plasma mass spectrometry. *J. Anal. At. Spectrom.* **5**(2), 125.

Viczian, M., Lasztity, A., and Barnes, R. M. (1990). Identification of potential environmental sources of childhood lead poisoning by inductively coupled plasma mass spectrometry. Verification and case studies. *J. Anal. At. Spectrom.* **5**(4), 293.

Vijayalakshmi, S., Prabhu, R. K., Mahalingam, T. R., and Mathews, C. K. (1992). A simple gas-liquid separator for continuous hydride introduction in ICP-MS. *At. Spectrosc.* **13**(1), 26–28.

Vijayalakshmi, S., Prabhu, R. K., Mahalingam, T. R., and Mathews, C. K. (1992). Application of ICP-MS in trace metal characterization of nuclear materials. *At. Spectrosc.* **13**(2), 61–66.

Vijayalakshmi, S., Prabhu, R. K., Mahalingam, T. R., and Mathews, C. K. (1992). Determination of trace metals in uranium oxide by inductively coupled plasma mass spectrometry combined with on-line solvent extraction. *J. Anal. At. Spectrom.* **7**(3), 565.

Vin Yi, Y., and Masuda, A. (1996). Simultaneous determination of ruthenium, palladium, iridium, and platinum at ultratrace levels by isotope dilution inductively coupled plasma mass spectrometry in geological samples. *Anal. Chem.* **68**(8), 1444.

Violante, N., Petrucci, F., Delle Femmine, P., and Caroli, S. (1998). Study of possible polyatomic interference in the determination of Cr in some environmental matrices by inductively coupled plasma mass spectrometry. *Microchem. J.* **59**(2), 269.

Vita, O. A., and Mayfield, K. C. (1995). Analysis of urine for U-235 and U-238 by ICP-MS. *ASTM Spec. Tech. Pub.*, p. 140.

Vlasankova, R., Otruba, V., Bendl, J., and Fisera, M. (1999). Preconcentration of platinum group metals on modified silicagel and their determination by inductively coupled plasma atomic emission spectrometry and inductively coupled plasma mass spectrometry in airborne particulates. *Talanta* **48**(4), 839.

Vogl, J., and Heumann, K. G. (1997). Determination of heavy metal complexes with humic substances by HPLC/ICP-MS coupling using on-line isotope dilution technique. *Fresenius' J. Anal. Chem.* **359**(4–5), 438.

Vollkopf, U., Klemm, K., and Pfluger, M. (1999). The analysis of high purity hydrogen peroxide by dynamic reaction cell ICP-MS. *At. Spectrosc.* **20**(2), 53.

Volosin, M. T. (1992). Quality control of high-purity process chemicals using ICP-MS. *Spectroscopy* **7**(4), 44–47.

Wagner, B., Garbos, S., Bulska, E., and Hulanicki, A. (1999). Determination of iron and copper in old manuscripts by slurry sampling graphite furnace atomic absorption spectrometry and laser ablation inductively coupled plasma mass spectrometry. *Spectrochim. Acta, Part B* **54**(5), 797.

Walder, A. J. (1997). Advanced isotope ratio mass spectrometry II: Isotope ratio measurement by multiple collector inductively coupled plasma mass spectrometry. *Chem. Anal.*, p. 83.

Walder, A. J., Koller, D., Reed, N. M., Hutton, R. C., and Freedman, P. A. (1993). Isotope ratio measurement by inductively coupled plasma multiple collector mass spectrometry incorporating a high efficiency nebulization system. *J. Anal. At. Spectrom.* **8**(7), 1037–1042.

Walker, G. S., Ridd, M. J., and Brunskill, G. J. (1996). A comparison of inductively coupled plasma atomic emission spectrometry and inductively coupled plasma mass spectrometry for determination of mercury in great barrier reef sediments. *Rapid Commun. Mass Spectrom.* **10**(1), 96.

Wang, C. F., Chen, W. H., Yang, M. H., and Chiang, P. C. (1995). Microwave decomposition for airborne particulate matter for the determination of trace elements by inductively coupled plasma mass spectrometry. *Analyst (London)* **120**(6), 1681.

Wang, C. F., Jeng, S. L., and Shieh, F. J. (1997). Determination of arsenic in airborne particulate matter by inductively coupled plasma mass spectrometry. *J. Anal. At. Spectrom.* **12**(1), 61.

Wang, C.-F., Jeng, S.-L., Lin, C. C., and Chiang, P.-C. (1998). Preparation of airborne particulate standards on PTFE-membrane filter for laser ablation inductively coupled plasma mass spectrometry. *Anal. Chim. Acta* **368**(1), 11.

Wang, C.-F., Chin, C.-J., Luo, S.-K., and Men, L.-C. (1999). Determination of chromium in airborne particulate matter by high resolution and laser ablation inductively coupled plasma mass spectrometry. *Anal. Chim. Acta* **389**(1), 257.

Wang, C.-F., Chang, C. Y., Chin, C. J., and Men, L. C. (1999). Determination of arsenic and vanadium in airborne related reference materials by inductively coupled plasma-mass spectrometry. *Anal. Chim. Acta* **392**(2), 299.

Wang, J. (1992). Investigation of matrix induced interferences in inductively coupled plasma mass spectrometry. Unpublished Ph.D. Thesis, University of Cincinnati, Cincinnati, OH.

Wang, J., Carey, J. M., and Caruso, J. A. (1994). Direct analysis of solid samples by electrothermal vaporization inductively coupled plasma mass spectrometry. *Spectrochim. Acta, Part B* **49B**(2), 193.

Wang, J., Tomlinson, M. J., and Caruso, J. A. (1995). Extraction of trace elements in coal fly ash and subsequent speciation by high-performance liquid chromatography with inductively coupled plasma mass spectrometry. *J. Anal. At. Spectrom.* **10**(9), 601.

Wang, J., Houk, R. S., Dreessen, D., and Wiederin, D. R. (1998). Identification of inorganic elements in proteins in human serum and in DNA fragments by size exclusion chromatography and inductively coupled plasma mass spectrometry with a magnetic sector mass spectrometer. *J. Am. Chem. Soc.* **120**(23), 5793.

Wang, T., Ge, Z., Wu, J., and Li, B. (1999). Determination of tungsten in bulk drug substance and intermediates by ICP-AES and ICP-MS. *J. Pharm. Biomed. Anal.* **19**(6), 937.

Wang, X., Zhuang, Z., Sun, D., Hong, J., Wu, X., Lee, F. S.-C., Yang, M. S., and Leung, H. W. (1999). Trace metals in traditional Chinese medicine: A preliminary study using ICP-MS for metal determination and as speciation. *At. Spectrosc.* **20**(3), 86.

Wangen, L. E., Bentley, G. E., Coffelt, K. P., Gallimore, D. L., and Phillips, M. V. (1991). Inductively coupled plasma-mass spectrometry drift correction based on generalized internal references identified by principal components factor analysis. *Chemometrics Intell. Lab. Syst.* **10**(3), 293.

Wangkarn, S., and Pergantis, S. A. (1999). Determination of arsenic in organic solvents and wines using microscale flow injection inductively coupled plasma mass spectrometry. *J. Anal. At. Spectrom.* **14**(4), 657.

Wanner, B., Moor, C., Richner, P., and Broennimann, R. (1999). Laser ablation inductively coupled plasma mass spectrometry (LA-ICP-MS) for spatially resolved trace element determination of solids using an autofocus system. *Spectrochim. Acta, Part B* **54**(2), 289.

Ward, N. I., Abou-Shakra, F. R., and Durrant, S. F. (1990). Trace element content of biological materials: A comparison of NAA and ICP-MS analysis. *Bio. Trace Elem. Res.*, p. 177.

Wardley, C. J., Cox, A., McCleod, C., and Morris, B. W. (1999). A longitudinal study of iodine excretion in normal pregnancy determined by inductively coupled plasma mass spectrometry. *J. Anal. At. Spectrom.* **14**(11), 1709.

Warmer, B., Richner, P., and Magyar, B. (1996). The role of modifiers in electrothermal vaporization inductively coupled plasma mass spectrometry (ETV-ICP-MS) for the determination of B, La and U. *Spectrochim. Acta, Part B* **51**(8), 817.

Warnken, K. W., Gill, G. A., Wen, L. S., and Griffin, L. L. (1999). Trace metal analysis of natural waters by ICP-MS with on-line preconcentration and ultrasonic nebulization. *J. Anal. At. Spectrom.* **14**(2), 247.

Warren, A. R. (1996). Simultaneous measurement of ion ratios by inductively coupled plasma—mass spectrometry with a twin quadrupole instrument. Unpublished Ph.D. Thesis, Iowa State University, Ames.

Warren, A. R., Allen, L. A., Pang, H. M., and Houk, R. S. (1994). Simultaneous measurement of ion ratios by inductively coupled plasma-mass spectrometry with a twin-quadrupole instrument. *Appl. Spectrosc.* **48**(11), 1360.

Watling, R. J. (1998). Sourcing the provenance of cannabis crops using inter-element association patterns fingerprinting' and laser ablation inductively coupled plasma mass spectrometry. *J. Anal. At. Spectrom.* **13**(9), 917.

Watling, R. J. (1998). In-line mass transport measurement cell for improving quantification in sulfide mineral analysis using laser ablation inductively coupled plasma mass spectrometry. *J. Anal. At. Spectrom.* **13**(9), 927.

Watling, R. J. (1999). Atomic spectroscopy perspectives: Novel application of laser ablation inductively coupled plasma mass spectrometry in forensic science and forensic archaeology. *Spectroscopy* **14**(6), 16.

Watling, R. J., Herbert, H. K., Delev, D., and Abell, I. D. (1994). Gold fingerprinting by laser ablation inductively coupled plasma mass spectrometry. *Spectrochim. Acta, Part B* **49B**(2), 205.

Watling, R. J., Lynch, B. F., and Herring, D. (1997). Use of laser ablation inductively coupled plasma mass spectrometry for fingerprinting scene of crime evidence. *J. Anal. At. Spectrom.* **12**(2), 195.

Watmough, S. A., and Hutchinson, T. C. (1996). Analysis of tree rings using inductively coupled plasma mass spectrometry to record fluctuations in a metal pollution episode. *Environ. Pollu.* **93**(1), 93.

Watmough, S. A., Hutchinson, T. C., and Evans, R. D. (1997). Application of laser ablation inductively coupled plasma-mass spectrometry in dendrochemical analysis. *Environ. Sci. Technol.* **31**(1), 114.

Watmough, S. A., Hutchinson, T. C., and Evans, R. D. (1998). Heavy metals in the environment—The quantitative analysis of sugar maple tree rings by laser ablation in conjunction with ICP-MS. *J. Environ. Qua.* **27**(5), 1087.

Watmough, S. A., Hutchinson, T. C., and Evans, R. D. (1998). Development of solid calibration standards for trace elemental analyses of tree rings by laser ablation inductively coupled plasma-mass spectrometry. *Environ. Sci. Technolo.* **32**(14), 2185.

Watters, R. L., Jr., Eberhardt, K. R., Beary, E. S., and Fassett, J. D. (1997). Protocol for isotope dilution using inductively coupled plasma- mass spectrometry (ICP-MS) for the determination of inorganic elements. *Metrologia* **34**(1), 87.

Wei, R., and Haraguchi, H. (1999). Multietement determination of major-to-ultratrace elements in river and marine sediment reference materials by inductively coupled plasma atomic emission spectrometry and inductively coupled plasma mass spectrometry. *Anal. Sci.: Int. J. Jpn. Soc. Anal. Chem.* **15**(8), 729.

Wei, M. T., and Jiang, S. J. (1999). Determination of thallium in sea-water by flow injection hydride generation isotope dilution inductively coupled plasma mass spectrometry. *J. Anal. At. Spectrom.* **14**(8), 1177.

Wei, W. C., Chen, C. J., and Yang, M. H. (1995). Determination of boron using mannitol-assisted electrothermal vaporization for sample introduction in inductively coupled plasma mass spectrometry. *J. Anal. At. Spectrom.* **10**(11), 955.

Weiss, D., Boyle, E. A., Chavagnac, V., Herwegh, M., and Wu, J. (2000). Determination of lead isotope ratios in seawater by quadrupole inductively coupled plasma mass spectrometry after Mg(OH)2 co- precipitation. *Spectrochim. Acta, Part B* **55**(4), 363–374.

Wells, M. L., and Bruland, K. W. (1999). An improved method for rapid preconcentration and determination of bioactive trace metals in seawater using solid phase extraction and high resolution inductively coupled plasma mass spectrometry. *Mar. Chem.* **63**(1), 145.

Wells, R. J., Skopec, S. V., Iavetz, R., and Robertson, J. (1995). Trace element analysis of heroin by ICP-MS. *Chem. Aust.* **62**(7), 14.

Wen, B., Shan, X. Q., Liu, R.-X., and Tang, H.-X. (1998). V. Recent developments in mass spectrometry - C - Inductively coupled plasma mass spectrometry. Preconcentration of trace elements in sea water with poly (acrylaminophosphonic - dithiocarbamate) chelating fiber for their determination by inductively coupled plasma mass spectrometry. *Mass Spectrom. Rev.* **17**(2), 116.

Werner, E., Roth, P., Wendler, I., Schramel, P., Hellmann, H., and Kratzel, U. (1997). Feasibility of ICP-MS for the assessment of uranium excretion in urine. *J. Radioanal. Nucl. Chem.* **226**(1–2), 201.

Westgate, J. A., Perkins, W. T., Fuge, R., Pearce, N. J. G., and Wintle, A. G. (1994). Trace-element analysis of volcanic glass shards by laser ablation inductively coupled plasma mass spectrometry: Application to tephrochronological studies. *Appl. Geochem.* **9**(3), 323–336.

Wickenheiser, E. B., Michalke, K., Drescher, C., Hirner, A. V., and Hensel, R. (1998). Development and application of liquid and gas-chromatographic speciation techniques with element specific (ICP-MS) detection to the study of anaerobic arsenic metabolism. *Fresenius' J. Anal. Chem.* **362**(5), 498.

Widmer, C. R., Krahenbuhl, U., Kramers, J., and Tobler, L. (2000). Lead isotope measurements on aerosol samples with ICP-MS. *Fresenius' J. Anal. Chem.* **366**(2), 171.

Wildhagen, D., Krivan, V., Gercken, B., and Pavel, J. (1996). Multi-element characterization of titanium(IV) oxide by electrothermal atomic absorption spectrometry, inductively coupled plasma atomic emission spectrometry, inductively coupled plasma mass spectrometry and total reflection x-ray fluorescence spectrometry after matrix-analyte separation. *J. Anal. At. Spectrom.* **11**(5), 371.

Wildner, H. (1998). Application of inductively coupled plasma sector field mass spectrometry for the fast and sensitive determination and isotope ratio measurement of non-metals in high-purity process chemicals. *J. Anal. At. Spectrom.* **13**(6), 573.

Wildner, H., and Wunsch, G. (1995). Comparison of nebulizer efficiencies with thermal trace-matrix- separation for the analysis of corrosive samples by ICP-MS on the semi-micro-scale. *J. Prakt. Chem.* **337**(7), 542.

Wilke, T., Wildner, H., and Wuensch, G. (1997). Ester generation for the determination of ultratrace amounts of boron in volatile high-purity process chemicals by inductively coupled plasma mass spectrometry. *J. Anal. At. Spectrom.* **12**(9), 1083.

Williams, C. A., Abou-Shakra, F. R., and Ward, N. I. (1995). Investigations into the use of inductively coupled plasma mass spectrometry for the determination of gold in plant materials. *Analyst (London)* **120**(2), 341.

Williams, J. G., and Jarvis, K. E. (1993). Preliminary assessment of laser ablation inductively coupled plasma mass spectrometry for quantitative multi-element determination in silicates. *J. Anal. At. Spectrom.* **8**(1), 25–34.

Willie, S. N., Grégoire, D. C., and Sturgeon, R. E. (1997). Determination of inorganic and total mercury in biological tissues by electrothermal vaporization inductively coupled plasma mass spectrometry. *Analyst (London)* **122**(8), 751.

Willie, S. N., Iida, Y., and McLaren, J. W. (1998). Determination of Cu, Ni, Zn, Mn, Co, Pb, Cd, and V in seawater using flow injection ICP-MS. *At. Spectrosc.* **19**(3), 67.

Willie, S. N., Tekgul, H., and Sturgeon, R. E. (1998). Immobilization of 8-hydroxyquinoline onto silicone tubing for the determination of trace elements in seawater using flow injection ICP-MS. *Talanta* **47**(2), 439.

Winge, R. K., Crain, J. S., and Houk, R. S. (1991). High speed photographic study of plasma fluctuations and intact aerosol particles or droplets in inductively coupled plasma mass spectrometry. *J. Anal. At. Spectrom.* **6**(8), 601–604.

Wolf, R. E., and Denoyer, E. R. (1995). Design and performance criteria for a new ICP-MS for environmental analysis. *At. Spectrosc.* **16**(1), 22.

Wolf, R. E., and Denoyer, E. R. (1996). Design and performance criteria for a new ICP-MS for environmental analysis. *Analusis* **24**(9–10), 19.

Wolf, R. E., and Grosser, Z. A. (1997). Overview and comparison of ICP-MS methods for environmental analyses. *At. Spectrosc.* **18**(5), 145.

Wolf, R. E., Thomas, C., and Bohlke, A. (1998). Analytical determination of metals in industrial polymers by laser ablation ICP-MS. *Appl. Surf. Sci.*, p. 299.

Wolf, S. F., and Bates, J. K. (1996). Radionuclide content of simulated and fully radioactive SRL waste glasses: Comparison of results from ICP-MS, gamma spectrometry, and alpha spectrometry. *Mater. Res. Soc. Symp. Procs.*, p. 107.

Wollenweber, D., Strassburg, S., and Wunsch, G. (1999). Determination of Li, Na, Mg, K, Ca and Fe with ICP-MS using cold plasma conditions. *Fresenius' J. Anal. Chem.* **364**(5), 433.

Wolnik, K. A., Heitkemper, D. T., Crowe, J. B., and Barnes, B. S. (1995). Application of inductively coupled plasma atomic emission and mass spectrometry to forensic analysis of sodium gamma hydroxy butyrate and ephedrine hydrochloride. *J. Anal. At. Spectrom.* **10**(3), 177.

Wondimu, T., Goessler, W., and Irgolic, K. J. (2000). Microwave digestion of "residual fuel oil" (NIST SRM 1634b) for the determination of trace elements by inductively coupled plasma-mass spectrometry. *Fresenius' J. Anal. Chem.* **367**(1), 35.

Woolard, D., Franks, R., and Smith, R. R. (1998). Inductively coupled plasma magnetic sector mass spectrometry method for stable lead isotope tracer studies. *J. Anal. At. Spectrom.* **13**(9), 1015.

Wu, J., and Boyle, E. A. (1997). Low blank preconcentration technique for the determination of lead, copper, and cadmium in small-volume seawater samples by isotope dilution ICPMS. *Anal. Chem.* **69**(13), 2464.

Wu, J., and Boyle, E. A. (1998). Determination of iron in seawater by high-resolution isotope dilution inductively coupled plasma mass spectrometry after Mg(OH)2 coprecipitation. *Anal. Chim. Acta* **367**(1), 183.

Wu, S., Zhao, Y. H., Feng, X., and Wittmeier, A. (1996). Application of inductively coupled plasma mass spectrometry for total metal determination in silicon-containing solid samples using the microwave-assisted nitric acid-hydrofluoric acid-hydrogen peroxide-boric acid digestion system. *J. Anal. At. Spectrom.* **11**(4), 287.

Wu, S., Feng, X., and Wittmeier, A. (1997). Microwave digestion of plant and grain reference materials in nitric acid or a mixture of nitric acid and hydrogen peroxide for the determination of multi-elements by inductively coupled plasma mass spectrometry. *J. Anal. At. Spectrom.* **12**(8), 797.

Wyse, E. J., and Fisher, D. R. (1994). Radionuclide bioassay by inductively coupled plasma mass spectrometry (ICP/MS). *Radia. Prot. Dosimetry* **55**(3), 199.

Xie, Q., and Kerrich, R. (1995). Application of isotope dilution for precise measurement of Zr and Hf in low-abundance samples and international reference materials by inductively coupled plasma mass spectrometry: Implications for Zr (Hf)/REE fractionations in komatiites. *Isot. Geosci.* **123**(1/4), 17.

Xie, Q., and Kerrich, R. (1995). Optimization of operating conditions for improved precision of zirconium and hafnium isotope ratio measurement by inductively coupled plasma mass spectrometry. *J. Anal. At. Spectrom.* **10**(2), 99.

Xie, Q., Jain, J., Sun, M., and Kerrich, R. (1994). ICP-MS analysis of basalt BIR-1 for trace elements. *Geostand. Newsl.* **18**(1), 53.

Xu, S. X., Stuhne-Sekalec, L., and Templeton, D. M. (1993). Determination of nickel in serum and urine by inductively coupled plasma mass spectrometry. *J. Anal. At. Spectrom.* **8**(3), 445–448.

Yabutani, T., Ji, S., Mouri, F., Sawatari, H., Itoh, A., Chiba, K., and Haraguchi, H. (1999). Multielement determination of trace elements in coastal seawater by inductively coupled plasma mass spectrometry with aid of chelating resin preconcentration. *Bull. Chem. Soc. Jpn.* **72**(10), 2253.

Yali, S., Xiyun, G., and Andao, D. (1998). Determination of platinum group elements by inductively coupled plasma-mass spectrometry combined with nickel sulfide fire assay and tellurium coprecipitation. *Spectrochim. Acta, Part B* **53**(10), 1463.

Yamada, H., Kiriyama, T., and Yonebayashi, K. (1996). Determination of total iodine in soils by inductively coupled plasma mass spectrometry. *Soil Sci. Plant Nut.* **42**(4), 859.

Yamamoto, M., Kofuji, H., Tsumura, A., and Yamasaki, S. (1994). Temporal feature of global fallout 020307Np deposition in paddy field through the measurement of low-level 020307Np by high resolution ICP-MS. *Radiochim. Acta* **64**(3/4), 217.

Yamamoto, M., Kofuji, S. K., Tsumura, A., Komura, K., Ueno, K., and Assinder, D. J. (1995). Determination of low-level 99Tc in environment31 samples by high resolution ICP-MS. *J. Radioan. Nucl. Chem.* **197**(1), 185.

Yamasaki, S. I. (1995). Total elemental analysis of soils by inductively coupled plasma-mass spectrometry (ICP-MS). *JARQ* **29**(1), 17.

Yan, X., Tanaka, T., and Kawaguchi, H. (1996). Reduced-pressure inductively coupled plasma mass spectrometry for nonmetallic elements. *Appl. Spectrosc.* **50**(2), 182.

Yan, X., Tanaka, T., and Kawaguchi, H. (1996). Electrothermal vaporization for the determination of halogens by reduced pressure inductively coupled plasma mass spectrometry. *Spectrochim. Acta, Part B* **51**(11), 1345.

Yan, X., Huang, B., Tanaka, T., and Kawaguchi, H. (1997). Langmuir probe potential measurements for reduced-pressure inductively coupled plasma spectrometry. *J. Anal. At. Spectrom.* **12**(7), 697.

Yan, X.-P., Kerrich, R., and Hendry, M. J. (1998). Determination of (ultra)trace amounts of arsenic(III) and arsenic(V) in water by inductively coupled plasma mass spectrometry coupled with flow injection on-line sorption preconcentration and separation in a knotted reactor. *Anal. Chem.* **70**(22), 4736.

Yan, X.-P., Hendry, M. J., and Kerrich, R. (2000). Speciation of dissolved iron(III) and iron(II) in water by on-line coupling of flow injection separation and preconcentration with inductively coupled plasma mass spectrometry. *Anal. Chem.* **72**(8), 1879.

Yang, K. L., Jiang, S. J., and Hwang, T. J. (1996). Determination of titanium and vanadium in water samples by inductively coupled plasma mass spectrometry with on-line preconcentration. *J. Anal. At. Spectrom.* **11**(2), 139.

Yang, K. X., Lonardo, R. F., Liang, Z., and Yuzefovsky, A. I. (1997). Determination of tin in nickel-based alloys by electrothermal laser-excited atomic fluorescence with confirmation of accuracy by inductively coupled plasma mass spectrometry and atomic absorption spectrometry. *J. Anal. At. Spectrom.* **12**(3), 369.

Yasuhara, H., Okano, T., and Matsumura, Y. (1992). Determination of trace elements in steel by laser ablation inductively coupled plasma mass spectrometry. *Analyst (London)* **117**(3), 395–400.

Yi, W., Halliday, A. N., Lee, D.-C., and Rehkamper, M. (1998). Precise determination of cadmium, indium and tellurium using multiple collector ICP-MS. *Geostand. Newsl.* **22**(2), 173.

Yi, Y. V., and Masuda, A. (1996). Isotopic homogenization of iridium for high sensitivity determination by isotope dilution inductively coupled plasma mass spectrometry. *Anal. Sci.: Int. J. Jpn. Soc. Anal. Chem.* **12**(1), 7.

Ying, H., Murphy, J., Tromp, J. W., Mermet, J. M., and Salin, E. D. (2000). Warning diagnostics for inductively coupled plasma-mass spectrometry. *Spectrochim. Acta, Part B* **55**(4), 311–326.

Yokoi, K., Alcock, N. W., and Sandstead, H. H. (1994). Co-occurrence of iron deficiency and mild zinc deficiency in the US women. Application of inductively coupled plasma-mass spectrometry for nutritional science. *Nippon Eiseigaku Zasshi* **49**(1), 8-04.

Yoshida, S., Muramatsu, Y., Tagami, K., and Uchida, S. (1996). Determination of major and trace elements in Japanese rock reference samples by ICP-MS. *Int. J. Environ. Anal. Chem.* **63**(3), 195.

Yoshinaga, J. (1996). Isotope ratio analysis of lead in biological materials by inductively coupled plasma mass spectrometry. *Tohoku J. Exp. Med.* **178**(1), 37.

Yoshinaga, J., and Morita, M. (1997). Determination of mercury in biological and environmental samples by inductively coupled plasma mass spectrometry with the isotope dilution technique. *J. Anal. At. Spectrom.* **12**(4), 417.

Yoshinaga, J., Matsuo, N., Imai, H., Nakazawa, M., Suzuki, T., and Morita, M. (1990). Application of inductively coupled plasma mass spectrometry (ICP-MS) to multi-element analysis of human organs. *Int. J. Environ. Ana. Chem.* 27.

Yoshinaga, J., Shibata, Y., and Morita, M. (1993). Trace elements determined along single strands of hair by inductively coupled plasma mass spectrometry. *Clin. Chem. (Winston-Salem, N.C.)* **39**(8), 1650.

Yoshinaga, J., Morita, M., and Edmonds, J. S. (1999). Determination of copper, zinc, cadmium and lead in a fish otolith certified reference material by isotope dilution inductively coupled plasma mass spectrometry using off-line solvent extraction. *J. Anal. At. Spectrom.* **14**(10), 1589.

Yoshinaga, J., Nakama, A., and Takata, K. (1999). Determination of total tin in sediment reference materials by isotope dilution inductively coupled plasma mass spectrometry after alkali fusion. *Analyst (London)* **124**(3), 257.

Young, S. M. M., Budd, P., Haggerty, R., and Pollard, A. M. (1997). Inductively coupled plasma-mass spectrometry for the analysis of ancient metals. *Archaeometry* **39**(2), 379.

Yu, L. (1994). Minimization of background interferences in isotope ratio analysis by inductively coupled plasma mass spectrometry. Unpublished Ph.D. Thesis, University of Missouri, Columbia.

Yu, L., Koirtyohann, S. R., Turk, G. C., and Salit, M. L. (1994). Selective laser-induced ionization in inductively coupled plasma mass spectrometry. *J. Anal. At. Spectrom.* **9**(9), 997.

Yukawa, M., Watanabe, Y., Nishimura, Y., Guo, Y., Yongru, Z., Lu, H., Zhang, W., Wei, L., and Tao, Z. (1999). Determination of U and Th in soil and plants obtained from a high natural radiation area in China using ICP-MS and g-counting. *Fresenius' J. Anal. Chem.* **363**(8), 760.

Yuzefovsky, A. I., and Miser, D. E. (1998). Analysis of discolored aluminum foil by laser ablation inductively coupled plasma mass spectrometry and scanning electron microscopy. *Appl. Spectrosc.* **52**(5), 629.

Zbinden, P., and Andrey, D. (1998). High pressure ashing—Determination of trace element contaminants in food matrices using a robust, routine analytical method for ICP-MS. *At. Spectrosc.* **19**(6), 214.

Zhang, H. (1998). Fundamental studies of helium inductively coupled plasma mass spectrometry for elemental analysis. Unpublished Ph.D. Thesis, George Washington University, Washington, DC.

Zhang, J., and Nozaki, Y. (1996). Rare earth elements and yttrium in seawater: ICP-MS determinations in the East Caroline, Coral Sea, and South Fiji basins of the western South Pacific Ocean. *Geochim. Cosmochim. Acta* **60**(23), 4631.

Zhang, L. S., and Combs, S. M. (1996). Using the installed spray chamber as a gas-liquid separator for the determination of germanium, arsenic, selenium, tin, antimony, tellurium and bismuth by hydride generation inductively coupled plasma mass spectrometry. *J. Anal. At. Spectrom.* **11**(11), 1043.

Zhang, L. S., and Combs, S. M. (1996). Determination of selenium and arsenic in plant and animal tissues by hydride generation inductively coupled plasma mass spectrometry. *J. Anal. At. Spectrom.* **11**(11), 1049.

Zhang, L.-S., and Combs, S. M. (1998). Double calibration (external calibration and stable isotope dilution) for determining selenium in plant tissue by hydride generation inductively coupled plasma mass spectrometry. *J. Assoc. Off. Anal. Chem. Int.* **81**(5), 1060.

Zhang, S., Shan, Z.-Q., Yan, X., and Zhang, H. (1997). The determination of rare earth elements in soil by ICP-MS. *At. Spectrosc.* **18**(5), 140.

Zhang, T.-H., Shan, X.-Q., Liu, R.-X., Tang, H.-X., and Zhang, S.-Z. (1998). Preconcentration of rare earth elements in seawater with poly(acrylaminophosphonic dithiocarbamate) chelating fiber prior determination by inductively coupled plasma mass spectrometry. *Anal. Chem.* **70**(18), 3964.

Zhang, X., and Koropchak, J. A. (1999). Direct chromium speciation using thermospray: Preliminary studies with inductively coupled plasma mass spectrometry. *Microchem. J.* **62**(1), 192.

Zhang, Z. W., Shimbo, S., Ochi, N., Eguchi, M., Watanabe, T., Moon, C. S., and Ikeda, M. (1997). Determination of lead and cadmium in food and blood by inductively coupled plasma mass spectrometry: A comparison with graphite furnace atomic absorption spectrometry. *Sci. Total Environ.* **205**(2–3), 179.

Zhang, J., Goessler, W., Geiszinger, A., Kosmus, W., Chen, B., Zhuang, G., Xu, K., and Sui, G. (1997). Multi-element determination in earthworms with instrumental neutron activation analysis and inductively coupled plasma mass spectrometry: A comparison. *J. Radioanal. Nucl. Chem.* **223**(1–2), 149.

Zheng, J., Goessler, W., and Kosmus, W. (1998). Speciation of arsenic compounds by coupling high-performance liquid chromatography with inductively coupled plasma mass spectrometry. *Mikrochim. Acta* **130**(1–2), 71.

Zheng, J., Ohata, M., and Furuta, N. (2000). Antimony speciation in environmental samples by using high-performance liquid-chromatography coupled to inductively coupled plasma mass spectrometry. *Anal. Sci.: Int. J. Jpn. Soc. Anal. Chem.* **16**(1), 75.

Zheng, J., Ohata, M., Furuta, N., and Kosmus, W. (2000). Speciation of selenium compounds with ion-pair reversed-phase liquid chromatography using inductively coupled plasma mass spectrometry as element-specific detection. *J. Chromatogr.* **874**(1), 55.